# About Island Press

Since 1984, the nonprofit organization Island Press has been stimulating, shaping, and communicating ideas that are essential for solving environmental problems worldwide. With more than 1,000 titles in print and some 30 new releases each year, we are the nation's leading publisher on environmental issues. We identify innovative thinkers and emerging trends in the environmental field. We work with world-renowned experts and authors to develop cross-disciplinary solutions to environmental challenges.

Island Press designs and executes educational campaigns in conjunction with our authors to communicate their critical messages in print, in person, and online using the latest technologies, innovative programs, and the media. Our goal is to reach targeted audiences—scientists, policymakers, environmental advocates, urban planners, the media, and concerned citizens—with information that can be used to create the framework for long-term ecological health and human well-being.

Island Press gratefully acknowledges major support from The Bobolink Foundation, Caldera Foundation, The Curtis and Edith Munson Foundation, The Forrest C. and Frances H. Lattner Foundation, The JPB Foundation, The Kresge Foundation, The Summit Charitable Foundation, Inc., and many other generous organizations and individuals.

Generous support for the publication of this book was provided by Daniel Hildreth.

The opinions expressed in this book are those of the author(s) and do not necessarily reflect the views of our supporters.

# ECOLOGY AND RECOVERY OF
# EASTERN OLD-GROWTH FORESTS

# Ecology and Recovery of Eastern Old-Growth Forests

*Edited by*

Andrew M. Barton and William S. Keeton

*Foreword by*

Thomas A. Spies

**ISLAND**PRESS

Washington | Covelo | London

Library of Congress Control Number: 2018946758

10 9 8 7 6 5 4 3 2 1

*Keywords*: Island Press, old-growth forests, forest ecology, natural disturbance, climate change, adaptation, silviculture, carbon, forest biodiversity, riparian, late-successional, invasive species, forest soils, temperate forests, dynamic ecosystems, restoration, sustainable forestry, conservation, resilience, landscape, stand development, succession, ecosystem services, flooding, hydroperiod, sea level rise, succession

# CONTENTS

My first encounter with an old-growth forest as a subject of scientific investigation was as an undergraduate ecology student visiting Hartwick Pines State Park in Michigan, a remnant patch of old-growth eastern white pine. I had known as a youth the story of how Michigan was once covered by old-growth pine and hardwood forests that were lost to logging in the late nineteenth and early twentieth centuries. I had seen the black-and-white pictures of loggers standing on piles of large boles and of landscapes of stumps. This story had special meaning for me because my great-great-grandfather founded a lumber company in 1880 in upper Michigan that was fed by logs from old-growth forests for almost 40 years until it closed when the timber ran out.

My scientific interest in old-growth forests really grew when I conducted my PhD research on vegetation and soil relationships in a remnant old-growth northern-hardwoods landscape in Sylvania Wilderness Area in Upper Michigan. I learned from my eastern old-growth experiences that old-growth forests were ecologically complex, extremely interesting scientifically, and rare. I also believed that old-growth forests could be an ecological model of a sort for how we might better design managed forests and landscapes to sustain forest values, including wood and native biodiversity.

My interest and experience in old growth was taken to another level when I began postdoctoral work in the Pacific Northwest in 1983 to study how old-growth Douglas fir-western hemlock forests differed in structure and composition of plant and animal communities from younger natural forests. With my move from Michigan to Oregon, old growth was suddenly no longer rare but relatively common (e.g., millions of acres), though history seemed to be repeating itself as I saw the log trucks with huge boles rolling down Oregon highways. Fortunately, federal forest policy changed during the 1990s as society recognized the high values of old-growth forests, and today the Pacific Northwest is home to extensive old-growth reserves, and a top priority for federal managers is to restore successional and disturbance processes in uniform plantations and dry forests where fire has been excluded.

Old growth in eastern North America may not get as much publicity as old growth on the west coast of the United States and Canada does, but it is no less interesting and important scientifically. This book is a timely and significant contribution to our evolving scientific knowledge of old-growth forest ecology in eastern North America and the world. It reveals the diversity of expressions of old-growth ecosystems in this region and helps to counter any notion that other types of old growth should be measured and valued against criteria used to define the massive old-growth coniferous forests of the West Coast. Old-growth forests in the eastern United States have their own interesting and awe-inspiring flavors, including old forests with massive hardwood trees and high tree-species richness, forests where frequent fire is an essential ecological process, and forests that develop in swamps and bottomlands and depend on periodic floods. The book confirms what we have learned in the Pacific Northwest: It is difficult to define old growth precisely, but that reality does not limit the scientific value of research to understand how forests with old trees develop and function ecologically and to reveal the different values they hold for society.

The authors of this book are preeminent forest ecologists. As accomplished as these scientists are, they are still a bit like blind men and women holding different parts of the elephant of complex old-growth ecosystems that once covered eastern North America. This book puts their collective knowledge together to give us a much better picture of the ecology and management issues of old growth in the eastern half of the continent.

The book covers a lot of ground, literally and figuratively, ranging from the boreal forests to the pine savannas of the southern United States and ranging from knowledge of old growth in terms of ecological history to forest-stream interactions, carbon dynamics, and silviculture. The geographic scope of the book and its emphasis on management and conservation in the Anthropocene make it especially valuable to today's managers. Eastern old-growth forests are not dusty museum pieces but are dynamic ecosystems that are changing as a result of climate change, invasive species, and altered disturbance regimes. These forests do not necessarily look and function in the same way as they did when they were first protected from logging. In many cases, management will be needed to conserve some of their most valuable characteristics. The remaining old forests are also not just scientific curiosities that are irrelevant to the larger landscapes in which they occur. Forests in the broader landscape around them have significant instrumental and intrinsic values to society. Many of these forests are now over 100 years old and are on their way to ecological maturity—some may even become the old growth of the future. We face many challenges to maintaining and increasing our forest values given climate change, invasive species, and changing social and economic systems. For example, how can we maintain the resilience of all our forests and facilitate their adaptation to rapid ecological change? The goals of most managers will not include growing old growth, but their goals typically include sustaining the values of forests over the long run. Old-growth forests are awesome in part because they reveal how forests can change and function when allowed to grow and develop over long periods—well beyond several human life spans. May the knowledge in this book help us to devise conservation and silvicultural approaches that sustain ecological and socioeconomic values of all our forests.

Thomas A. Spies
*Senior Scientist*
Pacific Northwest Research Station
USDA Forest Service
Corvallis, Oregon

# PREFACE

*Andrew M. Barton and William S. Keeton*

I (AMB) first set foot in old growth in the 1970s in Joyce Kilmer Memorial Forest in western North Carolina, backpacking with my father and a friend. I grew up not too far away in Asheville, and so I was familiar with the feel of thick air, the sight of dense woods, and the promise of abundant salamanders. But the towering tulip trees of Joyce Kilmer, far too big around for our three sets of arms, were a truly impressive sight, even for a teenager. Two decades earlier, Henry Oosting and Phillippe Bourdeau (1955) described the impressive ecological diversity of this old-growth cove forest—dozens of mesophytic tree species, tangles of rhododendrons, and a lushness of forbs, mosses, and ferns. The forest segregated into two heavily wooded zones— lower slopes and streamsides with a low rhododendron understory and upper slopes chock full of herbaceous plants. Even within each forest enclave, the vegetation differed considerably from place to place.

Old growth like that at Joyce Kilmer used to be thought of as the essence of stability and stasis. That's actually far from the truth. Investigating the natural disturbance dynamics of the forest in the 1970s and 1980s, Craig Lorimer (1980) and James Runkle (1982) documented a canopy punctured by small gaps where a tree or small group of trees had fallen, likely blown over by the wind. These sites turned out to be hot spots for the establishment of new tree seedlings and the growth of extant trees, which take advantage of the elevated light. About 1 percent of Joyce Kilmer was subject to gap production per year, creating a fine-grained forest mosaic of small patches of different ages—a constantly shifting quilt as old gaps filled and new ones were created. Still, on the scale of the entire forest, this was a relatively steady place with only modest fluctuations over time in forest structure and constituent species. Runkle (1996) wrote that, in forests like those at Joyce Kilmer Memorial Forest "major disturbances caused by the physical environment reach perhaps their lowest level of importance for any forest type."

More than 40 years after my first visit, Joyce Kilmer's 3,800 acres continue to receive protection. No trees have been harvested, no roads built, no developments proposed. Treefall gaps continue to form, and new tree seedlings strive for a place where the old have fallen. But the forest *has* changed. The hemlock woolly adelgid, a nonnative invasive insect, has killed many of the giant eastern hemlocks, piercing the canopy with new openings. A massive tree, hemlock is a foundation species in the forests of eastern North America, exerting unique control over the environment where it grows, providing conditions conducive to a diversity of species (Ellison et al. 2005). When hemlock goes, a cascade of ecological changes follows (Abella 2014).

---

Joyce Kilmer is but one tract in a rich and varied assemblage of old growth across eastern North America, remnants of the vast forest that existed before Euro-American settlement. This book is about those forests, more specifically about the science of eastern old growth. It is a sample of the remarkable research being carried out by ecologists from the pine savannas of the deep South to the Canadian boreal forest and west to the northern hardwood-hemlock forests of the Northern Lakes States region. "Old growth" is a scientific term, but it is also a powerful idea imbued with metaphor, spirituality, and politics. That old growth exists at all and receives some measure of protection owes a profound debt to the cultural resonance of the idea. Science is fundamental, however, to an understanding of the old-growth debate and to the future of these forests (Davis 1996; Spies and Duncan 2009; Wirth et al. 2009). Our mission here is to update and bring wider attention to the exciting advances in the science of old-growth forests in eastern North America over the past two decades and its application to the recovery and sustenance of these important ecosystems.

Despite a surge of research and conservation regarding these disparate patches across this vast landscape, old growth in the East continues to be overshadowed by old-growth forests elsewhere, especially in the Pacific Northwest and the tropics, where the tracts are larger and more ecologically intact and the surrounding issues more politically charged. Nevertheless, the old-growth forests of eastern North America are shaped by their own unique set of environmental forces and support a distinct constellation of ecosystems and species, and, as such, are equally important to the advance of ecological knowledge and the development of effective conservation strategies. Our geographic focus is a quasi-natural block of forestland bordered by the Atlantic Ocean to the east, the Gulf of Mexico to the south, prairie and deserts to the west, and tundra to the north. It is a swath of nature of sufficient size to explore the vital issue of ecoregional variation in the forces shaping old-growth forests—and the very notion of what constitutes old growth.

Our short tale about Joyce Kilmer Memorial Forest anticipates some of the main themes of this book. Old growth is complex. Old growth is heterogeneous across space. Old growth is dynamic, and natural disturbance is a chief driver of the amount and nature of old growth in any given area. Old growth is buffeted by human impact, even in sites remote from civilization. And, finally, old growth is a vital storehouse of biodiversity, ecological information, and ecosystem services such as carbon storage. The contributors to this book lay out cutting-edge science (in large measure advanced by their own research), explore the conservation challenges for the future, and, to a scientist, provide optimism that the sustenance of old-growth forests is possible.

More than twenty years after the first compendium of eastern old-growth ecology, edited by Mary Byrd Davis (1996), the world has changed. Symptoms of climate change are ubiquitous, invasive pests and pathogens spread apace, and land-use pressure continues to gobble up and fragment eastern landscapes. This book makes the case that old growth will, if anything, become even more relevant and vital in the face of these changing environmental conditions. Although they may look different than they did in the past, old-growth ecosystems will have an important place on the future landscape, absorbing carbon, harboring biological diversity, and helping humans and their fellow denizens of planet Earth adapt to those changes.

This book is the collective effort of many people. The idea for the book started several years ago with Daniel Hildreth, who has been a steadfast supporter of both old-growth science and this book. We are very grateful to Erin Johnson at Island Press, who was a remarkably patient, positive, insightful, and constructive editor. In fact, we greatly appreciate the efforts of the entire Island Press staff in bringing this book to fruition. Finally, we sincerely thank the many contributors to this book, who took time out of their busy schedules to synthesize their many years of research into the compelling reviews that comprise this book.

## References

Abella, S. R. 2014. "Impacts and management of hemlock woolly adelgid in national parks of the eastern United States." *Southeastern Naturalist* 13:16–45.

Davis, M. B., ed. 1996. *Eastern Old-Growth Forests: Prospects for Rediscovery and Recovery*. Washington, DC: Island Press.

Ellison, A. M., M. S. Bank, B. D. Clinton, E. A. Colburn, K. Elliott, C. R. Ford, D. R. Foster, et al. 2005. "Loss of foundation species: consequences for the structure and dynamics of forested ecosystems." *Frontiers in Ecology and the Environment 3*: 479–486.

Lorimer, C. G. 1980. "Age structure and disturbance history of a southern Appalachian virgin forest." *Ecology 61*: 1169–1184.

Oosting, H. J., and P. F. Boudreau. 1955. "Virgin hemlock forest segregates in the Joyce Kilmer Memorial Forest of western North Carolina." *Botanical Gazette 116*: 340–359.

Runkle, J. R. 1982. "Patterns of disturbance in some old-growth mesic forests of eastern North America." *Ecology 63*: 1533–1546.

Runkle, J. R. 1996. "Central mesophytic forests." In *Eastern Old-Growth Forests: Prospects for Rediscovery and Recovery*, edited by M. B. Davis, 161–177. Washington, DC: Island Press.

Spies, T. A., and S. L. Duncan, eds. 2009. *Old Growth in a New World: a Pacific Northwest Icon Reexamined*. Washington, DC: Island Press.

Wirth, C., G. Gleixner, and M. Heimann. 2009. *Old-Growth Forests: Function, Fate, and Value*. Ecological Studies 207. Berlin: Springer-Verlag.

# Chapter 1

# Introduction: Ecological and Historical Context

*Andrew M. Barton*

The science of old growth is multifarious, reflecting the complexity of these ecosystems and their considerable variation across the landscape and time. It is also tied to human agency, both the centuries of exploitation of forests and the evolution of the emotive ideas surrounding old growth. The goal of this introduction is to provide context for old-growth science and the diverse set of 14 chapters that follow.

We will start by addressing the perennial question *What is old growth?* Writing about old-growth ecosystems demands the penance of wrestling with the definition. It is a wickedly difficult but important exercise. We will then remind readers why people care about old-growth forests in the first place, in other words, why this book exists at all. Once we have established the what and why of this book, we will provide some ecological context, first geographically, and then temporally, by examining in detail how the sites supporting two extant eastern old-growth forests have changed over the past 20,000 years. In other words, we will address the questions of how old-growth ecosystems have reached their current state and how this should shape our expectations for the future in a rapidly changing world. Those two paleoecological narratives lead us directly to human history, first to the early North Americans, and then to modern times, in which old growth became an idea, a conservation goal, and a controversy. Although social and political aspects of old growth are not the focus of this book, they provide important background for the science, which affects and is affected by those currents. We will end by exploring different frameworks for making sense of and synthesizing the diversity of chapters to follow.

## What Is Old Growth?

Some readers might expect a sharp, clear definition that demarcates old growth from "not old growth." If so, what follows might prove to be a disappointment. Defining old growth has produced a cottage industry for scientists and forest managers striving for clarity and pragmatism. In a recent analysis of old-growth definitions and concepts, Pesklevits et al. (2011) cite more than 15 papers whose purpose is to define and circumscribe the subject. The US Forest Service toiled over definitions for a decade before publishing a guide to old-growth stands on national forests in the eastern United States (Tyrrell et al. 1998). The diversity of definitions, criteria, classifications, and confusions has been characterized vividly, as has the lack of consensus (Spies 2004; Wirth et al. 2009). Pesklevits et al. (2011) argue that defining old growth is a "wicked problem" (Rittel and Webber 1973) in the sense that it is "irreconcilably tricky or perpetually vexing." *Wicked* has also been used to describe the social problem of solving climate change (Grundmann 2016).

More recently, ecologists have embraced the idea that, "a consensus on the wording of an ecological definition of old growth will never be reached and may not be desirable, given the diversity of forests." (Spies 2004; see also Wirth et al. 2009). In fact, the discipline of ecology is full of terms, such as *community, ecosystem,* and *complexity* that are left vague but are operationalized for particular research, sites, or applications. Granted some of the confusion about what constitutes old growth emerges from the reality that "old growth is simultaneously an ecological state, a value-laden social concept, and a polarizing political phenomenon." (Pesklevits et al. 2011). Underlying the diversity of old-growth criteria, however, are real differences in these ecosystems across the local landscape and the continent. The challenge of any definition, therefore, is the tradeoff between generality and acknowledgement of complexity. We propose that the very act of grappling with old-growth forest definitions, as cumbersome as it might be, promotes an understanding of the variation in ecological patterns and processes (see chapter 4, for example).

Despite the variety of definitions, we can identify two intertwined attributes of old growth that are common to many conceptions and are particularly relevant for this book: forests with *old trees* that have been *largely undisturbed by people* since their origin. Hunter and White (1997) showed that these two axes should be thought of as continuous, that is, there is no objective threshold of either age or "naturalness" that separates old growth from somewhat old and natural forests. As forests age since their rebirth after the last disturbance, whether natural or anthropogenic, they slowly develop characteristics of old growth, and, by definition, the less intervention by humans, the more an ecosystem can be said to be under the control of nonhuman forces. As will become apparent in this book, the imprint of humans on nature, even old growth, increases apace regardless of remoteness and history, injecting new "wickedness" into defining and characterizing old growth.

Like all humans, scientists classify continuous phenomena into categories to better understand them. So, developing specific criteria that helps identify or characterize old-growth forests can be seen as an attempt to gain an understanding of the function, variability, and dynamics of forests that have developed over long periods with little human impact. Moreover, at least for some regions or specific forest types, old-growth criteria "may be useful for . . . inventorying stands . . . prioritizing sites for protection . . . determining whether or when forests . . . acquire an old-growth condition . . . " (Tyrrell et al. 1998).

Many criteria have been proposed, including the following:

- trees more than 50 percent of the maximum age of the dominant tree species;
- a variety of ages of dominant tree species;
- establishment of new individuals by gap-phase dynamics (i.e., the formation of small canopy gaps);
- the death of all members of the original cohort that established directly after the last major natural disturbance;
- and the presence of large snags and coarse woody debris (e.g., dead trunks and branches) on the forest floor.

Certainly, these criteria effectively describe the stereotypical old-growth forests of the Pacific Northwest, the cove forests of the southern Appalachians (chapter 4), and many of the mixed old-growth forests of the northeastern United States and southeastern Canada (chapter 6) and the northern Lake States region (chapter 7). As helpful as these detailed criteria might be, however, they would eliminate some forest types that are clearly very old and largely undisturbed by humans: centuries-old cedars perched on the cliff-face of the Niagara escarpment (Kelly and Larson 2007), for example, as well as some ecosystems described in this book (e.g., chapter 3). Therein lies the rub: The more specific the definition, the less it applies to the tremendous range of variation in forests and woodlands that meet a commonsense and ecologically important conception of "old growth."

Given that a goal of this book is to better understand the natural patterns and processes of old-growth forests and how these vary across the landscape, we embrace a permissive definition of what constitutes old-growth forest. In other words, we will let our contributors decide what old growth is for the systems in which they work, even if that stretches the boundaries of the most common conceptions. We are convinced that the diversity of ecological phenomena and ecosystems encompassed by this approach justifies its slackness.

## Why is Old Growth Important?

Why do individuals and societies care about old-growth forests? Clearly, the science, conservation, and controversy surrounding old growth arose and persist because these ecosystems are valued (Whitney 1987; Davis 1996; Spies and Duncan 2009; Wirth et al. 2009; Maloof 2011). Our goal is to provide a brief summary of the importance of old growth, divided into three categories of values: biodiversity (see Glossary), direct benefits to people, and moral standing. These are interrelated, but we separate them for convenience.

Old-growth forests harbor biodiversity at multiple levels. They provide a storehouse of genetic diversity that has evolved through eons and developed ecologically over centuries. A wide range of species thrive or even depend on the structures, resources, and long-term undisturbed nature of old-growth forests (chapter 11), and destruction would lead to a loss of biodiversity at the forest, regional, and planetary scales, the degree of such harm uncertain (Davis 1996; Spies and Duncan 2009; Wirth et al. 2009). The dependence of the Northern Spotted Owl on old-growth forests in the Pacific Northwest was, of course, the pivotal issue that launched old growth into international consciousness (Spies and Duncan 2009). Finally, old growth is a key stage in the successional processes of forests, and as such is a linchpin in the conservation goal of sustaining a diversity of habitats and ecosystem types (Spies 2004). Our knowledge of the importance of older forests for biodiversity and the complex species interactions therein continues to grow, especially regarding less charismatic, but still important, organisms (chapters 3, 9, 11, and 13).

There are multiple direct benefits of old growth to people, at individual and societal levels. Many derive great recreational pleasure, awe, and psychological and spiritual sustenance from old growth, which offers an experience apart from the heavy imprint of civilization, in a place where organisms have persisted for centuries (Leverett 1996; Moore 2007). Even if they do not actually venture into these places, some people derive well-being by simply knowing that old growth exists and is protected and that they or their descendants could visit them and enjoy them in the future (Loomis 2009). Put simply, most people love forests and trees, especially large, old ones. Old growth also provides essential indirect benefits to society, especially through ecosystem services such as the provision of clean water, as well as through more prosaic enhancements, such as boosting surrounding real estate values.

As described in chapter 14, ecosystem services can even accrue at a planetary level, for old-growth ecosystems store high levels of carbon, effectively sequestering it from the atmosphere where it traps heat. Recent research has overturned the conventional wisdom that old trees and forests are carbon neutral, revealing that they often continue to accrue additional carbon regardless of age (chapter 14). Finally, old-growth forests provide a unique research laboratory for scientists, allowing them to investigate the workings of nature with relatively few confounding human impacts compared to other ecosystems. Such research can inform our quest for improving management and conservation of all forests (Spies 2009).

Many religions, spiritual principles, and philosophies declare an ethical basis for the protection of old-growth forests, regardless of utilitarian human purpose. In other words, they give old-growth forests independent moral standing. These principles are based on connections to deities, sacredness, the special role of trees, and human stewardship. We refer the reader elsewhere for more in-depth discussions of these issues, which help explain the long-standing adoration of humans for old trees and forests (Whitney 1987; Albanese 1990; Proctor 2009).

## Geographic and Ecoregional Context

The old-growth forests discussed in this book occur across an enormous swath of North America, including the Deep South (chapters 2 and 3), southern Appalachians (chapter 4), central Appalachians (chapter 5), northeastern United States and southeastern Canada (chapter 6), northern Lake States region (chapter 7), and Canadian boreal zone (chapter 8). In the United States, this area encompasses east to west more than 2,500 kilometers and 28° longitude (67° to 95°) and north to south more than 3,500 kilometers and 23° latitude (25° to 48°). The boreal forest alone circumscribes an even larger area. Within these geographic boundaries are five Koppen-Trewartha climate zones (Belda et al. 2014): tropical (Aw), humid subtropical (Cf), temperate continental with a warm summer (Dca), temperate continental with a cool summer (Dcb), and boreal (E). Plant hardiness zones (based on minimum winter temperatures on an 11-point scale) range from 2 in southern Florida to 10 in the northern boreal forest (Daly et al. 2012; McKenney et al. 2014).

This geographical and climatic range supports a tremendous diversity of ecoregions. Ecoregional classifications attempt to take largely continuously varying patterns of ecosystems across the landscape and divide them into discrete units, which are organized in a hierarchical scheme. The goal is to simplify the complexity of nature in a way that facilitates our understanding and its management. As a reference for this book, we use an ecoregional classification for North America developed jointly by Mexico, the United States, and Canada (CEC 2006), which operates at a fine scale (Level III) nested into progressively more coarse scales (Levels II and I). The old-growth ecosystems in this book occur within Tropical Wet Forests, Eastern Temperate Forests, Northern Forests, the Hudson (Bay) Plain, and Taiga (Level I). These five are divided into 15 ecoregions at Level II and then further into 61 at Level III. Plate 1 provides a map of the ecoregional classification for all three levels, with Level III denoted with three digits.

## Paleoecological Context

The Nature Conservancy's 5,000-acre Big Reed Reserve in northern Maine is one of the largest tracts of continuous old growth in New England (Barton et al. 2012, 68). The forest is a mix of spruce-fir, northern hardwoods, and northern white-cedar swamps. Sugar maples (*Acer saccharum*; figure 1-1), yellow birches (*Betula alleghaniensis*), and cedars (*Thuja occidentalis*) one meter in diameter are common. Scattered white pines tower above the canopy. Moss and lichens cover the lower trunks of trees. The forest floor is shady and moist, crisscrossed with decomposing dead wood, and effusive with fungi. Because large natural disturbances are rare, early-successional species, such as aspen (*Populus tremuloides* and *P. grandidentata*), pin cherry (*Prunus pensylvanica*), and white birch (*Betula papyrifera*) are hard to find, which contrasts sharply with the vast acreage of early- to mid-successional timberlands surrounding the reserve. Although the trees are not enormous, one senses that Big Reed is every bit the "forest primeval," forever unchanging.

Of course, we would be wrong. Twenty-thousand years ago, this area was under a continental ice sheet nearly two kilometers thick that stretched

FIGURE 1-1. Large sugar maple (*Acer saccharum*) in Big Reed Reserve old growth, owned and managed by The Nature Conservancy in Maine. Photo credit: A. M. Barton.

from the Arctic to New Jersey (Borns et al. 2004). About that time, astronomical cycles tilted the climate system toward warming, and, by 12,000 to15,000 years ago, across the long span of Maine, the ice sheet retreated, raw land was exposed, life recolonized, and the gradual process of primary succession began to build new forests (Barton et al. 2012, 29). Pollen embedded in the sediment layers on the bottom of ponds and in small, damp hollows in and near Big Reed tell the story of vegetation change since deglaciation (Anderson 1986; Schauffler and Jacobson 2002; Dieffenbacher-Krall and Nurse 2005; Barton et al. 2012, 79). The first well-established vegetative cover was open tundra with grasses and herbs, succeeding to open taiga, and eventually closed spruce forests. From about 9,000 to 7,000 years ago, during an unusually warm period (mid-Holocene Hypsithermal), rapid warming and drier conditions led to a decline in spruce and predominance of pines and oaks. Charcoal deposited in pond sediments during this time suggests that fires were common during this era, which might have amplified favorable conditions for these species. Charcoal fragments are rare in sediments and soil at Big Reed since that time, suggesting that fire has played very little role for a long time.

Moderate cooling then shifted the forest to hemlock (*Tsuga canadensis*), yellow birch, and beech (*Fagus grandifolia*). Beech was a laggard of sorts, migrating from southern glacial refugia into Maine long after other tree species with which it freely mixes today. A temperate northern hardwood-hemlock forest persisted until about 1,500 years ago, with one notable exception. About 5,400 years ago, hemlock suffered an abrupt decline, not just in Maine but across its entire range, the result possibly of severe drought, or a pest outbreak, or both. After about 1,000 years, hemlock recovered to its predecline abundance. Surprisingly, not until about 1,000 years ago did spruce reassert its place as one of the most abundant trees in Big Reed Reserve and all of interior Maine. If we use 25 years (the average age of sexual maturity for common tree species in the reserve) as generation time, then, the current old-growth forest at Big Reed Reserve has been around for a mere 40 generations.

Nearly 2,500 kilometers to the southwest lies Big Spring Pines Natural Area in Carter County, Missouri, which supports a unique 345-acre tract of fire-associated, old-growth forest of white oak (*Quercus alba*), black oak (*Q. velutina*), scarlet oak (*Q. coccinea*), and shortleaf pine (*Pinus echinata*) (Stambaugh et al. 2005). This area was well south of the ice sheet, and sediments from nearby Cupola Pond document 20,000 years of vegetation change in the area (Jones et al. 2017).

Those oldest sediments, when Maine was under ice, are chock full of pine and spruce pollen. It is not a coincidence that these species turn up in Maine some 6,000 to 8,000 years later, after warming, deglaciation, and a continental-scale migration. About 15,000 years ago, oaks and ashes appeared, and, by 12,500 years ago, spruces and pines had declined, oaks dominated, and hickories (*Carya* spp.), hornbeams (*Ostrya virginiana*), and ironwoods (*Carpinus caroliniana*) were common. Vegetation shifted rapidly about 10,000 years ago, supporting an even more temperate vegetation, as spruces disappeared and pines, ashes, and hornbeams declined, leaving a mixed oak forest. The old-growth pine-oak forest found at Big Spring Pines today did not become fully established until less than 2,000 years ago, similar to the case for Big Reed Reserve. A surprising finding in this study is that the forest communities predominating from about 17,000 to 10,000 years ago do not occur anywhere today—the so-called "no-analog communities."

We have focused on two geographically distant old-growth tracts, one subject to past glaciation and one not. We could pick any number of other old-growth forest sites in eastern North America, and, although the actors might differ, the paleoecological story would be largely the same. Clear ecological lessons emerge from these deep histories. Forests are ever-changing, sometimes fast and sometimes drastically. These changes are driven largely by climate, which itself fluctuates as a result of astronomical cycles and accompanying shifts in greenhouse gases and albedo (Alley 2014). Other factors, especially fire, also play key roles in shaping natural communities (Poulos 2015). The evidence suggests that extant associations of species are, for the most part, ephemeral alliances, which freely transform over centuries and millennia in response to constant environmental change. Forests do not migrate as a unit as the climate changes, but instead, each species behaves independently, in an "individualistic manner," as ecologists say.

The dynamism of old growth and the individualistic nature of species do not diminish the importance of extant old-growth forests. These are rare habitats that represent a key stage in the successional development of forests, ecosystems upon which species depend, and forests where the actions of nature have accrued over centuries. Paleoecology does warn us, however, to expect substantial shifts in species composition, structure, and processes in old growth into the future. Some of these shifts may be within our predictive grasp; others, such as no-analog communities, might not. As discussed in multiple chapters to follow, long-term plans for sustaining old growth must take into consideration the intrinsic dynamism of nature, especially in a changing world (Millar et al. 2007).


## Human Context

The history of the migration of humans from Siberia into North America is hotly debated (Bourgeon et al. 2017; Holen et al. 2017). The first people crossed into North America across the Bering Land Bridge at least 14,000 years ago (Fagan 2011), but new evidence pushes that date back to 24,000 years or earlier (Bourgeon et al. 2017). Archeological evidence suggests the arrival of people in the East long after that date (as the land was freed from the grip of the continental ice sheet), eventually reaching New England and the Maritimes, for example, about 12,000 to 13,000 years ago (Chapdelaine 2012).

These early Americans developed sophisticated cultures. They felled trees, cleared areas, and burned where and when benefits accrued from these activities. There is evidence that hunting pressure was sufficiently high that humans drove dozens of genera of late Pleistocene megafauna, which did not evolve with humans, to extinction in as little as 1,000 years (e.g., Surovell et al. 2016). Plant domestication occurred as early as 10,000 years ago in present-day southern Mexico, and agricultural societies developed throughout North America, with the earliest evidence for eastern North America from 3,800 years before present in Illinois (Smith and Yarnell 2009). Acknowledging uncertainty, Thornton (2000) estimated an aboriginal population of more than 7 million in North America, north of present-day Mexico. The evidence suggests, then, that these first Americans influenced the trajectories of plant communities in some places, but on the whole, eastern North America was largely covered by what today we would call old-growth forests.

The arrival of Europeans brought a profoundly different level of anthropogenic impact. From 1500 to today, nearly all of the forests in the eastern United States and southeastern Canada have been harvested or cleared for agriculture. This was a result of technology, culture, and population size, which rose in North America (north of present-day Mexico) from 6 million in 1800 to 81 million in 1900 to 322 million in 2000. In a comprehensive analysis of landscape change in the eastern United States, Steyaert and Knox (2008) estimated 70 percent of the original old growth remained in 1850, 7 percent in 1920, and an insufficient amount to include as a land-use category in 1992. From 1650 to 1920, canopy height and leaf area index also declined, although that trend reversed from 1920 to 1992, as eastern forests regrew. Mary Byrd Davis (1996, 2003) estimated that, of the 381 million acres of forest in the eastern United States, about 2 million acres (0.5 percent) remains as original forest. Big Reed Reserve in northern Maine is instructive in this context, as it is a 5,000-acre island of old growth in a sea of frequently harvested private timberlands.

The negative impacts on forests of the activities of an increasingly large and materialistic human culture were first fully articulated in North America in the late 1800s (e.g., Marsh 1864), and eventually a wave of land and species conservation, such as laws and national parks and forests, ensued. Not to suggest that the first half of the twentieth century was silent on nature, but an upwelling of concern for the environment comparable to the late 1800s and early 1900s did not occur again until the 1960s and 1970s with the rise of environmentalism, modern ecological science, extensive land conservation, and legislation aimed at sustaining species and natural systems. Table 1 provides a timeline of events relevant to old growth.

TABLE 1-1. Timeline of important events relating to old-growth forests in eastern North America with a focus on social and policy events in the United States.

| Time | Event | Comments |
|------|-------|----------|
| 21,000–25,000 years ago (YBP) | Maximum extent of Wisconsin Ice Sheet | Stretched to just south of New York City |
| >24,000 YBP | Arrival of humans in North America | |
| ~14,000 YBP | Arrival of humans in eastern North America | Coincident with deglaciation |
| 11,000 YBP | Deglaciation to modern levels | Varied across North America |
| 5,000–9,000 YBP | Mid-Holocene warm period | Warmer and drier conditions than before or after this period |
| 1492 | Columbus arrives in the Americas | |
| 1744 | Preindustrial atmospheric carbon dioxide reaches 276 ppm | |
| 1800 | Population of North America (north of Mexico) reaches 6 million | Less than population estimated for early North Americans (7 + million) |
| 1854 | Henry David Thoreau: *Walden* | Inspiration for environmentalism |
| 1864 | George P. Marsh: *Man and Nature* | Inspiration for environmentalism |
| 1866 | Ernst Haeckel coins word "ecology" | Beginnings of the discipline of ecology |
| 1890s | Forestry begins in the United States | Imported from Germany |
| 1890 | Sequoia and Yellowstone national parks created | First national parks in the United States |
| 1891 | Forest Reserve Act | Creation of public lands in West |
| 1892 | John Muir founds the Sierra Club | Influential environmental organization |
| 1897 | Forest Management Act | National forests for timber and protection of forest and watersheds |
| 1900 | Society of American Foresters formed | By Gifford Pinchot, a founder of forestry in the United States |

*continued on next page*

TABLE 1-1. *continued*

| Time | Event | Comments |
| --- | --- | --- |
| 1900 | Population of North America (north of Mexico) reaches 81 million | |
| 1904 | Chestnut blight first detected | New York City; decimated American chestnut |
| 1905 | Creation of the US Forest Service | Led by Gifford Pinchot |
| 1911 | Weeks Act | Authorized national forests in the East |
| 1915 | Atmospheric carbon dioxide reaches 300 ppm | |
| 1916 | National Park Service established | |
| 1920 | Beech-bark disease first detected in North America | Nova Scotia; spread through most of American beech's range |
| 1947 | Aldo Leopold: *Sand County Almanac* | Influential in scientific disciplines and conservation movements |
| 1951 | The Nature Conservancy formed | Today, the largest private owner of nature reserves in the world |
| 1951 | Hemlock woolly adelgid first detected in North America | Virginia; rapidly spreading, decimating hemlocks |
| 1960 | Multiple-Use Sustained-Yield Act | Multiple-use of and sustained-yield in national forests |
| 1964 | Wilderness Act | Establishment of wilderness areas |
| 1970 | First Earth Day | Continues to today |
| 1970 | National Environmental Policy Act | Environmental impact statements |
| 1973 | Endangered Species Act | Recovery of endangered species |
| 1976 | National Forest Management Act | Protection of species and ecosystems |
| 1970s | Andrews Experimental Forest, OR | Intensified old-growth research |
| 1989 | Successful lawsuit on the Northern Spotted Owl | Brought by the Seattle Audubon in the Pacific Northwest |
| 1989 | USFS New Perspectives initiative | Ecosystem management, sustainability |
| 1994 | Northwest Forest Plan | Insured old growth, species conservation, and harvesting |

*continued on next page*

## The Rise of Old-Growth Science

The modern sense of old growth arose as a confluence of environmental consciousness, ecological science, and forest management (see Hays 2009; Johnson and Govatski 2013). When most people think of old-growth forests, the redwood groves of California or the Douglas-fir forests of the Pacific Northwest come to mind. That is in part due to their grandeur and antiquity, but also because the modern old-growth movement emerged there in the 1980s. Those events were the culmination, however, of decades of shifting social perspectives.

From not long after the establishment of the United States until the mid-1850s, the federal government generally acted to shift forest ownership from the public to the private domain, and logging rapidly cleared most of the lands in the East, eventually moving westward. In response to an outcry about the abuse of forests, the Forest Reserve (or Creative) Act of 1891, the Forest Management (or Organic) Act of 1897, and the Weeks Act of 1911 enabled the retention and creation of federal lands, including national forests, with the goal of ensuring a continuous supply of timber and protection of watersheds.

Before World War II, there was relatively little logging on these public

TABLE 1-1. *continued*

| Time | Event | Comments |
|------|-------|----------|
| 1996 | Davis: *Eastern Old-Growth Forests* | Milestone in eastern old-growth research |
| 1990s | Emerald ash borer and Asian long-horned beetle first detected | EAB-Canada and Michigan, ash decimated; ALB-New York City |
| 2000 | Population of North America (north of Mexico) reaches 322 million | |
| 2014 | Atmospheric carbon dioxide reaches 400 ppm | Unprecedented in the past 10 million years |

SOURCES: National Park Service. 2017. "Conservation Timeline: 1800-2000." Accessed January 10, 2018. https://www.nps.gov/mabi/learn/historyculture/conservation-timeline-1801-1900.htm & 1901-2000.htm; Neftel, A., H. Friedli, E. Moor, H. Lotscher, H. Oeschger, U. Siegenthaler, and B. Stauffer. 1994. "Historical carbon dioxide record from the Siple Station ice core." Carbon Dioxide Information Analysis Center, University of Bern, Bern, Switzerland. Accessed January 20, 2018. http://cdiac.esd.ornl.gov/ftp/trends/co2/siple2.013; NOAA. 2018. "Trends in Atmospheric CO2, Mauna Loa, Hawaii." Earth System Research Laboratory, Global Monitoring Division. Accessed January 10, 2018. https://www.esrl.noaa.gov/gmd/ccgg/trends/full.html; see also references in the text.

lands, as timber companies could (and wished to) supply the demand from their own lands. After World War II, a rapidly growing economy and programs promoting the purchase of homes for the many returning GIs led to a surge in demand for wood. Forests on federal lands were quickly recruited to help meet that need, and the US Forest Service developed a sophisticated system of staff and methods to harvest wood, often using clear-cutting. In response to these new pressures, the Multiple-Use Sustained-Yield Act was passed in 1960, stipulating sustainable harvesting but also spelling out multiple allowed uses in national forests. The Wilderness Act of 1964 enabled one of those uses on any federal lands, as described in the following famous passage (Public Law 88-577 (16 U.S. C. 1131-1136)): "A wilderness, in contrast with those areas where man and his own works dominate the landscape, is hereby recognized as an area where the earth and its community of life are untrammeled by man, where man himself is a visitor who does not remain."

Three pieces of legislation in the 1970s directly set the stage for the rise of old growth as a national issue as well as the attendant science (Wilkinson and Anderson 1987).

- The National Environmental Policy Act of 1970 mandated the examination of environmental impacts for all federal land actions.
- The Endangered Species Act of 1973 legislated the evaluation and recovery of species in danger of extinction.
- The National Forest Management Act of 1976 required efforts to maintain the viability of species and the natural communities that support them on national forests.

These major laws went hand in hand with the growth of underlying ecological and forest sciences. Whole new ecological disciplines (and attendant societies and journals) focusing on these issues arose, including conservation biology in 1985 (Society for Conservation 2018) and landscape ecology in 1986 (Barrett et al. 2015). These legislative, management, and scientific developments each have their own evolutionary pathway, but they should be viewed as mutually reinforcing and an expression of changes in public views of nature and its management in the United States.

These roiling currents (table 1-1) came to a head in the Pacific Northwest in the 1980s. Employing new perspectives and techniques, researchers in national forests in that region had begun to overturn conventional wisdom, documenting the complexity of old-growth forests and the wildlife supported by these habitats. A successful lawsuit in 1989 required the US Forest Service to develop a comprehensive plan for protecting the Northern Spotted Owls and the old-growth habitat on which they depended. This issue was soon to attract the attention of the entire country, including President Clinton, culminating in the Northwest Forest Plan of 1994, a strategy for ensuring the protection of biodiversity as well as sustainable levels of harvesting. The plan continues to govern management in the region, with mixed success, but with the result that more old growth likely remains than would without the agreement (Spies and Duncan 2009).

Julie Wondolleck (2009) wrote that, "although understandably frustrating for all involved, the old-growth conflict has nonetheless . . . attracted widespread attention and illuminated the full array of values at stake . . . identified new questions and spotlighted the imperative for ongoing research." In fact, that imperative sparked a passion for discovery, conservation, and research on old-growth forests elsewhere, including in the eastern United States. Mary Byrd Davis published the first survey of old-growth forests for the eastern United States in 1993. The first conference on eastern old growth was held in the same year at the University of North Carolina at Asheville, followed by a symposium at Harvard Forest in Petersham, Massachusetts, and then a wider conference in 1994 in Williamstown, Massachusetts (Leverett 1996). An outgrowth of those conferences, *Eastern Old-Growth Forests: Prospects for Rediscovery and Recovery*, edited by Mary Byrd Davis, published in 1996, disseminated the groundbreaking work of 34 authors, who addressed the identification, values, preservation, and restoration of old-growth forests. This iconic book can be viewed as both an expression and an amplifier of ideas and initiatives started years earlier. The eastern region of the US Forest Service devoted significant resources toward old growth, culminating in major publications (e.g., Tyrrell et al. 1998). Similar initiatives regarding old growth occurred in Canada (Mosseler et al. 2003). The development of the science of old-growth forests during that time is reflected in the increase from 46 papers using the phrase "old-growth" in the ecological science literature between 1970 and 1980 to 2,089 between 1995 and 2005 (Wirth et al. 2009).

## This Book

The mission of this book is to disseminate a diverse sample of cutting-edge, old-growth research in eastern North America pursued over the past several decades, especially since the publication of *Eastern Old-Growth Forests* in 1996. Ecology as a science progressed dramatically during that time in many ways—methods, statistics, multidisciplinarity, commitment to the mission of addressing environmental problems, and more. New tracts of old growth have been discovered and protected, but the anthropogenic threats, such as climate change and invasive organisms, have only intensified. The contributions to this book offer an opportunity to take stock regarding how the science of old growth has advanced and responded to new opportunities and challenges.

The book begins geographically, moving south to north: bottomland hardwood forests (chapter 2) and pine savannas (chapter 3) in the South, a diversity of ecosystems in the southern Appalachians (chapter 4), mixed mesophytic and more xeric forests in the central Appalachians (chapter 5), natural forest dynamics and their application to silviculture in the northeast United States and southeastern Canada (chapter 6), northern hardwood-hemlock forests in the northern Lake States region (chapter 7), and the boreal forest in central Canada (chapter 8). These chapters examine specific old-growth forest types, reserves, or regions, integrating multiple aspects of the ecology of those forests and strategies for conservation.

The remainder of the chapters address issues relevant to most, if not all, old-growth forests in the East and beyond. Three chapters examine the interactions of the above-ground forest with other ecosystem components: streams (chapter 9), underground biota and processes (chapter 10), and biodiversity (chapter 11). The next three chapters explore old-growth forests in the context of global environmental change: the threat of nonnative, invasive organisms (chapter 12), the potential and challenges of employing silviculture to create old-growth structures (chapter 13), and the role of old growth in climate change mitigation (chapter 14). Finally, chapter 15 synthesizes key patterns, advances, and challenges, and proposes future directions for old-growth science in eastern North America.

This book is not meant to be comprehensive in terms of geography, forest types, anthropogenic stressors, or conservation strategies. Space allowed only a representative slice of the full range of those elements. Each chapter tells a compelling story that can be read independently without reference to the others. Collectively, however, the chapters reveal a great deal about old-growth forest ecology, conservation, and recovery in a world dominated by humans. Moreover, they raise compelling general questions about old-growth forests beyond the individual forest or region. To what extent can we make sense out of the tremendous ecological heterogeneity of old growth? To what extent are the many definitions and delineations of old growth compatible despite differences in underlying ecology, land-use history, forest-management constraints, and intellectual perspectives? Likewise, there are similarities, but also striking contrasts, in conservation challenges and potential solutions. To what extent can the issues facing one system provide guidance in another? Finally, what is the role of old-growth forests in a rapidly changing, human-dominated world? These sets of questions raise a central conundrum: Can we bridge the diversity among old-growth regions and ecosystems in such a vast region and develop a framework that illuminates underlying conceptions, ecological patterns, and conservation pathways, or does each ecosystem require independent consideration? We will return to these questions in the concluding chapter.

## Acknowledgements

Sincere thanks to Helen Poulos, whose comments greatly improved this chapter.

## References

Albanese, C. L. 1990. *Nature Religion in America: from the Algonkian Indians to the New Age*. Chicago: University of Chicago Press.

Alley, R. B. 2014. *The Two-Mile Time Machine: Ice Cores, Abrupt Climate Change, and Our Future*. Updated Edition. Princeton: Princeton University Press.

Anderson, R. S., R. B. Davis, N. G. Miller, and R. Stuckenrath. 1986. "History of late- and post-glacial vegetation and disturbance around Upper South Branch Pond, northern Maine." *Canadian Journal of Botany 64*: 1977–1986.

Barrett, G. W., T. L. Barrett, and J. Wu. 2015. *History of Landscape Ecology in the United States*. New York: Springer-Verlag.

Barton, A. M., A. S. White, and C. V. Cogbill. 2012. *The Changing Nature of the Maine Woods*. Durham, NH: University of New Hampshire Press.

Belda, M., E. Holtanová, T. Halenka, and J. Kalvová. 2014. "Climate classification revisited: from Köppen to Trewartha." *Climate Research 59*: 1–13.

Borns, Jr. H. W., L. A. Doner, C. C. Dorion, G. L. Jacobson Jr., M. R. Kaplan, K. J. Kreutz, T. V. Lowell, W. B. Thompson, and T. K. Weddle. 2004. "The deglaciation of Maine, U.S.A." In *Quaternary Glaciations—Extent and Chronology, Part II: North America*, edited by J. Ehlers, and P. L. Gibbard, 89–109. Amsterdam, The Netherlands: Elsevier.

Bourgeon, L., A. Burke, and T. Higham. 2017. "Earliest human presence in North America dated to the last glacial maximum: new radiocarbon dates from Bluefish Caves, Canada." *PLoS ONE 12*(1): e0169486. https://doi.org/10.1371/journal.pone.0169486.

Chapdelaine, C. 2012. *Late Pleistocene Archaeology and Ecology in the Far Northeast*. College Station, TX: Texas A&M Press.

Commission on Environmental Cooperation (CEC). 2006. *Ecological Regions of North America, Levels I-III*. Montreal, Canada: Commission for Environmental Cooperation.

Daly, C., M. P. Widrlechner, M. D. Halbleib, J. I. Smith, and W. P. Gibson. 2012. "Development of a new USDA plant hardiness zone map for the United States." *Journal of Applied Meteorology and Climatology 51*: 242–264.

Davis, M. B. 1993. *Old Growth in the East: A Survey*. Richmond, VT: Cenozoic Society.

Davis, M. B., ed. 1996. *Eastern Old-Growth Forests: Prospects for Rediscovery and Recovery*. Washington, DC: Island Press.

Davis, M. B. 2003. *Old Growth in the East: A Survey*. Revised edition. Georgetown, KY: Appalachia-Science in the Public Interest.

Dieffenbacher-Krall, A. M., and A. M. Nurse. 2005. "Late-glacial and Holocene record of lake levels of Mathews Pond and Whitehead Lake, northern Maine, USA." *Journal of Paleolimnology 34*: 283–310.

Fagan, B. M. 2011. *The First North Americans: an Archeological Journey*. New York: Thames and Hudson.

Grundmann, R. 2016. "Climate change as a wicked social problem." *Nature Geoscience 9*: 562–563. doi:10.1038/ngeo2780.

Hays, S. P. 2009. *The American People and the National Forests: the First Century of the U.S. Forest Service*. Pittsburg, PA: University of Pittsburg Press.

Holen, S. R., T. A. Deméré, D. C. Fisher, R. Fullagar, J. B. Paces, G. T. Jefferson, J. M. Beeton, et al. 2017. "A 130,000-year-old archaeological site in southern California, USA." *Nature 544*: 479–483.

Hunter M. L., and A. S. White. 1997. "Ecological thresholds and the definition of old-growth forest stands." *Natural Areas Journal 17*: 292–296.

Johnson, C., and D. Govatski. 2013. *Forests for the People: the Story of America's Eastern National Forests*. Washington, DC: Island Press.

Jones, R. A., J. W. Williams, and S. T. Jackson. 2017. "Vegetation history since the last glacial maximum in the Ozark highlands (USA): a new record from Cupola Pond, Missouri." *Quaternary Science Reviews 170*: 174–187.

Kelly, P. E., and D. W. Larson. 2007. *The Last Stand: a Journey Through the Ancient Cliff-face Forest of the Niagara Escarpment*. Toronto: Dundurn.

Leverett, R. 1996. "Extent and location." In *Eastern Old-Growth Forests: Prospects for Rediscovery and Recovery*, edited by M. B. Davis, 3–17. Washington, DC: Island Press.

Loomis, J., 2009. "Nontimber economic values of old-growth forests: what are they, and how do we preserve them?" In *Old Growth in a New World: a Pacific Northwest Icon Reexamined*, edited by T. A. Spies, and S. L. Duncan, 211–221. Washington, DC: Island Press.

Maloof, J. 2011. *Among the Ancients: Adventures in the Eastern Old-Growth Forests*. Washington, DC: Ruka Press.

Marsh, G. P. 1864. *Man and Nature, Or, Physical Geography as Modified by Human Action*. New York: Charles Scribner.

McKenney, D. W., J. H. Pedlar, K. Lawrence, P. Papadopol, K. Campbell, and M. F. Hutchinson. 2014. "Change and evolution in the plant hardiness zones of Canada." *BioScience 64*: 341–350. doi:10.1093/biosci/biu016.

Millar, C. I., N. L. Stephenson, and S. L. Stephens. 2007. "Climate change and forests of the future: managing in the face of uncertainty." *Ecological Applications 17*: 2145–2151.

Moore, K. D. 2007. "In the Shadow of the Cedars: the Spiritual Values of Old–Growth Forests." *Conservation Biology 21*: 1120–1123. https://doi.org/10.1111/j.1523-1739.2007.00669.x.

Mosseler, A., I. Thompson, and B. A. Pendrel. 2003. "Overview of old-growth forests in Canada from a science perspective." *Environmental Reviews 11* (S1): S1, S7. https://doi.org/10.1139/a03-018.

Pesklevits, A., P. Duinker, and P. Bush. 2011. "Old-growth forests: anatomy of a wicked problem." *Forests 2*: 343-356. doi:10.3390/f2010343

Poulos, H. 2015. "Fire in the northeast: learning from the past, planning for the future." *Journal of Sustainable Forestry 34*: 6–29.

Proctor, J. 2009. "Old growth and a new nature: ambivalence of science and religion." In *Old Growth in a New World: a Pacific Northwest Icon Reexamined*, edited by T. A. Spies, and S. L. Duncan, 104–115. Washington, DC: Island Press.

Rittel, H. W. J., and M. M. Webber. 1973. Dilemmas in a general theory of planning. *Policy Sciences 4*: 155–169. doi:10.1007/bf01405730.

Schauffler, M., and G. Jacobson. 2002. "Persistence of coastal spruce refugia during the Holocene in northern New England, USA, detected by stand-scale pollen stratigraphies." *Journal of Ecology 90*: 235–250.

Smith, B. D., and R. A. Yarnell. 2009. "Initial formation of an indigenous crop complex in eastern North America at 3800 B.P." *Proceedings of the National Academy of Sciences USA 106*: 6561–6566.

Society for Conservation Biology. "History." Accessed January 8, 2018. https://conbio.org /about-scb/who-we-are/scb-history.

Spies, T. A. 2004. "Ecological concepts and diversity of old-growth forests." *Journal of Forestry 102*: 14–20.

Spies, T. A., and S. L. Duncan., eds. 2009. *Old Growth in a New World: a Pacific Northwest Icon Reexamined*. Washington, DC: Island Press.

Stambaugh, M., R. Guyette, and C. Putnam. 2005. "Fire in the pines: a 341-year fire and human history at Big Spring Pines Natural Area, Ozark National Scenic Waterways." *Park Science 23*: 43–47.

Steyaert, L. T., and R. G. Knox. 2008. "Reconstructed historical land cover and biophysical parameters for studies of land-atmosphere interactions within the eastern United States." *Journal of Geophysical Research 113*. D02101, doi:10.1029/2006JD008277.

Surovell, T. A., S. R. Pelton, R. Anderson-Sprecher, and A. D. Myers. 2016. "Test of Martin's overkill hypothesis using radiocarbon dates on extinct megafauna." *PNAS 113*: 886–891. doi: 10.1073/pnas.1504020112.

Thornton, R. 2000. "Population history of North American Indians." In *A Population History of North America*, edited by M. R. Haines, and R. H. Steckel, Chapter 2. Cambridge, UK: Cambridge University Press.

Tyrrell, L. E., G. J. Nowacki, T. R. Crow, D. S. Buckley, E. A. Nauertz, J. N. Niese, J. L. Rollinger, and J. C. Zasada. 1998. *Information about Old Growth for Selected Forest Type Groups in the Eastern United States*. General Technical Report NC-197. North Central Forest Experiment Station. St. Paul, MN: USDA Forest Service.

Whitney, G. G. 1987. "Some reflections on the value of old-growth forests, scientific and otherwise." *Natural Areas Journal 7*: 92–99.

Wilkinson, C. F., and H. M. Anderson. 1987. *Land and Resource Planning in the National Forests*. Washington, DC: Island Press.

Wirth, C., G. Gleixner, and M. Heimann, eds. 2009. *Old-growth Forests: Function, Fate, Value*. Ecological Studies 207. Berlin: Springer-Verlag.

Wondolleck, J. 2009. "Old growth: evolution of an intractable conflict." In *Old Growth in a New World: a Pacific Northwest Icon Reexamined*, edited by T. A. Spies and S. L. Duncan, 176–188. Washington, DC: Island Press.

Chapter 2

# Old-Growth and Mature Remnant Floodplain Forests of the Southeastern United States

*Loretta L. Battaglia and William H. Conner*

Bottomland hardwood ecosystems (BLH) occupy the floodplains of low-gradient streams and rivers in the Atlantic and Gulf Coastal Plain Provinces of the southeastern United States (figure 2-1; King et al. 2012). Although they make up a relatively small portion of the landscape compared to their upland counterparts, healthy BLH forests, particularly ones with seasonal flooding and flow, have high productivity (Conner and Day 1976; Megonigal et al. 1997). They provide necessary habitat for wildlife and numerous other species and are thus critical for supporting biodiversity in the region (Sharitz and Mitsch 1993; Conner and Sharitz 2005). These forests are also responsible for the provisioning of many ecosystem goods and services, including flood control, maintenance of water quality, recreation, and nursery habitat for commercial fisheries, to name a few (King et al. 2009).

## Hydrology

Hydrology is the primary factor that drives BLH forest structure and function (Wharton et al. 1982). The timing, depth, and duration of flooding depend on inputs of snowmelt, rainfall, and groundwater, which can vary considerably across the historic geographical range of BLH ecosystems. Degree of tolerance to the stress of flooding and associated anaerobic conditions determines, in part, which species can regenerate, grow, survive, and reproduce in the floodplain (McKnight et al. 1981; Battaglia et al. 2004). Thus, flooding is a constraining driver of community composition, but at the same time, flood pulses also provide intimate ecological connections be-

FIGURE 2-1. Map illustrating the historical extent of bottomland hardwood forest flanking rivers and streams in the southeastern United States. The Fall Line delineates the northern boundary of the Coastal Plain. The largest contiguous expanse of bottomland hardwood forests occurred in the Mississippi Alluvial Valley. Figure created by William Sipek.

tween floodplains and rivers that maintain transport of propagules; govern exchange of sediment, organic matter, and nutrients; and create ephemeral nursery habitat necessary for many aquatic and semi-aquatic species (Junk et al. 1989). Where these connections are severed through anthropogenic activities, such as flood control structures, the floodplains become degraded and exhibit shifts in composition that reflect hydrologic changes.

Hydroperiods can also differ among and within communities at smaller scales as floodwaters create, destroy, and interact with fluvial geomorphological features, such as terraces, creating spatiotemporal heterogeneity in habitat characteristics and patterning across floodplain landscapes (Hupp 2000). Differences in hydroperiod are underlain by the resultant slight elevation gradient that produces much of the variation in hydrologic conditions and, in turn, patterns in plant composition. Indeed, there can be an almost complete turnover of species composition with elevation changes often less than a meter (Wharton et al. 1982; Sharitz and Mitsch 1993). The elevation gradient may unfold over broader spatial scales across the floodplain and/or at small scales in the form of pit and mound topography, elevated pneumatophores, or downed logs. The microtopographic variation creates differences in flooding depth, frequency, and duration, which in turn influence regeneration, scale of seedling recruitment patterns, and community composition (Huenneke and Sharitz 1986; Battaglia et al. 2000; Collins and Battaglia 2002).

## Canopy Disturbances

Common in bottomland forests, canopy disturbances range in size from single treefalls (King and Antrobus 2001) to largescale blowdowns from tornadoes (Nelson et al. 2008) and hurricanes (Putz and Sharitz 1991). They generate microsite variation on the forest floor and simultaneously open the canopy, modifying the amount and pattern of light availability (Battaglia et al. 1999). Light availability is often secondary to hydrology as a driver but, nevertheless, a pivotal determinant of successful seedling regeneration and trends in forest dynamics (Hall and Harcombe 1998). Canopy gaps and blowdowns are often sites that support the establishment of shade-intolerant species, such as loblolly pine (*Pinus taeda*) and sweetgum (*Liquidambar styraciflua*) (Battaglia and Sharitz 2006), but higher light availability in these areas can also promote recruitment of individuals in the advance regeneration layer, such as oaks (*Quercus* spp.), into upper forest strata (Collins and Battaglia 2008).

## History

These forests once occupied vast areas along the floodplains of streams and rivers in the southeastern United States (Sharitz and Mitsch 1993), but less than 25% of the original four million hectares remain (Lockaby 2009). Early European settlers viewed these forests as impenetrable barriers to expansion. This dark, foreboding wilderness was unlike anything they had ever seen, considering that the forests in Europe had been cleared centuries before (Leverett 1996). During the 1600s and 1700s, millions of hectares of forestland were cut and burned. During the 1800s, timber barons realized there was a huge profit to be made from the remaining virgin forests, and by the 1920s, most of the original forests had been cut (Leverett 1996; Fredrickson et al. 2005). Much of what remains is fragmented and degraded, and the few remaining undisturbed tracts are of great ecological and societal significance. Massive habitat loss of this ecosystem contributed to the extinction of the Carolina Parakeet and likely to that of the Ivory-billed Woodpecker, as well as to the near extinction of many other animal and plant species.

## Vegetation Description

Today, BLH forests occupy approximately 12 million hectares and contain about one-third of the hardwood volume in the southern region. Early successional habitats near rivers include species such as cottonwood (*Populus deltoides*) and black willow (*Salix nigra*.), which often grade into stands of sweetgum, water oak (*Quercus nigra*), and green ash (*Fraxinus pennsylvanica*). In mixed stands on the ridges—slight rises of a few centimeters to one meter above a surrounding flat—species such as cherrybark oak (*Quercus pagoda*), hackberry (*Celtis laevigata*), American beech (*Fagus grandifolia*), water oak, hickories (*Carya* spp.), and sweetgum are typical. Blackgum (*Nyssa sylvatica*), white oak (*Quercus alba*) and other upland hardwoods sometimes occur in the highest and driest areas.

Southern baldcypress (*Taxodium distichum*) is often found along with hardwoods in bottomlands where water is too deep for competitive species (Dennis 1988). This species generally occurs in pure, dense, even-aged stands. In some stands, especially where cypress was historically removed, it may codominate with water tupelo (*Nyssa aquatica*).

Where isolated baldcypress trees are found mixed with hardwood species, a change in the hydrology of the site, such as channel alteration during the life of the stand, has probably occurred.

## Old-Growth Bottomland Hardwood Forests: Congaree National Park

The Congaree National Park is the largest and one of the best examples of old-growth BLH forest in the southeastern United States (Davis 1996). Tributaries of the Congaree River, which nourish this floodplain, originate in the Piedmont physiographic province, and, thus, it is a red river system, as opposed to a blackwater one that begins in the Coastal Plain. Approximately 4,450 hectares of the park are regarded as old-growth forest that escaped intensive or extensive land-use activities by humans (Doyle 2009; Gaddy 2012), and they harbor several state and national champion-sized trees, several of which exceed 40 meters in height (Jones 1997; https://www.nps.gov/cong/index.htm.). The old-growth forest itself is quite diverse structurally, with uneven-aged stands, treefalls, misshapen tree canopies, and standing dead snags that reflect the forest's disturbance history (Battaglia et al. 1999) and senescence of old trees. Vegetation on the floodplain comprises a mosaic of successional states but also closely tracks hydrologic features. The drier end of the elevation gradient that supports bottomland hardwoods is where the expanse of old growth occurs. With less frequent flooding, here the forest is compositionally diverse and contains several oak species, loblolly pine, hackberry, red maple (*Acer rubrum*), green ash, and sweetgum in the canopy. American holly (*Ilex opaca*) and pawpaw (*Asimina triloba*) are common in the subcanopy and shrub strata. Within the park, baldcypress was selectively logged between 1895 and 1910 (Rikard 1988). Older baldcypress trees remain, but they are rare and isolated.

The operation of the Saluda Dam, 78 kilometers upstream of the park, has reduced the magnitude, frequency, and duration of high flood events since its construction in 1930 (Patterson et al.1985). Abrupt changes in the flood regime of a swamp environment—drier or wetter—can lead to abrupt change in growth responses of flood-tolerant trees (Conner and Day 1992; Young et al. 1995) and, over time, shifts toward species composition more aligned with the new hydroperiod. Unfortunately, long-term studies within floodplain forest ecosystems have been limited in duration and scope (Jones et al. 1994; Bell 1997; Hodges 1997; Harcombe et al. 2002; Conner et al. 2011; Conner et al. 2014), and, therefore, it remains unclear how hydrologic changes have and will impact long-term forest dynamics.

Hurricanes have played a prominent role in structuring the floodplain forest of the Congaree (Putz and Sharitz 1991; Allen et al. 1997; Battaglia et al. 1999). The dominant canopy species in this forest are sweetgum, loblolly pine, and bottomland oaks. Because these species are relatively shade-intolerant, their reestablishment primarily occurs in areas heavily disturbed by hurricanes, such as Hurricane Hugo in 1989 (Zhao et al. 2006). Hurricane Hugo disproportionately removed some late successional species, such as cherrybark (*Q. pagoda*) and Shumard oaks (*Q. shumardii*), changing the forest from a "cathedral-like park appearance" (R. R. Sharitz, pers. comm.) to one with patches of heavy damage and a tangle of early successional vegetation (e.g., southern dewberry [*Rubus trivialis*]) in the years following the storm (Battaglia and Sharitz 2006).

## Old-Growth Baldcypress Forests

Baldcypress swamps with trees 500 to over 1,500 years old are among the most notable old-growth forests left in the southeastern United States (Stahle 1996). Most surviving old-growth baldcypress stands have been selectively cut, and the remaining forests are dominated by trees deemed architecturally or otherwise defective and, therefore, not harvested. The oldest documented baldcypress stand in the southern United States is located along the Black River, a tributary of the Cape Fear River in southeastern North Carolina. Living baldcypress trees ranging up to 1,700 years old have been discovered at this site (Stahle et al. 1988). In southern Louisiana and Mississippi, trees up to 1,300 years old have been found along the Pearl River drainage. Such ancient trees are rare, but it is not uncommon to find small baldcypress stands that are hundreds of years old (Stahle et al. 1992).

Only three virgin baldcypress forests are known to remain in the southeastern United States: (1) the National Audubon Society Sanctuary at Four Holes Swamp, South Carolina, (2) Corkscrew Swamp, Florida, and (3) a private parcel of old growth baldcypress in southwestern Arkansas. These three tracts only total approximately 2,400 hectares out of about 16 million hectares of virgin baldcypress forest estimated to have originally existed in the South (Stahle et al. 2006). Fortunately, a few other stands of ancient baldcypress survived the era of massive timber cutting and agricultural land clearing. Two important areas are the old-growth baldcypress woodlands along the Black River in North Carolina, and the selectively logged old-growth baldcypress-tupelo forests of Bayou DeView, Arkansas, which is the 2005 "rediscovery" site (though never confirmed) of the critically endangered and most likely extinct Ivory-billed Woodpecker (Fitzpatrick et al. 2005).

The Four Holes Swamp site is the largest of the remaining old-growth baldcypress-tupelo forests (Porcher 1981). This roughly 720-hectare fragment is located in the Coastal Plain of South Carolina and is part of a moving "brown water" system nourished by small creeks and streams that meet and split on the floodplain without a discernible main channel. The old-growth forest canopy is dominated by baldcypress trees, some of which exceed 36 meters in height. Water tupelo makes up the roughly 25-meter tall forest subcanopy. Small trees (e.g., planera tree [*Planera aquatica*]), shrubs (e.g., Virginia sweetspire [*Itea virginica*]), and a rich herbaceous flora occur but are relegated to slightly elevated, drier microsites. The canopy and subcanopy species occur in all size classes, from regenerating seedlings to saplings and large trees, suggesting uneven aged stands.

Proximity to the coast means this forest is often in the path of tropical storms such as Hurricane Hugo, a powerful Category 4 storm when it made landfall nearby in coastal South Carolina. As in other wetland forests of this type, damage in swamp forests was mostly limited to small branch loss and defoliation (Putz and Sharitz 1991), whereas pines and bottomland oaks in higher and drier parts of the floodplain suffered greater damage and mortality. In addition, fire scars and some charring indicate occasional fire disturbances, presumably when the swamp dries during droughts.

## Notable Mature Bottomland Hardwood Forests

In southern Illinois at the northern terminus of the BLH range (Robertson et al. 1978), there are abundant floodplains that flank major rivers and streams. Many have been highly degraded, but there remain some remarkable examples of BLH forest. The Cache River floodplains are not old growth but do support internationally recognized wetlands, exemplary forests that have been conserved, and others that are actively being restored. This system harbors restored stands of different ages, fragments of mature BLH stands (McLane et al. 2012) and isolated old (and perhaps ancient) baldcypress trees in sloughs and along the Cache River and her tributaries.

Many of the forests in this system have been cut several times and converted to agriculture. In the interest of restoring forest cover and functionality to the system, as well as conserving soil, restoration of this system has been ongoing for several decades. Despite the disturbances and degradation, this system is a wonderful regional example of a floodplain landscape with elements of old stands, young stands, and others in between. Today, there are many parcels of land along the Cache River in the broader floodplain where restoration activities have now woven together a connected system. The Cache River and associated wetlands are arguably the best of what is left in a landscape where there have been largescale losses of floodplain wetlands and severe hydrologic changes (figure 2-2).

FIGURE 2-2. *Top*: Cache River wetlands in southern Illinois. Baldcypress trees dominate the canopy of Heron Pond. Native rose shrubs are also visible in the understory. *Right*: Mature baldcypress forest in one of the swamps along the Cache River. Photo credit: L. L. Battaglia.

Heron Pond is an iconic part of southern Illinois and a piece of a wetland complex recognized by the Ramsar Convention on wetlands as one of international significance (figure 2-2). The forest occupies a shallow depression and supports baldcypress forest in the canopy, with small shrubs such as native swamp rose (*Rosa palustris*) occurring on elevated microsites within the swamp. There are also several floating aquatic species, including duckweed (*Lemna minor*), that form the floating understory in this forest. Nearby ridges are dominated by typical BLH species such as sweetgum, ash, hackberry, and perhaps most notably, cherrybark oak. The state champion of the latter is on a ridge near Heron Pond.

The baldcypress state champion can be seen while paddling out into

one of the swamps along the Cache River (figure 2-2). Some of the isolated trees in these sloughs and backswamps are old, even though most of this species was harvested. The pockets that escaped are old, but it is difficult to get an accurate age of these trees because their centers are rotten. Nevertheless, this system is a regional standout and an important site for wetland conservation and restoration.

The Big Woods area in the Barataria Preserve Unit of Jean Lafitte National Historical Park and Preserve is south of New Orleans, Louisiana. This preserve contains an exceptional example of coastal BLH forest in the Mississippi Alluvial Valley (figure 2-1). Although not technically old growth, this forest is mature, relatively old (about 175 to 200 years old, with some isolated trees that may be older) and important to include here because it is in the southernmost part of the Lower Mississippi Alluvial Valley (LMAV). With all the characteristics described earlier for other BLH, it is also subject to rapid relative sea level rise (Denslow and Battaglia 2002), a common stressor in the LMAV. As forests in these areas disassemble, they may convert to open water or alternatively transition more slowly to marsh. These coastal floodplains also experience frequent tropical storms and hurricanes, including associated saline storm surges.

Community composition is closely aligned with elevation. The drier and higher portion of the floodplain is the natural levee of Bayou des Familles (figure 2-3a). The species that are least flood tolerant are located here, including the iconic live oak (*Quercus virginiana*), which is not regenerating (Denslow and Battaglia 2002). Moving into wetter areas, (i.e., down the elevation gradient), there is a blend of bottomland species, including red maple (*Acer rubrum*), sweetgum, green ash, hackberry, American elm (*Ulmus americana*) and several others. In the flooded backswamp, there is a mixture of baldcypress and pumpkin ash (*Fraxinus profunda*).

Overall, seedling regeneration is quite limited in this system, which is thought to be due in part to increased duration of flooding (Denslow and Battaglia 2002). Another factor may be the dominant dwarf palmetto (*Sabal minor*) that occurs in great abundance throughout most of the gradient (figure 2-3a). In the wetter end, the floating understory is dominated by water hyacinth (*Eichhornia crassipes*) and *Salvinia* spp., two noxious species native to South America (figure 2-3b).

Despite vulnerability to the inexorable change due to relative sea level rise (Shirley and Battaglia 2006) and the invasive species problems, this forest supports a wide array of biodiversity, with numerous plant (Wall and Darwin 1999) and wildlife species and abundant habitat for resident and migratory birds. The forest has great structural diversity and strange architecture in some of the tree canopies, which is undoubtedly a testament to branch loss and canopy damage during previous storm events. There are many trees that have been toppled and a number of pit and mound regeneration sites that provide isolated, protected pools for various amphibian species to reproduce without fish predators and elevated safe sites for seedling regeneration. Coarse woody debris is common on the forest floor, but it decomposes more readily in the drier end of the gradient that experiences periodic flooding. However, in the semi-permanently to permanently flooded backswamp that only dries out during extreme regional droughts, debris often remains submerged and does not decompose due to anaerobic conditions.

FIGURE 2-3. Forest at Barataria Preserve Unit, Jean Lafitte National Historical Park and Preserve, Jefferson Parish, Louisiana. (a) Taken in the forest on the natural levee of Bayou des Familles where dwarf palmetto (*Sabal minor*) is pervasive in the understory. Large live oaks (*Quercus virginiana*) like the one in the photo are flood intolerant and limited to the highest point on the floodplain. (b) The backswamp is heavily invaded by water hyacinth (*Eichhornia crassipes*) and *Salvinia* spp., two noxious floating invaders. Photo credit: L. L. Battaglia.

## Young Versus Old Bottomland Hardwood Forests

Younger secondary- and tertiary-growth bottomland forests often differ substantially in comparison to mature or old-growth BLH forests. For example, many of these younger forests are even-aged because they developed following cessation of agricultural activity (Battaglia et al. 1995) or largescale clearcutting. These even-aged cohorts typically lack structural diversity and other features essential as habitat for wildlife and their prey, such as standing dead snags and downed coarse woody debris. Also, these forests are generally depauperate in species richness where the landscape is highly fragmented and the species pool is limited by distance to seed source. These forests are also usually devoid of features common in old-growth forests, such as pit and mound microsites, that promote diversity and provide safe sites for seedling regeneration as well as amphibian reproduction (chapters 10 and 11).

Some of these younger forests that have developed after years of agricultural activity may have long-lived seedbanks containing agricultural weeds that emerge and compete with germinating or planted bottomland hardwood seedlings. Moreover, these disturbed sites are susceptible to invasions of exotic species.

Younger forests may not provide the same complement of ecosystem goods and services, such as nutrient cycling, recreation, and hunting (chapters 1, 10, and 14). Old-growth and mature BLH forests generally provide more structural and compositional diversity and, consequently, habitat heterogeneity, for a wider array of wildlife and other species (chapter 11). Other elements that often take time to form and may be lacking include mycorrhizal associations and microbial networks (chapter 10). Where feasible, connecting young fragments to mature forests on the landscape (hydrologically and/or spatially) will be beneficial to promote structural and functional restoration.

## Conclusion

Fragments of old-growth and immature BLH forests are vulnerable to many threats in the surrounding modern landscape. Surrounding land use can influence inputs of nutrients, sediments, and pollutants, particularly in agricultural watersheds where runoff is a problem. Although BLH forests are good at filtering excess nutrients and other contaminants, there is a point of saturation beyond which they can no longer effectively filter or uptake. Beyond that point, these systems can become degraded and experience shifts in composition and loss of function, which can have a big impact on areas farther downstream. A prime example is the hypoxic zone in the Gulf of Mexico, which is an ever-growing reminder of the linkage between land use, eutrophication, water quality, loss of wetlands, and river flood-plain connectivity (Rabalais et al. 2010).

Bottomland forests have high productivity and biodiversity, and the very conditions that support these attributes make them also suitable for establishment and spread of exotic species (chapter 12). There are numer-ous invasive species that pose a threat to remaining forests and fragments. On the drier end of the elevation gradient, species such as Chinese tallow (*Triadica sebifera*) and Chinese privet (*Ligustrum sinense*) can establish (Denslow and Battaglia 2002). Where disturbances open the canopy, these species easily invade, displacing native bottomland species, and, then, it is difficult to eradicate or even control their populations. In swamp commu-nities that are semi-permanently or permanently flooded there are a num-ber of noxious floating and submerged aquatic species that have become problematic, including water hyacinth and *Salvinia* spp.

Historically, many of these bottomlands were drained and converted to agriculture. While this type of conversion has slowed, if it does not cease, many parts of the BLH will be further degraded. Because these forests are so fragmented, small ones often have a lot of edge surrounding them. Edge habitat is vulnerable to exotic invasions and other negative aspects of the urban-wildland interface, and therefore often represents a degraded form of the more intact core habitat (chapter 12). The smaller the fragment, the more susceptible it is to surrounding landscape influences. In addition, hydrologic changes associated with channel straightening, flood control structures, and levees can quickly change and further degrade these flood-plain ecosystems.

BLH forests are also threatened by climate change. In coastal land-scapes, sea level rise and tropical storm surges present the additional stress of rising waters and saline intrusions into freshwater systems not adapted to these pulses (Visser et al. 2012; Krauss et al. 2009). More broadly, changes in precipitation regimes and temperature are likely to have many conse-quences for the composition and functioning of these ecologically valuable forests across their entire range.

Although restoring hydrologic regimes is not necessarily feasible in many cases, restoring hydrologic *connectivity*, where that linkage has been severed, may be possible and is likely to improve functionality. Within the constraints of a particular site, it may be possible to restore some semblance of a natural hydroperiod, including flood pulsing, which is essential for many species to regenerate. For example, artificially flooding BLH stands all the time may lead to regeneration failure over time and loss of forest diversity.

Restoration of species that are known to be dispersal-limited should be high on the list for reestablishment through seeding or planting. Passive afforestation often does not work for heavy-seeded, late-successional species if a nearby seed source is lacking (Battaglia et al. 2002; Battaglia et al. 2008). Hickories (*Carya* spp.) and oaks, important for many wildlife species, are often lacking in these early successional bottomlands (Battaglia et al. 1995) and potentially could be restored through such active restoration efforts.

With respect to climate change, connecting forest fragments together with corridors may be useful for facilitating migration of species responding to shifting climate conditions. Finally, conservation and protection of these precious old-growth and mature forests, which will likely provide ecological resilience for a wide range of species, will be essential to preserve biodiversity and ecosystem function.

## Acknowledgements

The authors wish to thank Sammy King for comments on a previous draft of this chapter. This material is based upon work supported in part by NIFA/USDA, under project number SC-1700531. Technical Contribution No. 6632 of the Clemson University Experiment Station.

## References

Allen, B. P., E. F. Pauley, and R. R. Sharitz. 1997. "Hurricane impacts on liana populations in a southeastern bottomland forest." *Journal of the Torrey Botanical Society 124*: 34–42.

Battaglia, L. L., B. S. Collins, and R. R. Sharitz. 2004. "Do published tolerance ratings and dispersal factors predict species distributions in bottomland hardwoods?" *Forest Ecology and Management 198*: 15–30.

Battaglia, L. L., S. A. Foré, and R. R. Sharitz. 2000. "Seedling emergence, survival and size in relation to light and water availability in two bottomland hardwood species." *Journal of Ecology 88*: 1041–1050.

Battaglia, L. L., J. R. Keough, and D. W. Pritchett. 1995. "Early secondary succession in a southeastern U.S. alluvial floodplain." *Journal of Vegetation Science 6*: 769–776.

Battaglia, L. L., P. R. Minchin, and D. W. Pritchett. 2002. "Sixteen years of old-field succession and reestablishment of a bottomland hardwood forest in the Lower Mississippi Alluvial Valley." *Wetlands 22*: 1–17.

Battaglia, L. L., P. R. Minchin, and D. W. Pritchett. 2008. "Evaluating dispersal limitation in passive bottomland forest restoration." *Restoration Ecology 16*: 417–424.

Battaglia, L. L., and R. R. Sharitz. 2006. "Response of floodplain forest species to spatially condensed gradients: a test of the flood-shade tolerance tradeoff hypothesis." *Oecologia 147*: 108–118.

Battaglia, L. L., R. R. Sharitz, and P. R. Minchin. 1999. "Patterns of seedling and overstory composition along a gradient of hurricane disturbance in an old-growth bottomland hardwood community." *Canadian Journal of Forest Research 29*: 144–156.

Bell, D. T. 1997. "Eighteen years of change in an Illinois streamside deciduous forest." *Journal of the Torrey Botanical Society 124*: 174–188.

Collins, B. S., and L. L. Battaglia. 2002. "Microenvironmental heterogeneity and *Quercus michauxii* regeneration in experimental canopy gaps." *Forest Ecology and Management 155*: 281–292.

Collins, B. S., and L. L. Battaglia. 2008. "Oak regeneration in bottomland hardwood forests." *Forest Ecology and Management 255*: 3026–3034.

Conner, W. H., and J. W. Day, Jr. 1976. "Productivity and composition of a baldcypress-water tupelo site and a bottomland hardwood site in a Louisiana swamp." *American Journal of Botany 63*: 1354–1364.

Conner, W. H., and J. W. Day, Jr. 1992. "Diameter growth of *Taxodium distichum* (L.) Rich. and *Nyssa aquatica* L. from 1979-1985 in four Louisiana swamp stands." *American Midland Naturalist 127*: 290–299.

Conner, W. H., J. A. Duberstein, J. W. Day, Jr., and S. Hutchinson. 2014. "Impacts of changing hydrology and hurricanes along a flooding/elevation gradient in a south Louisiana forested wetland from 1986-2009." *Wetlands 34*: 803–814. 10.1007/s13157-014-0543-0.

Conner, W. H., and R. R. Sharitz. 2005. "Forest communities of bottomlands." In *Ecology and Management of Bottomland Hardwood Systems: The State of Our Understanding*, edited by L. H. Fredrickson, S. L. King, and R. M. Kaminski. Gaylord Memorial Laboratory Special Publication No. 10. Puxico, MO: University of Missouri-Columbia.

Conner, W. H., B. Song, T. M. Williams, and J. T. Vernon. 2011. "Community structure and aboveground productivity in a longleaf pine-swamp blackgum forest drainage, South Carolina, USA." *Journal of Plant Ecology 4*: 67–76.

Davis, M. B., ed., 1996. *Eastern Old-Growth Forests: Prospects for Rediscovery and Recovery*, 383. Washington, DC: Island Press.

Dennis, J. V. 1988. *The Great Cypress Swamps*, 142. Baton Rouge, LA: Louisiana State University Press.

Denslow, J. S., and L. L. Battaglia. 2002. "Stand composition and structure across a changing hydrologic gradient: Jean Lafitte National Park." *Wetlands 22*: 738–752.

Doyle, T. W. 2009. "Modeling floodplain hydrology and forest productivity of Congaree Swamp, South Carolina." US Geological Survey Scientific Investigations Report 2009-5130: 46.

Fitzpatrick, J. W., M. Lamertink, M. D. Luneau, Jr., T. W. Gallagher, B. R. Harrison, G. M. Sparling, K. V. Rosenberg, et al. 2005. "Ivory-billed Woodpecker (*C. principalis*) persists in continental North America." *Science 308*: 1460–1462.

Fredrickson, L. H., S. L. King, and R. M. Kaminski, eds. 2005. *Proceedings of the Symposium on Ecology and Management of Bottomland Hardwood Systems: the State of our Understanding*. Columbia, MO: University of Missouri Press.

Gaddy, L. L. 2012. *The Natural History of Congaree Swamp*, 108. Columbia, SC: Terra Incognita Books.

Hall, R. B. W., and P. A. Harcombe. 1998. "Flooding alters apparent position of floodplain saplings on a light gradient." *Ecology 79*: 847–855.

Harcombe, P. A., C. J. Bill, M. Fulton, J. S. Glitzenstein, P. L. Marks, and I. S. Elsik. 2002. "Stand dynamics over 18 years in a southern mixed hardwood forest, Texas, USA." *Journal of Ecology 90*: 947–957.

Hodges, J. D. 1997. "Development and ecology of bottomland hardwood sites." *Forest Ecology and Management 90*: 117–125.

Huenneke, L. F., and R. R. Sharitz. 1986. "Microsite abundance and distribution of woody seedlings in a South Carolina cypress-tupelo swamp." *American Midland Naturalist 115*: 328–335.

Hupp, C. R. 2000. "Hydrology, geomorphology, and vegetation of Coastal Plain rivers in the southeastern United States." *Hydrological Processes 14*: 2991–3010.

Jones, R. H. 1997. "Status and habitat of big trees in Congaree Swamp National Monument." *Castanea 62*: 22–31.

Jones, R. H., R. R. Sharitz, S. M. James, and P. M. Dixon. 1994. "Tree population dynamics in seven South Carolina mixed-species forests." *Bulletin of the Torrey Botanical Club 121*: 360–368.

Junk, W. J., P. B. Bayley, and R. E. Sparks. 1989. "The flood pulse concept in river-floodplain systems." In *Proceedings of the International Large River Symposium*, edited by D. P. Dodge. *Can. Spec. Publ. Fish. Aquat. Sci. 106*: 110–127. Ottawa: Department of Fisheries and Oceans.

King, S. L., and T. J. Antrobus. 2001. "Canopy disturbance patterns in a bottomland hardwood forest in northeast Arkansas, USA." *Wetlands 21*: 543–553.

King, S. L., L. L. Battaglia, C. R. Hupp, R. F. Keim, and B. G. Lockaby. 2012. "Floodplain wetlands of the southeastern Coastal Plain." In *Wetland Habitats of North America: Ecology and Conservation Concerns*, edited by D. Batzer, and A. Baldwin, 253–266. Berkeley, CA: University of California Press.

King, S. L., R. R. Sharitz, J. W. Groninger, and L. L. Battaglia. 2009. "Ecology, restoration, and management of floodplains: a synthesis." *Wetlands 29*: 624–634.

Krauss, K. W., J. A. Duberstein, T. W. Doyle, W. H. Conner, R. H. Day, L. W. Inabinette, and J. L. Whitbeck. 2009. "Site condition, structure, and growth of baldcypress along tidal/non-tidal salinity gradients." *Wetlands 29*: 505–519.

Leverett, R. 1996. "Definitions and history." In *Eastern Old-Growth Forests: Prospects for Rediscovery and Recovery*, edited by M. B. Davis, 3–17. Washington, DC: Island Press.

Lockaby, B. G. 2009. "Floodplain ecosystems of the Southeast: linkages between forests and people." *Wetlands 29*: 407–412.

McKnight, J. S., D. D. Hook, O. G. Langdon, and R. L. Johnson. 1981. "Flood tolerance and related characteristics of trees of the bottomland forests of the southern United States." In *Wetlands of Bottomland Hard-wood Forests*, edited by J. R. Clark, and J. Benforado, 29–69. New York: Elsevier.

McLane, C. R., L. L. Battaglia, D. J. Gibson, and J. G. Groninger. 2012. "Succession of exotic and native species assemblages across a chronosequence of restored floodplain forests: a test of the parallel dynamics hypothesis." *Restoration Ecology 20*: 202–210.

Megonigal, J. P., W. H. Conner, S. Kroeger, and R. R. Sharitz. 1997. "Aboveground production in southeastern floodplain forests: a test of the subsidy-stress hypothesis." *Ecology 78*: 370–384.

Nelson, J. L., J. W. Groninger, L. L. Battaglia, and C. M. Ruffner. 2008. "Bottomland hardwood forest vegetation and soils recovery following a tornado and salvage logging." *Forest Ecology and Management 256*: 388–395.

Patterson, G. G., G. K. Speiran, and B. H. Whetstone. 1985. "Hydrology and its effect on distribution of vegetation in Congaree Swamp National Monument, South Carolina." US Geological Survey Water-Resources Investigations Report 85-4256.

Porcher, R. D. 1981. "The vascular flora of the Francis Beidler Forest in Four Holes Swamp, Berkeley and Dorchester Counties, South Carolina." *Castanea 46*: 248–280.

Putz, F. E., and R. R. Sharitz. 1991. "Hurricane damage to old-growth forest in Congaree Swamp National Monument, South Carolina, U.S.A." *Canadian Journal of Forest Research 21*: 1765–1770.

Rabalais, N. N., R. J. Díaz, L. A. Levin, R. E. Turner, D. Gilbert, and J. Zhang. 2010. "Dynamics and distribution of natural and human-caused hypoxia." *Biogeosciences 7*: 585–619.

Rikard, M. 1988. "Hydrologic and vegetation relationships of the Congaree." National Park Service. Clemson Cooperative Park Study Unit.

Robertson, P. A., G. T. Weaver, and J. A. Cavanaugh. 1978. "Vegetation and tree species patterns near the northern terminus of the southern floodplain forest." *Ecological Monographs 48*: 249–267.

Sharitz, R. R., and W. J. Mitsch. 1993. "Southern floodplain forests." In *Biodiversity of the Southeastern United States, Lowland Terrestrial Communities*, edited by W. H. Martin, S. G. Boyce, and A. C. Echtemacht, 311–372. New York: John Wiley.

Shirley, L. J., and L. L. Battaglia. 2006. "Assessing vegetation change in coastal marsh-forest transitions along the northern Gulf of Mexico using National Wetlands Inventory data." *Wetlands 26*: 1057–1070.

Stahle, D. W. 1996. "Tree rings and ancient forest history." In *Eastern Old-Growth Forests: Prospects for Rediscovery and Recovery*, edited by M. B. Davis, 321–343. Washington, DC: Island Press.

Stahle, D. W., M. K. Cleaveland, R. D. Griffin, M. D. Spond, F. K. Fye, R. B. Culpepper, and D. Patton. 2006. "Decadal drought effects on endangered woodpecker habitat." *Eos, Transactions of the American Geophysical Union 87*: 121–125.

Stahle, D. W., M. K. Cleaveland, and J. G. Hehr.1988. "North Carolina climate changes reconstructed from tree rings: A.D. 372 to 1985." *Science 240*: 1517–1519.

Stahle, D. W., R. B. VanArsdale, and M. K. Cleaveland. 1992. "Tectonic signal in baldcypress trees at Reelfoot Lake, Tennessee." *Seismological Research Letters 63*: 439–447.

Visser, J. M., J. W. Day, Jr., L. L. Battaglia, G. P. Shaffer, and M. W. Hester. 2012. "Mississippi River Delta wetlands." In *Wetland Habitats of North America: Ecology and Conservation Concerns*, edited by D. Batzer and A. Baldwin, 63–74. Berkeley, CA: University of California Press.

Wall, D. P., and S. P. Darwin. 1999. "Vegetation and elevational gradients within a bottomland hardwood forest of Southeastern Louisiana." *The American Midland Naturalist 142*: 17–30.

Wharton, C. H., W. M. Kitchens, E. C. Pendleton, and T. W. Sipe. 1982. "Ecology of bottomland hardwood swamps of the southeast: a community profile." Biological Services Prog., FWS/OBS-81/37. Washington, DC: U.S. Dept. of the Interior, Fish and Wildlife Service. https://www.nwrc.usgs.gov/techrpt/81-37.pdf.

Young, P. J., B. D. Keeland, and R. R. Sharitz. 1995. "Growth response of baldcypress to an altered hydrologic regime." *American Midland Naturalist 133*: 206–212.

Zhao, D., B. P. Allen, and R. R. Sharitz. 2006. "Twelve year response of old-growth southeastern bottomland hardwood forests to disturbance from Hurricane Hugo." *Canadian Journal of Forest Research 36*: 3136–3147.

# Chapter 3

## Fire-Maintained Pine Savannas and Woodlands of the Southeastern United States Coastal Plain

*Robert K. Peet, William J. Platt, and Jennifer K. Costanza*

Descriptions by early naturalists provide glimpses of presettlement southeastern coastal plain landscapes. Means (1996) quoted Bartram (1791): "This plain is mostly . . . the great long-leaved pine (*P. palustris* Linn.), the earth covered with grass, interspersed with an infinite variety of herbaceous plants, and embellished with extensive savannas, always green, sparkling with ponds of water. . . . We left the magnificent savanna . . . , passing through a level, open, airy pine forest, the stately trees scatteringly planted by nature, arising straight and erect from the green carpet, embellished with various grasses and flowering plants, and gradually ascending the sand hills . . . "

Bartram's experience cannot be repeated today. Noss et al. (2015) estimated that more than 96 percent of pine savannas and woodlands (hereafter savannas), central to the North American Coastal Plain (NACP) biodiversity hot spot, have been converted to other uses by human actions or severely degraded by fire suppression. Most remaining savannas have been degraded further by logging, open-range grazing, and fire suppression. Such degraded systems have produced a history of misguided ecological ideas that often conflict with recent concepts emerging from the small number of remaining savannas that do have old-growth attributes and have been burned regularly using prescribed fires that mimic natural lightning-fire regimes (Platt 1999). Here, we present a concept of old growth for pine savannas based on three interrelated components: tree populations and ground-layer vegetation, together with feedbacks produced by both layers in the context of long-ongoing evolutionary relationships with fire.

Pine savannas with old-growth attributes exist. They do not, however, resemble old-growth forests (see chapter 1). Intact old-growth savannas

39

have a characteristic two-layer physiognomy: large, often scattered over-story trees (mostly pines) above a continuous ground-layer vegetation in which grasses are aspect dominants (see figures 3-1, 3-2, 3-3). The pine populations (fire-tolerators, sensu Pausas 2017) are multiaged, ranging from recently recruited grass-stage individuals to overstory trees several centuries old. Such populations persist in an intact state only on a few high-quality sites (Varner and Kush 2004). These trees produce needles that are flame-retardant when green but incendiary when shed and dried, and, thus, they increase fire intensity and generate pyrodiversity throughout their lifespans (Platt et al. 2016). Such influence of pines on the ecosystem con-tinues after death as large upright snags not only are repeatedly struck by lightning but also provide animal homesites. Once on the ground, logs, which can be common in nonlogged sites managed with frequent fires, provide microhabitats for animals and serve as local firebreaks, generating pyrodiversity for years before they are eventually consumed by fire (Her-mann 1993). Pine savannas also contain greater ground-layer biodiversity and endemism than is typical of eastern forests, in large part resulting from a long evolutionary history of association with frequent fire in the context of relatively stable climate (Hoctor et al. 2006; Noss et al. 2015). Similar

FIGURE 3-1. Old-growth xeric sand ridge pine savanna with longleaf pine (*Pinus palustris*) in overstory and wiregrass (*Aristida stricta*) as aspect dominant in the ground layer. Croatan National Forest, North Carolina. Photo credit: R. K. Peet.

FIGURE 3-2. Old-growth flatwoods pine savanna with longleaf pines (*Pinus palustris*) and South Florida slash pines (*Pinus densa*) in the overstory. Cutthroat grass (*Panicum abcissum*) and saw palmetto (*Serenoa repens*) are aspect dominants in the ground layer. Tomlin Gully, Avon Park Air Force Base, Florida. Photo credit: W. J. Platt.

FIGURE 3-3. Old-growth upland clayhill pine savanna-woodland with longleaf pines (*Pinus palustris*) in the overstory. Wiregrass (*Aristida beyrichiana*), Indian grass (*Sorghastrum spp.*), and little blue stem (*Schizachyrium scoparium*) are aspect dominants in the ground layer. Wade Tract, Georgia. Photo credit: W. J. Platt.

high-diversity ground-layer vegetation occurs worldwide in savannas and grassland ecosystems (Parr et al. 2014; Veldman et al. 2015). The physiognomy of pine savannas also is crucial for integrity of numerous endemic animal populations known to depend on old trees and/or fire-maintained groundcover (e.g., gopher tortoises, Red-Cockaded Woodpeckers, reticulated flatwoods salamanders; Hermann 1993).

We propose that "near-natural" fire regimes are integral old-growth attributes. Frequent surface fires during late spring and early summer, when lightning fires occur most frequently (the "lightning-fire season"), maintain pine savannas with old-growth attributes (Platt 1999). Savanna vegetation is quickly altered following anthropogenic changes in fire regimes; lengthening fire-return intervals or shifting seasonal timing of fires result in changes in structure and loss of biodiversity (e.g., Palmquist et al. 2014, 2015; Platt et al. 2015). Further, with fire exclusion, regionally abundant rainfall results in closed-canopy forests (Wahlenberg 1946; Platt 1999; Beckage et al. 2009). Over the past halfcentury, studies of sites with old-growth attributes have shown that recurrent fires during the lightning-fire season maintain savanna pines, the diverse ground-layer vegetation, and the characteristic physiognomy (Fill et al. 2015; Noss et al. 2015).

Tree populations and ground-layer vegetation are not readily reconstituted once disrupted. Flammable vegetation (sensu Mutch 1970) and vegetation feedbacks on fuels that influence characteristics of fires in the context of distinct seasonal environments have molded these fire-influenced landscapes over millions of years (Noss et al. 2015). Reintroducing fire per se is not sufficient in that fire characteristics are altered in reconstructed habitats (Robertson and Ostertag 2007). The most useful approaches to reconstruction should be ones that enhance all old-growth attributes as part of a planned program aimed at reconstituting vegetation-fire feedbacks (Fill et al. 2015).

## Southeastern Pine Savannas and Woodlands in a Global Context

Pine savannas share physiognomy and frequent occurrence of fire with other savanna ecosystems (Archibald et al. 2013). Moreover, southeastern savannas share nutrient impoverishment and climatic buffering with other tropical and subtropical ecosystems designated as "Old, Climatically Buffered, Infertile Landscapes" (OCBILs; Hopper 2009, 2016). Mucina and Wardell-Johnson (2011) redefined climatic buffering as climatic stability, identified soil impoverishment as a function of landscape age, and specified predictable fire regimes (and hence, adaptation to fire) as an important aspect of OCBILs. Similar concepts regarding adaptation to frequent fire regimes have emerged in different habitats worldwide (Keeley et al. 2011; Ratnam et al. 2011), increasing recognition of ecological differences between fiery ecosystems and closed-canopy forests (Parr et al. 2014). Pine savannas exemplify how environmentally induced aspects of OCBILs can influence emergent attributes, including high biodiversity and endemism. Including the NACP, half of the 36 terrestrial biodiversity hot spots occur in OCBILs (Noss et al. 2015; Hopper et al. 2016).

Pine savannas of the southeastern US coastal plain generally occur within a relatively flat landscape on highly leached marine sediments, typically with low levels of phosphorus. The occurrence of these savanna ecosystems south of the zone of glaciation and near the climatic buffering influences of the Atlantic and/or Gulf Coast facilitated species persistence during periods of glaciation when much of the eastern North American flora was subject to significant climatic stress and consequent movement and species loss (Sorrie and Weakley 2006; Noss et al. 2015). The flora of these systems, as with other OCBILs, has ancient origins and has evolved in the context of infertile soil conditions and frequent fires. Such conditions favored dominance by a diverse set of flammable grasses and sedges, often with flat terrain and seasonal environments (sensu Platt et al. 2015), increasing the likelihood of fires spreading across herbaceous landscapes once ignited.

Globally, fiery ecosystems are disappearing. Declines are attributable primarily to decreasing occurrence of fires in fragmented landscapes (Parr et al. 2014; Noss et al. 2015). For example, Andela et al. (2017) reported a global decline of 24 percent in annually burned area over a period of less than two decades. Certainly, fire suppression has resulted in increases in woody plants (Platt et al. 2015) and decreases in diversity of the endemic-rich herb layer of many OCBILs, including pine savannas (Palmquist et al. 2014, 2015). More broadly, Pausas and Ribeiro (2017) proposed that fire regime is a primary factor explaining plant diversity around the globe, even after accounting for productivity. They suggest that fires delay competitive exclusion, increase landscape heterogeneity and generate new niches. Peet et al. (2014) made similar points for pine savannas: Fire reduces aboveground biomass, shifting from the largely asymmetric competition for light (larger plants aboveground have a disproportionate advantage) to more symmetric belowground competition where resource use is proportional to plant belowground investment. Hence conservation of OCBILs, such as pine savannas, needs to involve active management, using scientifically based prescribed fires (Fill et al. 2015).

## Environmental and Biogeographic Patterns of Biodiversity

In presettlement times, pine savannas dominated the coastal plain land-scape from southern Virginia to southern Florida and west to eastern Texas. While within that region there was considerable biogeographic variation (figure 3-4), longleaf pine (*P. palustris*) was the dominant tree over the vast majority of the landscape (Wahlenberg 1946; Frost 2006; Peet 2006). After review of historical documents, Frost (2006) reported that longleaf pine had dominated 52 percent of the uplands and codominated another 33.2 percent across the original range of the species. In addition, South Florida slash pine (*P. densa*) was the sole pine on sandy uplands and in what are locally called *rocklands* (savannas on limestone outcroppings) in the Florida Everglades and Keys (Schmitz et al. 2002).

We also recognize transition areas (figure 3-4) where longleaf pine shared dominance with other pines and various hardwoods. At the north-eastern end of the range, most sites are relatively moist, owing to extremely low topographic relief. There, pond pine (*P. serotina*) typically was codominant. Over the inland portion of the range, including parts of the Piedmont, the southern tip of the Appalachians, and the interior coastal plain of Mississippi and Alabama, longleaf pine often shared dominance with short-leaf pine (*P. echinata*) and various oaks. At the western end of the range of longleaf pine in Louisiana and Texas, shortleaf and loblolly pine (*P. taeda*) co-occurred in upland regions; we recognize a core area of longleaf savannas with bluestem understory, transitional areas to the north with a mix of pines and hardwoods, and transitional areas to coastal prairies in the south. In central Florida, longleaf pine shared dominance with south Florida slash pine (*P. densa*). Further, oaks (*Quercus* spp.), pond cypress (*Taxodium ascendens*), and northern slash and pond pine (*P. elliottii, P. serotina*) were trees associated with savanna habitat inclusions, commonly at either the dry or wet ends of the moisture gradient, which resulted in additional heterogeneity within the broader pine savanna matrix (Peet 2006).

Despite the relatively low diversity of the tree layer of southeastern pine savannas, there is considerable diversity in the composition of the ground layer, both across the region and at the local scale. Variation in the aspect-dominant grasses has been used to distinguish ground-layer divisions of pine savannas (figure 3-4). The commonly recognized core concept has been one of a ground layer with wire grass (*Aristida* spp.) as the most abundant grass. This type occupies much of the central part of the range, except for a gap in South Carolina where wire grass is absent, likely because of Pleistocene range contraction. In fact, the wire grass of the southern

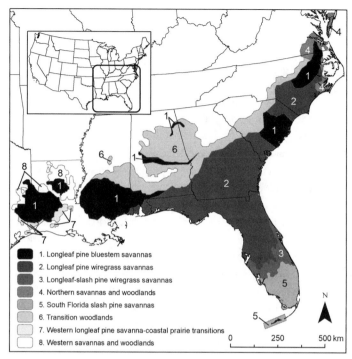

FIGURE 3-4. The ranges of different types of southeastern pine savannas in the southeastern United States. The six divisions are based on abundance of overstory trees and ground-layer grasses as described in the text. Modified from Frost (1993, 2006) and Platt (1999).

portion is morphologically distinct from the more northern wire grass and is sometimes recognized as a separate species (*Aristida beyrichiana* versus *A. stricta*). Where wire grass is not present, bluestems (*Andropogon* spp., *Schizachyrium* spp.) tend to be aspect-dominants, as in the north-central portion of North Carolina, central South Carolina, and southwest Alabama westward to Texas, as well as in the southern Florida rocklands.

Recognition of geographic regions based on grass aspect dominance does not do justice to the diversity and geographic turnover of ground-layer floristic composition in pine savannas. In contrast to near monodominance of the overstory, the ground layer commonly contains a plethora of species, including many endemics. In fact, roughly 85 percent of the many endemic plant species of the NACP occur in fire-maintained savannas or their associated inclusions (Noss et al. 2015). Most of these endemic species are narrowly distributed (Sorrie and Weakley 2006), and, as inhabitants largely of fire-maintained pine savannas, contribute to the considerable turnover in species composition, both among habitats and geographically (Carr et al. 2009; Palmquist et al. 2015).

At small scales, ground-layer diversity can be substantial. Moist, frequently burned pine savannas on silty soils typically have around 90 species per 1,000 square meters, with values often exceeding 120 (Peet et al. 2014). Similarly, well-drained silty soils can support high richness at the 1,000-square-meters scale, typically over 80 species, whereas sandy soils support less diverse longleaf types. At a scale of 1 square meter, species richness can be even more impressive. The current United States record value of 52 species in a square meter was observed in a North Carolina longleaf stand on moist, silty soils that had experienced almost annual fires for at least 75 years. After a change in management practices to somewhat lower fire frequency, smallscale richness dropped dramatically, and richness at the 1,000-square-meters scale dropped modestly (Palmquist et al. 2014). A parallel study over a 20-year period across many longleaf sites in the Carolinas found significant increases and decreases in species richness associated with increased and decreased fire frequency (Palmquist et al. 2015). These studies suggest that both local microhabitat and fire frequency influence species richness (also see Schmitz et al. 2002 for the south Florida rocklands).

The dramatically reduced extent of southeastern pine savannas since European settlement, coupled with the apparent physiognomic simplicity of scattered pines over grass, resulted in a long period during which the vegetation of these systems received less attention than other vegetation types in eastern North America. However, in recent years, our understanding of the compositional variation of these systems has greatly expanded. This has been driven largely by the development of a comprehensive US National Vegetation Classification (USNVC), now widely employed by state and federal agencies and conservation organizations (Franklin et al. 2012). In the USNVC, associations constitute the lowest level of an eight-level hierarchy and are combined hierarchically into alliances, groups, and macrogroups. The Longleaf Pine Macrogroup contains 131 associations (table 3-1). This comes close to the 156 associations in the Appalachian-Northeast Oak-Hickory-Pine Forest and Woodland Macrogroup with many different tree species dominating the various associations across a topographically and edaphically much more complex and equally large geographic region. To provide a logical structure within which to conceptualize these many USNVC associations, Peet (2006) recognized categories that could be arranged within a two-dimensional framework of soil texture and moisture, which were subsequently incorporated as groups within the USNVC. These groups included sand barrens and xeric sandy uplands over entisols (figure 3-1), moist flatwoods over sandy spodosols (figure 3-2), silty uplands over well-drained ultisols (figure 3-3), and savannas (sensu stricto) over moist ultisols. These groups with their current official names and component alliances are shown in table 3-1, along with a group containing montane and piedmont longleaf types and a group containing thin-soil South Florida slash pine rockland savannas.

The high local diversity and large number of USNVC associations results from local co-occurrence of a great variety of small-sized, different life forms and their turnover with both geographic distance and local microhabitats (Peet 2006). Different species occupy different strata within the ground layer, such that ground-hugging rosettes co-occur with upper layer grasses and shrubs, as well as with smaller upright forbs and lianas. Different species flower at different times, so the flowering phenology presents an ever-changing display during the year, from sunbonnets (*Chaptalia tomentosa*) in January–February to the peak flowering of $C_4$ grasses and composites in the fall, to gentians (*Gentiana catesbaei*) in November–December (Platt et al. 1988). Different functional and phylogenetic groups characterize different portions of the longleaf savannas. Moist, silty savannas host a spectacular diversity of charismatic species such as orchids (*Calopogon* spp., *Cleistesiopsis* spp., *Plantanthera* spp., *Pogonia ophioglossoides*, *Spiranthes* spp.) and insectivorous species (*Dionaea muscipula*, *Drosera* spp., *Pinguicula* spp., *Sarracenia* spp., *Utricularia* spp.), along with a great diversity of sedges (*Rhynchospora* spp., *Scleria* spp.). Drier, silty sites support a great diversity of Fabaceae. Moist, sandy sites are renowned for their diversity of Ericaceae. Asteraceae are abundant across all these types, often with basal rosettes that facilitate winter photosynthesis. Owing to the predominance of infertile soils, annual plants are uncommon, except for a few legumes, such as partridge pea (*Chamaecrista* sp.), and a group of hemiparasites (e.g., *Agalinus* spp., *Buchnera* spp., *Schwalbea americana*, *Seymeria* spp.). In the end, no description can do justice to the often bewildering array of species and life forms in the pine savanna ground layer or to the high turnover of species along local soil moisture and texture gradients, as well as across the geographic region.

## Role of Fire in Pine Savannas

Fire regimes across the coastal plain have been altered directly by humans and indirectly by habitat fragmentation (Wahlenberg 1946; Frost 2006). The most reliable data on presettlement fire frequency and seasonality come from scars produced in annual growth rings of old pines

TABLE 3-1. Southeastern pine savanna associations recognized in the United States National Vegetation Classification.

| Alliance | Assoc. | States | Provinces |
| --- | --- | --- | --- |
| **Dry-Mesic Loamy Longleaf Pine Woodland** | | | |
| West Gulf Coastal Plain Longleaf Pine / Blackjack Oak / Bluestem Woodland | 6 | LA, TX | Coastal Plain |
| West Gulf Coastal Plain Upland Longleaf Pine / Bluestem Woodland | 6 | LA, TX | Coastal Plain |
| Southeastern Coastal Plain Longleaf Pine / Sand Post Oak / Wiregrass Woodland | 6 | NC, SC, GA, FL, AL, MS | Coastal Plain |
| Southeastern Coastal Plain Longleaf Pine / Blackjack Oak Clayhill Woodland | 4 | NC, SC, GA, FL, AL, MS | Coastal Plain |
| Southeastern Coastal Plain Upland Longleaf Pine / Wiregrass Woodland | 14 | NC, SC, GA, FL, AL, MS, LA | Coastal Plain |
| **Xeric Longleaf Pine Woodland** | | | |
| Longleaf Pine / Bluejack Oak Sandhill Woodland | 7 | LA, TX | Coastal Plain |
| Longleaf Pine / Turkey Oak / Pineland Three-awn Woodland | 9 | VA, NC, SC | Coastal Plain |
| Longleaf Pine / Turkey Oak / Little Bluestem Woodland | 9 | SC, GA, FL | Coastal Plain |
| Longleaf Pine / Turkey Oak - Sand Live Oak Woodland | 9 | GA, FL, AL | Coastal Plain |
| Longleaf Pine / Turkey Oak / Three-awn Woodland | 3 | GA, FL, AL, MS | Coastal Plain |
| **Wet-Mesic Longleaf Pine Open Woodland** | | | |
| Atlantic Coastal Plain Wet Longleaf Pine Savanna | 14 | NC, SC, GA, FL | Coastal Plain |
| West Gulf Coastal Plain Longleaf Pine Wet Savanna | 3 | LA, TX | Coastal Plain |
| East Gulf Coastal Plain Wet Pine Open Woodland | 8 | GA, FL, AL, MS, LA | Coastal Plain |
| **Mesic Longleaf Pine Flatwoods - Spodosol Woodland** | | | |
| Southern Coastal Plain Mesic Longleaf Pine Flatwoods | 16 | SC, GA, FL, AL, MS, LA | Coastal Plain |

*continued on next page*

and stumps (Rother et al. 2018). In coastal savannas, presettlement return intervals averaged 1 to 3 years with occasionally longer intervals (Wahlenberg 1946; Frost 2006), and fire scars occurred overwhelmingly during transitions from early to latewood within annual rings (Rother et al. 2018). These transitions in type of wood production occur during the lightning-fire season, at the time (May–June) when lightning-ignited fires occur most frequently throughout the coastal plain (Platt et al. 2015). These studies, together with models based on recurrent fire conditions (e.g., Guyette et al. 2012), indicate high frequencies of fires, especially during the fire season. Thus, historical fire regimes were characterized by rather predictable attributes related to frequency and season, a characteristic of OCBIL habitats more broadly.

High frequencies of prescribed fires during fire seasons in sites with old-growth attributes entrain aspects of fuels and, hence, fire characteristics. For example, short-return intervals (1 to 2 years) result in limited fuel biomass. Thus, low intensity fires consume primarily live and dead fine fuels in the ground layer. Hence, variation in fine fuel biomass and flammability of fuels have measurable effects on fire characteristics at ground level where plants have protected regenerative buds (Ellair and Platt 2013). Nonetheless, sensitivity of fire characteristics to local variation in fuel characteristics and variation in synoptic weather conditions, especially during fire seasons (Platt et al. 2015), results in a range of characteristics within and among fires (Platt et al. 2016; Dell et al. 2017).

TABLE 3-1. *continued*

| Alliance | Assoc. | States | Provinces |
|---|---|---|---|
| **Mesic Longleaf Pine Flatwoods - Spodosol Woodland** | | | |
| Atlantic Coastal Plain Mesic Longleaf Pine Flatwoods | 5 | VA, NC, SC | Coastal Plain |
| **South Florida Slash Pine Rockland** | | | |
| Florida Slash Pine Rockland Woodland | 4 | FL | Coastal Plain |
| **Shortleaf Pine - Oak Forest and Woodland** | | | |
| Montane Longleaf Pine - Shortleaf Pine Woodland | 8 | NC, SC, GA, AL | Piedmont, Mountains |

Vegetation-fire feedbacks modify fire regimes. Plants that contribute substantial rapidly drying biomass or increase flammability of fine fuels increase likelihoods of fires. For example, $C_4$ grasses grow rapidly after fire-season fires and produce aboveground biomass that dries readily, resulting in a matrix of flammable fuels (Platt 1999; Simpson et al. 2016). Leaves of woody shrubs (e.g., savanna oaks) are retained on plants, then released near the onset of the fire season, adding flammable fuels to the ground layer (Kane et al. 2008). During the fire season, this matrix of fine fuels has low ignition temperatures and combusts rapidly, producing sufficient heat to combust adjacent fuels and spread fires across landscapes under dry weather conditions (Slocum et al. 2003, 2010).

Superimposed on feedbacks produced by ground-layer plants are those produced by pines. Needles of overstory pines like longleaf pine are flame-resistant when green, but incendiary when dry (Platt 1999). Because needles are shed continually, the ground layer becomes laced with pyrogenic fuels that increase maximum temperatures and durations of heating throughout the ground layer. These fire-feedbacks generate pyrodiversity at scales capable of influencing species composition and dynamics of ground-layer vegetation (Platt et al. 2016). Pervasive fire feedbacks, especially those involving pines, have likely generated fire regimes with enhanced likelihoods of frequent (almost annual) fire-season fires.

## Dynamics of Pine Savanna Plant Populations and Communities

Studies in savannas with old-growth attributes have shown that pine populations are broadly influenced by two types of disturbance. Tropical cyclones, which recur at intervals averaging about a decade along the Gulf of Mexico coast, drive stands away from demographic equilibria, shortening tree replacement cycles via differential mortality of older, larger trees, and opening space for recruitment (Platt and Rathbun 1993; Platt et al. 2000). Recurrent fires, noncatastrophic for savanna pines with thick bark and protected buds, block recruitment of hardwood trees, maintaining open space eventually colonized by pines (Platt 1999; Beckage et al. 2009). Recruitment does not occur in stands with older trees that are not burned for long periods; as pines die, open spaces become filled with hardwoods (e.g., Glitzenstein et al. 1995; Johnson et al. 2018).

In frequently burned old-growth stands, recruitment of pines involves patch dynamics. Seedlings germinate throughout stands in periodic mast years. Juvenile pines are killed by pine needle-fueled fires and thus mostly survive in open patches large enough to be devoid of shed pine needles and in which root competition from surrounding large trees is reduced (Platt and Rathbun 1993; Platt et al. 2016). In addition, fallen pines (as for example, after hurricanes) provide firebreaks that may create unburned patches in which seedlings survive (Platt and Rathbun 1993). Nonetheless, recruitment does not necessarily occur once open patches reach sufficient size for recruits to survive fires (Platt et al. 1988) because growth and survival are influenced by ground-layer vegetation (Grace and Platt 1995). Consequently, open patches often are maintained for multiple decades (Platt and Rathbun 1993; Platt 1999) before colonization and formation of patches of grass-stage pines. Once present and released from the grass stage, pines slowly grow and thin, resulting in a few trees reaching the overstory in perhaps a century. As trees grow and thinning occurs, recruitment patches merge into a matrix of randomly dispersed large trees (Noel et al. 1998).

Old-growth pine populations persist but cannot be viewed as in demographic equilibria. Projections for population changes over time based on individual-tree demography in sites with old-growth attributes suggest rapidly growing populations (Platt et al. 1988), but projections for population changes based in patch-based demography suggest spatially heterogeneous populations whose dynamics are driven by rates of disturbance, such as hurricane frequency (Platt and Rathbun 1993). Indeed, dynamics of old-growth pine populations most likely varied in spatial and temporal structure, with tree replacement cycles much shorter along coasts with associated frequent lightning storms and tropical cyclones than in inland areas more protected from such disturbance.

Pine savannas with old-growth attributes also contain hardwood trees and shrubs that occur mainly in the ground layer (Platt 1999). Established plants commonly resprout after fire (e.g., Schafer and Just 2014), forming clusters of clonal stems (Platt et al. 2015). For example, oak populations (e.g., *Quercus laevis*, *Q. margarettae*) often contain genetic individuals with aggregated clonal stems. Some species (e.g., *Quercus minima*, *Q. elliottii*, *Castanea pumila*, *Morella pumila*, *Ilex glabra*, *Vaccinium stamineum*) do not have arboreal stages as large as congeners in surrounding forests. These species appear to have lost tree stages and gained underground geoxylic structures as evolutionary responses to frequent fire, as hypothesized for some woody species in other savanna habitats (e.g., Simon et al. 2009; O'Donnell et al. 2015).

Populations of hardwoods have enigmatic status in frequently burned old-growth pine savannas. Although the presence of species without tree stages suggests long evolutionary relationships with fire, increases in abundance have been designated as shrub encroachment attributed to human disturbance (see review in Platt et al. 2015). Past logging and fire suppression are proposed to have released populations of native hardwood trees once fire intensity was reduced (e.g., Glitzenstein et al. 1995; Gilliam and Platt 1999). As a result, reduction of hardwoods has been a perennial focus of both silvicultural and restoration efforts (e.g., Platt et al. 2015), though a limited amount of research has pointed to concerns about conservation practices that aim to reduce hardwoods (Hiers et al. 2014). Hypotheses regarding shrub-tree population structure and dynamics have emerged from studies of sites with old-growth attributes (e.g., Grady and Hoffman 2012). These hypotheses have been based on the idea that hardwood trees are unlikely to survive repeated fires fueled by pyrogenic pine needles, but may persist in open areas away from pines (Platt et al. 2016). Thus, spatial heterogeneity of hardwood tree populations may have depended on the spatial distribution of pyrogenic pines (Ellair and Platt 2013). This concept may provide an approach useful in conservation of hardwood tree populations in sites with old-growth attributes.

Dynamics of populations and relations to fire are less well-known for ground-layer species than trees. Meristematic tissues of most grasses and forbs (and seeds in some monocarpic species) are protected by location at or below ground level. Ground-layer plants, especially $C_4$ grasses, tend to respond postfire with rapid aboveground growth and reproduction, as well as often by clonal growth (e.g., Platt et al. 1988; Fill et al. 2012; Peet et al. 2014). These studies indicate that ground-layer plant populations in sites with old-growth attributes, as in many savannas (Lamont et al. 2011; Clarke et al. 2013), tend to be persistent over time despite recurrent fires. The resulting direct regeneration involves only minor changes in species composition of pine savannas following fires. In contrast, even a modest reduction in fire frequency can result in a layer of dead grasses shading the ground, significantly reducing the smallscale diversity of herbaceous plants. Smallscale disturbances, such as hot spots resulting from the burning of downed trees or localized animal disruptions (e.g., gopher tortoises), also may be important in regeneration by both longleaf pine and ground-layer plants (Hermann 1993; Grace and Platt 1995).

## Remaining Old-Growth Pine Savannas and Woodlands

We identified old-growth southeastern pine savanna sites based on the three components of old growth: tree populations, intact ground-layer vegetation, and a prescribed fire regime with attributes producing ecological conditions postulated to have occurred historically. A fourth important attribute of an old-growth landscape is sufficient extent and habitat heterogeneity for viable populations of plants and animals to have persisted in a fiery landscape. Employing all four of these criteria, remaining old-growth pine savanna occurs only as clusters of sites in the Red Hills region of southern Georgia (Wade Tract and surrounding and nearby conservation easements that maintain high-quality pine savanna landscapes) and on Eglin Air Force Base (Patterson Natural Area and nearby high-quality tracts).

To better capture the range of diversity in persisting old-growth southeastern pine savannas, we compiled a list of tracts with more than one old-growth attribute. We started with a list compiled by Varner and Kush (2004), wherein they identified 15 old-growth sites covering 5,094 hectares. Natural ground-layer vegetation was maintained on only 9 sites; the rest were undergoing ground-layer restoration. In our list (table 3-2), we indicate dominant vegetation types, and for tracts containing old-growth tree populations, the current estimate of acreage. Sites with old tree populations and a long-term history of frequent fire are important for understanding natural fire dynamics of these ecosystems. We identified seven sites that have both old-growth tree populations and a long history of frequent fires. Five occur in Florida and one each in North Carolina and Georgia. In a second list, we identify sites with old-growth trees but with degraded or uncertain ground-layer conditions (also see Johnson et al. 2018). It is possible that a few of these belong in our first list or will shift to that list with recurrent prescribed fires. These sites are spread across Alabama, Florida, Georgia, and North Carolina, and none is particularly large. Finally, we compiled a list of sites with relatively intact ground-layer vegetation, regularly managed with fire and an overall extent of over 10,000 hectares. For this list, we include 14 longleaf and 2 slash pine sites. These all have the potential to regain old-growth tree populations if properly managed, using frequent, fire-season fires. All these large sites are on a military base, national forest, or national wildlife refuge. Additional areas of extensive but more degraded longleaf vegetation are identified in the Range-Wide Conservation Plan for Longleaf Pine compiled by the Regional Working Group for America's Longleaf.

TABLE 3-2. Examples of nearly old-growth pine savanna vegetation of the southeastern Coastal Plain and adjacent areas. Vegetation types are defined at the bottom of the table.

| Site | State | Area-ha | Veg Type |
|---|---|---|---|
| **Sites with old-growth trees and ground-layer vegetation** | | | |
| Croatan National Forest, Pringle Road | NC | 20 | SH |
| Wade Tract / Red Hills Plantations | GA | 3200 | CL |
| Eglin Air Force Base | FL | 3650 | SH & FW |
| Goethe State Forest | FL | 75 | FW or SH |
| Platt Branch | FL | 160 | FW |
| Venus Flatwoods | FL | 40 | FW |
| Tomlin Gully (Avon Park Air Force Base) | FL | 200 | FW |
| **Sites with old-growth trees but uncertain or degraded ground-layer vegetation** | | | |
| Arnett Branch | NC | 46 | MT |
| Boyd Tract | NC | 24 | CL |
| Bonnie Doone Tract | NC | 65 | CL |
| Camp LeJeune Tract | NC | 20 | SH |
| Appling Tract / Moody Preserve | GA | 120 | CL, SH |
| Brooksville / Big Pine Tract | FL | 170 | SH, CL |
| Tiger Creek | FL | 2 | SH |
| Mountain Longleaf National Wildlife Refuge | AL | 45 | MT |
| Horn Mountain | AL | 20 | CL |
| **Longleaf sites with old-growth ground-layer vegetation and area cumulatively >10,000 ha** | | | |
| Frances Marion National Forest | SC | | FW, SH |
| Camp LeJeune | NC | | SH, FW, SV |
| Fort Bragg | NC | | SH |
| Fort Gordon | GA | | SH |
| Fort Stewart | GA | | SH, FW, SV |
| Fort Benning | GA | | SH, CL |
| Eglin Air Force Base | FL | | SH, FW |
| Avon Park Air Force Base | FL | | FW, SV |

*continued on next page*

## Maintenance of Southeastern Pine Savanna Ecosystems in the Future

Pervasive human-induced changes in fire regimes have now broadly disrupted the long vegetation-fire coevolution of pine savannas. As a result, nearly all savannas have shifted away from old-growth conditions (Fill et al. 2015). Because of wildfire exclusion, remaining tracts with even some characteristics of old growth have only prescribed fire regimes, and those often differ markedly from lightning-fire regimes,

TABLE 3-2. *continued*

| Site | State | Area-ha | Veg Type |
|---|---|---|---|
| **Longleaf sites with old-growth ground-layer vegetation and area cumulatively >10,000 ha** | | | |
| Apalachicola National Forest / St. Marks National Wildlife Refuge | FL | | FW, SH, SV |
| Ocala National Forest | FL | | SH, FW |
| Osceola National Forest / Okefenokee National Wildlife Refuge | FL | | SH, FW |
| Kisatchie National Forest / Fort Polk | LA | | CL |
| De Soto National Forest | MS | | SH, CL |
| Conecuh National Forest | AL | | SH, CL |
| **Slash Pine Sites with >10,000 ha old-growth ground-layer vegetation** | | | |
| Everglades National Park | FL | | SH, FW, SV, RK |
| Big Cypress National Park | FL | | SH, FW, SV |
| **Veg Type Key** | | | |
| CL: Clay/Loam | Dry-Mesic Loamy Longleaf Pine Woodland (NVC G009) | | |
| FW: Flatwoods | Mesic Longleaf Pine Flatwoods - Spodosol Woodland (NVC G596) | | |
| MT: Montane | Shortleaf Pine - Oak Forest & Woodland (NVC G012) | | |
| RK: Rockland | South Florida Slash Pine Rockland (NVC G005) | | |
| SH: Sandhill | Xeric Longleaf Pine Woodland (NVC G154) | | |
| SV: Savanna | Wet-Mesic Longleaf Pine Open Woodland (NVC G190) | | |

with shifts in seasonality and frequency (cf. Platt et al. 2015). The small size of most old-growth sites makes it difficult to represent the full range of natural heterogeneity (Hoctor et al. 2006; Sorrie and Weakley 2006). This fragmentation, as well as increasing urbanization surrounding old-growth sites, affects the ability of managers to conduct prescribed fires. In the future, further land-use changes may threaten maintenance of old-growth conditions and restoration of degraded sites. Urbanization in the region is expected to increase markedly, leading to increased fragmentation (Terando et al. 2014) and the possibility of cascading effects that result in future prescribed fire becoming increasingly difficult due to a combination of fuel accumulation and fragmentation in a human-dominated matrix.

Climate change also is likely to impact all aspects of old-growth characteristics of pine savannas. Increases in temperature are likely in the region by the end of the century (USGCRP 2017). Changes in precipitation are more uncertain but include the possibility of increases in average precipitation along the Atlantic Coast and decreases in the western part of the coastal plain (USGCRP 2017). Nonetheless, even increased precipitation, when combined with temperature increases, could result in increased potential evapotranspiration and drought. Growth and survival of pines should be less sensitive to drought than for other tree species. Increased cloud-ground lightning strikes and more intense tropical storms could concurrently shift age-size class distributions of pines and facilitate tree replacement cycles (Romps et al. 2014), but seedling establishment could become hampered by moisture stress, especially on xeric sites (Loudermilk et al. 2016). The likelihood of altered water cycles and water availability in the future means that small wetlands within these ecosystems may be particularly vulnerable to climate change. The importance of these wetland habitats for many endemic species, their current rarity as part of frequently burned landscapes, and their vulnerability to climate change make them a major priority for conservation of old growth in the future. Decreased water availability may also pose substantial constraints on fire prescriptions, especially in drier conditions, as on xeric soils and in the west gulf portion of the coastal plain.

Despite the potential for increasing threats and vulnerability, we suggest a high likelihood that savannas with old-growth characteristics will be maintained, and even expanded, across the region. We now have a solid base of data from sites with old-growth attributes upon which to base restoration and management. Several types of conservation partnerships that focus on pine savanna restoration have begun in the southeastern United

States. Regional partnerships between federal agencies and a range of public and private stakeholders have identified restoration of pine ecosystems and prescribed fire regimes, mimicking natural fires as key strategies for conservation in the region. Examples include Landscape Conservation Cooperatives and the Southeast Conservation Adaptation Strategy. Region-wide plans for conservation have identified priority landscapes for restoring pine savannas that span all major geographic regions of southeastern pineland. And indeed, implementation teams have begun restoring pine savannas in every state in the range.

Research on the ecology and management of the remaining old-growth savannas can provide a wealth of knowledge to guide management and conservation efforts based on prescribed fire regimes that restore old-growth attributes produced by vegetation-fire feedbacks. For example, research on the effects of fire on old-growth sites suggests that restoration of savanna characteristics cannot be done simply by reestablishing fire. When fires alone (and especially if not conducted during the fire season) are reestablished in long-unburned sites, they do not restore the savanna species composition, especially in the ground layer. Moreover, they can have other unintended consequences on community characteristics such as increased mortality of overstory trees or proliferation of woody shrubs, thus leading to de novo ecosystems (Varner et al. 2005; Platt et al. 2015b). Full restoration of savannas often should include thinning dense overstory trees (Noel et al. 1998) and reestablishment of ground-layer savanna species, especially those that produce important vegetation-fire feedbacks (Fill et al. 2015; Platt et al. 2016).

Moving pine savanna restoration forward requires careful integration of knowledge of past evolution with uncertainties about the future. Overall, recognition of the long evolutionary history of these ecosystems in association with fire, coupled with knowledge of the extreme heterogeneity of species and community composition and dynamics at multiple scales derived from research in existing old-growth sites, should be useful in developing concepts for restoration despite the uncertainties of future environmental conditions. Acceptance of variability in environmental conditions over space and time, while also guaranteeing that responses to specific environmental conditions molded by past evolution are maintained, should be important if remaining and restored sites are to be resilient to future change (Hiers et al. 2014; Fill et al. 2015). Maintaining and restoring fires, and especially vegetation-fire feedbacks integral to the evolutionary history of pine savannas, will be especially important for the future of pine savanna ecosystems.

## Conclusion

Contemporary old-growth concepts emerged from study of closed-canopy temperate forests. Generally, "old growth" denotes forested areas dominated by populations of trees with limited anthropogenic mortality. Forests containing such older trees are hypothesized to resemble those present historically and to persist if situated in a spatial context that supports natural patterns of regeneration driven by ongoing gap dynamics punctuated by relatively infrequent disturbances. In southeastern pine savannas, old-growth attributes provide a framework of core concepts integral to the ecology and evolution of the species, ecosystems, and landscapes comprising the NACP hot spot of biodiversity. Study of old-growth pine savannas has been instrumental in shifting ecological paradigms in those ecosystems from pine savannas as early successional ecosystems maintained by natural and human disturbances to persistent ecosystems molded over millions of years of coevolution with fire as an endogenous ecological process. These ecosystems with high biodiversity and endemism persist after natural fires, while key species exert feedbacks, continually modifying the natural ecological process of fire, resulting in a high diversity of species indigenous to these unique fiery ecosystems. These emergent concepts provide scientifically based guides for ongoing conservation initiatives to increase and maintain substantial areas of old-growth pine savannas and woodlands, supporting biodiversity and endemic species in a sustainable way.

To achieve these goals, it will be necessary to preserve and manage large areas with careful restoration of fire regimes that mimic the natural fire regimes that occurred under presettlement conditions. The substantial geographic turnover in composition across the coastal plain and the large number of local endemic species means that a large and dispersed network of preserves managed in accordance with locally modified fire regimes will be needed. If these goals are accomplished, then old-growth trees, ground-layer vegetation and the indigenous animals should be able to maintain persistent populations. The major challenges are likely to be the fragmented nature of most of the landscape, the relatively modest number of large tracts with an intact ground layer that has burned frequently, and the increasing difficulty of using fire on wildlands imbedded in human development. But, we emphasize it is possible, given the foundation laid by our increased understanding of the old-growth attributes of pine savannas and woodlands, and that on-going initiatives are leading us in the right direction.

# References

Andela, N., D. C. Morton, L. Giglio, Y. Chen, R. G. van der Werf, P. S. Kasibhatla, R. S. DeFries, et al. 2017. "A human-driven decline in global burned areas." *Science 356*: 1356–1362.

Archibald, S., C. E. R. Lehmann, J. L. Gomez-Dans, and R. A. Bradstock. 2013. "Defining pyromes and global syndromes of fire regimes." *Proceedings of the National Academy of Sciences of the United States of America 110*: 6442–6447.

Bartram, W. 1791. *Travels through North & South Carolina, Georgia, East & West Florida*. Philadelphia, PA: James and Johnson.

Beckage, B., W. J. Platt, and L. Gross. 2009. "Vegetation, fire and feedbacks: A disturbance-mediated model of savannas." *The American Naturalist 174*: 805–818.

Carr, S. C., K. M. Robertson, W. J. Platt, and R. K. Peet. 2009. "A model of geographic, environmental and regional variation in vegetation composition of pyrogenic pinelands of Florida." *Journal of Biogeography 36*: 1600–1612.

Clarke, P. J., M. J. Lawes, J. J. Midgley, B. B. Lamont, F. Ojeda, G. E. Burrows, N. J. Enright, et al. 2013. "Resprouting as a key functional trait: how buds, protection and resources drive persistence after fire." *New Phytologist 197*: 19–35.

Dell, J. E., L. A. Richards, J. J. O'Brien, E. L. Loudermilk, A. T. Hudak, S. M. Pokswinski, B. C. Bright, et al. 2017. "Overstory-derived surface fuels mediate plant species diversity in frequently burned longleaf pine forests." *Ecosphere 8(10): e01964*. doi: 10.1002/ecs2.1964.

Ellair, D. P., and W. J. Platt. 2013. "Fuel composition influences fire characteristics and understory hardwoods in pine savannas." *Journal of Ecology 101*: 192–201.

Fill, J. M., S. M. Welch, J. L. Waldron, and T. A. Mousseau. 2012. "The reproductive response of an endemic bunchgrass indicates historical timing of a keystone process." *Ecosphere 3*, art61. http://dx.doi.org/10.1890/ES12-00044.1.

Fill, J. M., W. J. Platt, S. M. Welch, J. L. Waldron, and T. A. Mousseau. 2015. "Updating models for restoration and management of fiery ecosystems." *Forest Ecology and Management 356*: 54–63.

Franklin, S., D. Faber-Langendoen, M. Jennings, T. Keeler-Wolf, O. Loucks, A. McKerrow, R. K. Peet, et al. 2012. "Building the United States National Vegetation Classification." *Annali di Botanica 2*: 1–9.

Frost, C. C. 1993. "Four centuries of changing landscape patterns in the longleaf pine ecosystem." *Proceedings Tall Timbers Fire Ecology Conference 18*: 17–44.

Frost, C. C. 2006. "History and future of the longleaf pine ecosystem." In *The Longleaf Pine Ecosystem: Ecology, Silviculture, and Restoration*, edited by S. Jose, E. J. Jokela, and D .L. Miller, 9–42. New York: Springer-Verlag.

Gilliam, F. S., and W. J. Platt. 1999. "Effects of long-term fire exclusion on tree species composition and stand structure in an old-growth longleaf pine forest." *Plant Ecology 140*: 15–26.

Glitzenstein, J. S., W. J. Platt, and D. R. Streng. 1995. "Effects of fire regime and habitat on tree dynamics in north Florida longleaf pine savannas." *Ecological Monographs 65*: 441–476.

Grace, S. L., and W. J. Platt. 1995. "Effects of adult tree density and fire on the demography of pre-grass stage juvenile longleaf pine (*Pinus palustris* Mill.)." *Journal of Ecology 95*: 75–86.

Grady, J. M., and W. A. Hoffmann. 2012. "Caught in a fire trap: recurring fire creates stable size equilibria in woody resprouters." *Ecology 93*: 2052–2060.

Guyette, R. P., M. C. Stambaugh, D. C. Dey, and R. Muzika. 2012. "Predicting fire frequency with chemistry and climate." *Ecosystems 15*: 322–335.

Hermann, S. M. 1993. "Small-scale disturbances in longleaf pine forests." *Proceedings of the Tall Timbers Fire Ecology Conference 18*: 263-274.

Hiers, J. K., J. R. Walters, R. J. Mitchell, J. M. Varner, L. M. Conner, L. A. Blanc, and J. Stowe. 2014. "Ecological value of retaining pyrophytic oaks in longleaf pine ecosystems." *Journal of Wildlife Management 78*: 383–393.

Hoctor, T. S., L. D. Harris, R. F. Noss, and K. A. Whitney. 2006. "Spatial ecology and restoration of the longleaf pine evosystem." In *The Longleaf Pine Ecosystem: Ecology, Silviculture, and Restoration*, edited by S. Jose, E. J. Jokela, and D.L. Miller, 337-402. New York: Springer-Verlag.

Hopper, S. D. 2009. "OCBIL theory: towards an integrated understanding of the evolution, ecology and conservation of biodiversity on old, climatically buffered, infertile landscapes." *Plant and Soil 322*: 49–86.

Hopper, S. D., F. A. O. Silveira, and P. L. Fiedler. 2016. "Biodiversity hotspots and Ocbil theory." *Plant and Soil 403*: 167–216.

Johnson, E. D., T. Spector, J. K. Hiers, D. Pearson, J. M. Varner, and J. Bente. 2018. "Defining old-growth stand characteristics in fragmented natural landscapes: a case study of old-growth pine in Florida (USA) state parks." *Natural Areas Journal 38*: 88–98.

Kane, J. M., J. M. Varner, and J. K. Hiers. 2008. "The burning characteristics in southeastern oaks: discriminating fire facilitators from fire impeders." *Forest Ecology and Management 256*: 2039-2045.

Keeley, J. E., J. G. Pausas, P. W. Rundel, W. J. Bond, and R. A. Bradstock. 2011. "Fire as an evolutionary pressure shaping plant traits." *Trends in Plant Science 16*: 406–411.

Lamont, B. B., N. J. Enright, and T. He. 2011. "Fitness and evolution of resprouters in relation to fire." *Plant Ecology 212*: 1945–1957.

Loudermilk, E. L., J. K. Hiers, S. Pokswinski, J. J. O'Brien, A. Barnett, and R. J. Mitchell. 2016. "The path back: oaks (*Quercus* spp.) facilitate longleaf pine (*Pinus palustris*) seedling establishment in xeric sites." *Ecosphere 7*: e01361. doi: 10.1002/ecs2.1361.

Means, D. B. 1996. "Longleaf Pine Forest, Going, Going, . . ." In *Eastern Old-Growth Forests: Prospects for Rediscovery and Recovery*, edited by M. B. Davis, 210–229. Washington, DC: Island Press.

Mucina, L., and G. W. Wardell-Johnson. 2011. "Landscape age and soil fertility, climatic stability, and fire: beyond the OCBIL framework." *Plant and Soil 341*: 1–23.

Mutch, R. W. 1970. "Wildland fires and ecosystems—A hypothesis." *Ecology 51*: 1046-1051.

Noel, J. M., W. J. Platt, and E. B. Moser. 1998. "Structural characteristics of old- and second-growth stands of longleaf pine (*Pinus palustris*) in the Gulf coastal region of the U.S.A." *Conservation Biology 12*: 533–548.

Noss, R. F., W. J. Platt, B. A. Sorrie, A. S. Weakley, D. B. Means, J. Costanza, and R. K. Peet. 2015. "How global biodiversity hotspots may go unrecognized: Lessons from the North American coastal plain." *Diversity and Distributions 21*: 236–244.

O'Donnell, F. C., K. K. Caylor, A. Bhattachan, K. Dintwe, P. D'Odorico, and G. S. Okin. 2015. "A quantitative description of the interspecies diversity of belowground structure in savanna woody plants." *Ecosphere 6(9):* 154. http://dx.doi.org/10.1890/ES14-00310.1.

Palmquist, K. A., R. K. Peet, and A. S. Weakley. 2014. "Changes in plant species richness following reduced fire frequency and drought in one of the most species-rich savannas in North America." *Journal of Vegetation Science 25*: 1426–1437.

Palmquist, K. A., R. K. Peet, and S. R. Mitchell. 2015. "Scale-dependent responses of longleaf pine vegetation to fire frequency and environmental context across two decades." *Journal of Ecology 103*: 998–1008.

Parr, C. L., C. E. R. Lehmann, W. J. Bond, W. A. Hoffmann, and A. N. Anderson. 2014. "Tropical grassy biomes: misunderstood, neglected, and under threat." *Trends in Ecology and Evolution 29*: 205–213.

Pausas, J. 2017. "Evolutionary fire ecology: lessons learned from pines." *Trends in Plant Science 20*: 318-324.

Pausas, J. G., and E. Ribeiro. 2017. "Fire and plant diversity at the global scale." *Global Ecology and Biogeography 26*: 889–897.

Peet, R. K. 2006. "Ecological Classification of Longleaf Pine Woodlands." In *The Longleaf Pine Ecosystem: Ecology, Silviculture, and Restoration*, edited by S. Jose, E. J. Jokela, and D. L. Miller, 51–93. New York: Springer-Verlag.

Peet, R. K., K. A. Palmquist, and S. M. Tessel. 2014. "Herbaceous layer species richness of southeastern forests and woodlands: patterns and causes." In *The Herbaceous Layer in Forests of Eastern North America*, 2nd edition, edited by F. S. Gilliam and M. R. Roberts, 255–276. Oxford, England: Oxford University Press.

Platt, W. J. 1999. Southeastern pine savannas. In *The Savanna, Barren, and Rock Outcrop Communities of North America*, edited by R. C. Anderson, J. S. Fralish, and J. Baskin, 23–51. Cambridge, England: Cambridge University Press.

Platt, W. J., R. F. Doren, and T. Armentano. 2000. "Effects of Hurricane Andrew on stands of slash pine (*Pinus elliottii var. densa*) in the everglades region of south Florida (USA)." *Plant Ecology 146*: 43–60.

Platt, W. J., D. P. Ellair, J. M. Huffman, S. E. Potts, and B. Beckage. 2016. "Pyrogenic fuels produced by savanna trees can engineer humid savannas." *Ecological Monographs 86*: 352–372.

Platt, W. J., A. K. Entrup, E. K. Babl, C. Coryell-Turpin, V. Dao, J. A. Hebert, C. D. LaBarbera, et al. 2015. "Short-term effects of herbicides and a prescribed fire on restoration of a shrub-encroached pine savanna." *Restoration Ecology 23*: 909–917.

Platt, W. J., G. W. Evans, and M. M. Davis, 1988. "Effects of fire season on flowering of forbs and shrubs in longleaf pine forests." *Oecologia 76*: 353–363.

Platt, W. J., G. W. Evans, and S. L. Rathbun. 1988. "The population dynamics of a long-lived conifer (*Pinus palustris*)." *American Naturalist 131*: 491–525.

Platt, W. J., S. J. Orzell, and M. G. Slocum. 2015. "Seasonality of fire weather strongly influences fire regimes in south Florida savanna-grassland landscapes." *PLoS ONE 10(1)*: e0116952.

Platt, W. J., and S. L. Rathbun. 1993. "Dynamics of an old-growth longleaf pine population." *Proceedings Tall Timbers Fire Ecology Conference 18*: 275–297.

Ratnam J., W. J. Bond, R. J. Fensham, W. A. Hoffmann, S. Archibald, C. E. R. Lehmann, M. T. Anderson, et al. 2011. "When is a 'forest' a savanna, and why does it matter?" *Global Ecology and Biogeography 20*: 653–660.

Robertson, K. M., and T. E. Ostertag. 2007. "Effects of land use on fuel characteristics and fire behavior in pinelands of Southwest Georgia." *Proceedings of the Tall Timbers Fire Ecology Conference 23*: 181–191.

Romps, D. M., J. T. Seeley, D. Vollaro, and J. Molinari. 2014. "Projected increase in lightning strikes in the United States due to global warming." *Science 346*: 851–854.

Rother, M.T., J.M Huffman, G.L Harley, W.J Platt, N. Jones, K.M. Robertson, and S.L. Orzell. 2018. Cambial phenology informs tree-ring analysis of fire seasonality in Coastal Plain pine savannas. *Fire Ecology 14(1)*: 164–185. doi: 10.4996/fireecology.140116418

Schafer, J. L., and M. G. Just. 2014. "Size dependency of post-disturbance recovery of multi-stemmed resprouting trees." *PLoS ONE 9*: e105600.

Schmitz, M., W. J. Platt, and J. DeCoster. 2002. "Substrate heterogeneity and numbers of plant species in Everglades savannas (Florida, USA)." *Plant Ecology 160*: 137–148.

Simon, M. F., R. Grether, L. P. de Quelroz, C. Skema, R. T. Pennington, and C. E. Hughes. 2009. "Recent assembly of the Cerrado, a neotropical plant diversity hotspot, by in situ evolution of adaptations to fire." *Proceedings of the National Academy of Sciences 106*: 20359–20364.

Simpson, K. J., B. S. Ripley, P. Christin, C. M. Belcher, C. E. R. Lehmann, G. H. Thomas, C. P. Osborne, et al. 2016. "Determinants of flammability in savanna grass species." *Journal of Ecology 104*: 138–148.

Slocum, M. G., B. Beckage, W. J. Platt, S. L. Orzell, and W. Taylor. 2010. "Effect of climate on wildfire size: A cross-scale analysis." *Ecosystems 13*: 828–840.

Slocum, M. G., W. J. Platt, and H. C. Cooley. 2003. "Effects of differences in prescribed fire regimes on patchiness and intensity of fires in subtropical savannas of Everglades National Park, Florida." *Restoration Ecology 11*: 91–102.

Sorrie, B. A., and A. S. Weakley. 2006. "Conservation of the endangered *Pinus palustris* ecosystem based on Coastal Plain cenrtes of plant endemism." *Applied Vegetation Science 9*: 59–66.

Terando, A., J. K. Costanza, C. Belyea, R. R. Dunn, A. J. McKerrow, and J. A. Collazo. 2014. "The southern megalopolis: using the past to predict the future of urban sprawl in the Southeast U.S." *PLoS ONE 9*: e102261.

USGCRP, 2017. D. J. Wuebbles, D. W. Fahey, K. A. Hibbard, D. J. Dokken, B. C. Stewart, and T. K. Maycock, eds. *Climate Science Special Report: Fourth National Climate Assessment, Volume I.* Washington, DC: U.S. Global Change Research Program. doi: 10.7930/J0J964J6.

Varner, J. M., D. R. Gordon, F. E. Putz, and J. K. Hiers. 2005. "Restoring fire to longunburned *Pinus palustris* ecosystems: novel fire effects and consequences for longunburned ecosystems." *Restoration Ecology 13*: 536–544.

Varner, J. M. III, and J. S. Kush. 2004. "Remnant old-growth longleaf pine (*Pinus palustris* Mill.) savannas and forests of the southeastern USA: Status and threats." *Natural Areas Journal 24*: 141–149.

Veldman, J. W., E. Buisson, G. Durigan, G. W. Fernandes, S. Le Stradic, G. Mahy, D. Negreiros, et al. 2015. "Toward an old-growth concept for grasslands, savannas, and woodlands." *Frontiers in Ecology and the Environment 13*: 154–162.

Wahlenberg, W. G. 1946. *Longleaf Pine. Its Use, Ecology, Regeneration, Protection, Growth, and Management.* Washington, DC: Charles Lathrop Pack Forestry Foundation.

# Chapter 4

# Old-Growth Forests in the Southern Appalachians: Dynamics and Conservation Frameworks

*Peter S. White, Julie P. Tuttle, and Beverly S. Collins*

In the southern Appalachian Mountains, compositions, structures, and dynamics of forest communities vary across steep topographic gradients, such as elevation and slope aspect, position on slope, steepness, and slope shape (Whittaker 1956). For instance, in mesic sites, the forest transitions across elevations from lower elevation cove hardwoods and hemlock forests to higher elevation northern hardwoods and, where the mountains surpass approximately 1,680 meters, spruce-fir forests. At mid and low elevations, cove forests on protected sites transition to oak-dominated forests on drier soils and finally to pine and xeric hardwood forests on exposed south- to west-facing sites. Despite a long history of human influence, remnant old-growth forests have survived across these landscape gradients and now comprise, in aggregate, one of the largest totals for old-growth acreage in eastern North America (Davis 1996).

We have three goals in this chapter: to describe the natural and human disturbances that have shaped old-growth forests in this region; to describe the structural variation of old-growth forest in Great Smoky Mountains National Park, using remarkable datasets from the 1930s and 1990s; and to present a framework for evaluating, delineating, and conserving these forests. The 1930s dataset shows forests at the start of chestnut blight and just prior to the time when fire suppression was becoming effective (MacKenzie and White 1998). The 1990s dataset is after balsam woolly adelgid and beech bark disease introduction, but before arrival of the hemlock woolly adelgid. In a concluding section, we discuss the implications of these dynamics for the delineation and future of old-growth forests in the high mountain southern Appalachians or southern Blue Ridge region (Davis 2000).

## Old Growth and Human and Natural Disturbances

Historically, the southern Appalachians supported diverse forest communities with trees of impressive age and size. On mesic mid- and low-elevation sites, trees sometimes exceeded 400 years, 3-meter diameters, and 60-meter heights. Naturalist William Bartram explored the mountains in the late 1700s. The tone of his account displayed astonishment at the magnificence of the forests (Bartram 1791). Bartram also described human impacts, including Native American villages and agriculture fields on the productive soils along rivers and streams and in open forests, plains, and cane breaks that Native Americans likely managed with fire. Though old-growth forests covered much of the landscape outside settled areas, charred wood from fire pits shows that firewood gathering could deplete the highest quality hardwood species from forests surrounding Native American villages (Delcourt 1987).

Beginning in earnest about 1800, European settlers replaced declining Native American populations and expanded or reestablished agriculture on valley flats and lower mountain slopes. These settlers also introduced cattle, sheep, and pigs, increased harvest for timber and fuel, and continued the use of fire. However, the most dramatic transformation of southern Appalachian forests was caused by widespread and unregulated corporate logging that began in the late 1800s and continued until about 1930, even as advocates of conservation and more scientific forestry practices were instituting protective measures. The most exploitive phase of logging extended from the lowest to the highest elevations and sometimes was followed by severe slash fires and soil erosion. Some sites with severe soil erosion still lack complete tree cover 80 to 100 years after logging.

Just before the most intense logging, Ayers and Ashe (1905) completed a forest survey to assess timber in western North Carolina and eastern Tennessee. They reported average tree ages between 104 and 221 years, thus dating these forests to before European settlement of these lands. However, the landscape was quite varied. At one end of the spectrum were forests around settled areas, where logging, land clearing, and fire were extensive. At the other end of the spectrum, in areas far from railroads (at the time), logging impacts were absent or targeted only high-value trees, and "some remarkable fine timber trees" (Ayers and Ashe 1905) still made up the relatively unbroken forest. Areas of older forest included the north slope of the Great Smoky Mountains, where old growth remains today. Logging impacts eventually slowed and became less severe with creation of six national forests, an effort made possible by passage of the Weeks Act in 1911, as well as the establishment of Great Smoky Mountains National Park, authorized by Congress for land purchases in 1926 and established as a park in 1934. The legacy of logging means that today's landscape is largely covered by second-growth forests of relatively uniform ages, with a dearth of both old-growth and early successional habitats (Greenberg et al. 2011). Remnant sites escaped logging and settlement for a variety of reasons, including presence on steep, high elevation, less accessible, and less commercially viable sites. Thus, remnant old-growth forests are a nonrandom sample of the original landscape.

## Remaining Old Growth

How much old growth remains in the southern Appalachians? This question is hard to answer with precision for three reasons: First, it depends on the criteria used to delineate old growth (a subject we develop more fully below); second, there are problems in accurately documenting past human disturbance on some sites; and, third, the forests, even when not subject to logging or settlement, are continuing to change because of natural and indirect human disturbances. Despite these issues, it is estimated that more than 100,000 hectares of old-growth forest remain in the southern Appalachians (Pyle 1985; Johnson 1995; Messick 2004).

Great Smoky Mountains National Park protects about 43,900 hectares of old-growth forest defined as areas that were not logged or farmed (Pyle 1985; as modified in Tuttle and White 2016). Pyle (1985) mapped an additional 59,400 hectares of "diffuse disturbance" areas with milder human impacts such as selective logging, fire, and grazing. Outside Great Smoky Mountains National Park, Messick (2004; Yost et al. 1994; Johnson 1995) used lack of human disturbance and tree age to define three categories for old-growth delineation: A. no sign of human disturbance, trees older than 150 years; B. many trees older than 150 years, but evidence of such disturbance as chestnut blight (one might add other invasive forest pests—see below) or culling more than 50 years ago OR no sign of human disturbance but trees younger than 150 years, probably due to natural disturbance; and B+. larger tracts that had a mix of characteristics of categories A and B. In these three categories, Messick (2004) compiled documentation for 43,500 hectares of old-growth forest on six national forests, comprising 4.5 percent of those public lands, with an additional 190 hectares on private lands. Messick (pers. comm.) has also noted that additional old-growth forests have been documented since his 2004 compilation and that an updated old-growth catalog is much needed to delineate old growth on national forest land.

Some of the forests in Great Smoky Mountains National Park labeled as "diffuse disturbance" (Pyle 1985) contain areas that would qualify as old growth under Messick's definition—areas with minor logging effects ("culling") but no settlement. In one attempt to bridge the gap between Messick's and Pyle's categories, Johnson (1995; see also Yost et al. 1994) estimated 60,700 hectares of old growth in Great Smoky Mountains National Park (versus Pyle's 43,900 hectares without logging or settlement), which would bring the regional total to 105,900 hectares. Clearly, a consistent evaluation of old growth in the southern Appalachians is needed, given the importance, uniqueness, and future impacts to these forests.

## Ongoing Natural and Human Disturbances in Old-Growth Forests

The remnant old-growth forests in the southern Appalachians are now—and have always been—subject to a range of natural disturbances (Harmon et al. 1983; White et al. 2011). Smaller disturbances, such as single tree falls, usually lead to minor if any changes in forest composition, but large ones can initiate succession, thereby producing structural and compositional variation. In the case of smallscale gap dynamics, dendrochronological studies have shown that old-growth forests have repeated patterns of suppression and release of tree growth (Di Filippo et al. 2017). Kincaid and Parker (2008) reported that release events that caused at least a 50-percent increase in growth in old-growth hemlock forests in Great Smoky Mountains National Park occurred with a frequency of 0.48 per year per tree. Busing (2005) reported mortality rates for trees of less than 10 centimeters in diameter at breast height of 0.5 to1.4 percent per year. Runkle (1985) reported turnover rates in cove forest canopies of about 1 to 2 percent per year, with 9.5 percent of the canopy in recent gaps (Runkle 1985).

These forests also experienced larger disturbances. For example, an 1840s recruitment peak in Joyce Kilmer, an old-growth cove forest in western North Carolina, coincides with historical accounts of a hurricane in 1835, and a second recruitment peak coincides with an ice storm in 1915 (Butler et al. 2014). Patterns of larger-scale natural disturbances in the southern Appalachians reflect the region's susceptibility to tropical storm remnants, wind and lightning storms, ice storms, debris avalanches and flood scour on steep creek valleys after intense rain storms, periodic and severe drought, and fire (White et al. 2011). Natural disturbances resulting in stand replacement occur once every several hundred years at most on individual sites (Butler et al. 2014; Pederson et al. 2014).

Although natural fires do occur in the southern Appalachians (Cohen et al. 2007), most fires in the last century have resulted from human ignitions. After suppression became effective in the 1940s, human-caused and lightning fires continued, but suppression reduced fire sizes and fire became much less frequent. The topographic and moisture gradients that control forest composition and structure also control fire behavior in the humid southern Appalachian landscape, and even in extreme conditions, fires leave behind a mosaic of unburned, low-severity, moderate-severity, and high-severity patches. Although mesic forests have a low incidence of fire, old-growth submesic to xeric stands were shaped by fires in the presuppression era. During the suppression era, species favored by fire, such as *Pinus rigida* and *Pinus pungens*, have declined. Fire, with wind and other disturbances, plays an important role in oak regeneration, as well (Lafon et al. 2017).

A second disturbance agent, and one that interacts with fire in xeric pine stands, is the southern pine beetle, a native insect. When older susceptible pines are aggregated, as they can be if regenerating in older burned patches, beetle outbreaks result in patches of fuel load that can increase the frequency and severity of fire. Without fire, southern pine beetle-caused mortality accelerates succession to hardwoods.

Beginning with chestnut blight in the 1920s, a series of nonnative, invasive pathogens and herbivores have influenced southern Appalachian forest types (Harmon et al. 1983; chapter 12). Chestnut canopy trees, eliminated by about 1950, were once among the largest and most dominant trees, particularly in submesic to subxeric forests. Recruitment peaks in old-growth cove stands in the Joyce Kilmer Memorial Forest between 1920 and 1940 may have been due to a combination of ice storms, severe drought, late spring freezes, and chestnut mortality (Butler et al. 2014). Balsam woolly adelgid began spreading through high elevation spruce-fir forests in 1956, causing widespread mortality to mature Fraser firs (*Abies fraser*; Kaylor et al. 2017). Hemlock woolly adelgid entered the southern Appalachians in the early-2000s and is causing the loss of a large and dominant tree of the old-growth forest remnants. Other invasive pests and diseases include beech bark disease, butternut canker, dogwood anthracnose (Jenkins et al. 2007), and the currently invading emerald ash borer. Other indirect human influences include the European wild boar; locally large deer populations; atmospheric pollutant deposition, including acid rain, nitrogen deposition, and near-ground ozone exposure (Sullivan 2017); and the projected effects of climate warming.

There is an ongoing conservation response to many of these threats, with regional efforts most focused on invasive species and pollutant deposition. For example, researchers are seeking to create blight-resistant chestnut trees, while chemical and biocontrol trials are underway for the hemlock woolly adelgid. Fraser fir reproduced prolifically in stands affected by the first wave of the balsam woolly adelgid invasion from 1960 through 1980. Consequently, there is a possibility that some of these stands will be able to contribute another regeneration cohort before the next infestation (Kaylor et al. 2017). In addition, conservation efforts are aimed at preventing the movement of the emerald ash borer by restricting the movement of firewood. Finally, because Great Smoky Mountain National Park is rated a Class I Airshed under the Clean Air Act, substantial research has been carried out on pollutant deposition and near-ground ozone exposure in the park to provide evidence in regulatory efforts to increase air quality (e.g., Fakhraei et al. 2016).

## Structure of Old Growth in Great Smoky Mountains National Park in the 1930s and 1990s

The distribution and variability of old-growth forests of Great Smoky Mountains National Park (plate 3) reflect the natural and human disturbances we have reviewed above. In the park, forests not disturbed by logging or settlement occupy an estimated 21 percent of the park landscape (figure 4-1a). These forests are not a random sample of the original landscape but are biased towards higher elevations, with settlement and diffuse disturbances covering lower to middle elevations and logging covering most of the elevation gradient (figure 4-1b).

The National Park Service employed forest ecologist Frank Miller to carry out a vegetation survey in the 1930s, just as the park was completing land purchases. This dataset forms a unique quantitative view of the park, just after chestnut blight was entering the area (decades before other invasive pests) and just as fire suppression was becoming effective. We used this dataset in two ways, as described below, to illustrate the complexity of historic and ongoing disturbances in the southern Appalachian landscape.

To estimate the importance of indirect human disturbance across this landscape, we used species distribution models developed by Tuttle and based on the 1930s data for five dominant tree species (American chestnut, Fraser fir, hemlock, beech, and pitch pine). These models use plot occurrences to predict species frequency from an elevation and landform index (Tuttle unpublished; figure 4-2). The models indicate the potential impact

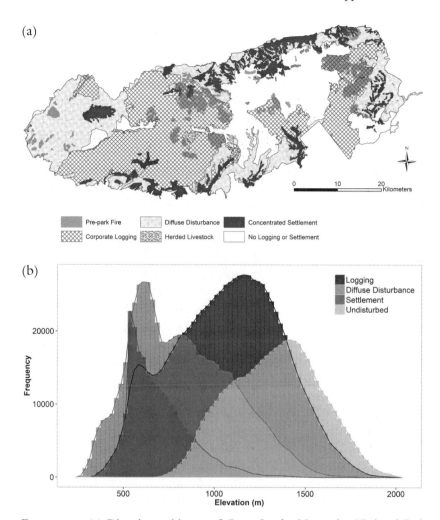

FIGURE 4-1. (a) Disturbance history of Great Smoky Mountains National Park (Pyle 1985, as modified by Tuttle). (b) Distribution of disturbance categories by elevation showing that remnant undisturbed forests are biased towards higher elevations. (The spike in settlement and diffuse disturbance at an elevation of about 550 meters shows the effect of Cades Cove, an unusual limestone window in the western part of the park that has relatively flat topography.)

of chestnut blight, balsam woolly adelgid, hemlock woolly adelgid, and beech bark disease, with the pitch pine model used as surrogate for the potential effects of fire suppression on the most xeric slope positions. These models show that the disturbances affect, with varying degrees, all parts of the Smokies landscape. The disturbances began at different times, causing a pattern of overlap in time as well. Chestnut blight caused near complete mortality of canopy chestnut trees between 1930 and 1950. The heaviest pulse of mortality from balsam woolly adelgid occurred between 1975 and 1990. The invasions of hemlock woolly adelgid and beech bark disease began in the early-2000s and are ongoing in 2017.

Tuttle also compiled 1990s vegetation data for the park and used these data to show both the temporal and spatial variability of old-growth forests and to illustrate the effects of the natural and human disturbances reviewed above (Tuttle and White 2016). Pyle's (1985) disturbance map (with corrections by Tuttle) was used to classify plots. We have extracted summaries for two of Pyle's categories (figure 4-1a): "undisturbed" (no prepark logging or settlement) and "diffuse disturbance" (not subject to largescale logging or settlement but with diffuse direct human disturbance like grazing, fire, and selective cutting). After matching the 1930s and 1990s datasets for environmental conditions (see Tuttle and White 2016 for methods), clear structural differences emerge. The forests of the 1930s that were not disturbed by prepark logging or settlement had about 4 percent higher basal area, 15 percent lower density, and 14 percent higher quadratic mean diameters (a measure of average tree size that weights large trees more than small trees) than the forests of the 1990s (table 4-1). The data also show that the plots with diffuse disturbance changed between the 1930s and 1990s, but this comparison shows a different pattern: Basal area was 35 percent lower in the 1930s than in the 1990s, and quadratic mean diameter did not differ between the two periods. For the diffuse disturbance plots, densities were much lower in the 1930s, by 53 percent, than in the 1990s.

Compared to undisturbed plots in the 1990s, the diffuse disturbance plots in the 1990s had about 11 percent lower basal area, 5 percent higher density, and 10 percent lower quadratic mean diameter (table 4-1). Thus, the plots with diffuse disturbance show the legacy of the generally higher level of disturbance they sustained, compared to the undisturbed plots.

The increase in density in both disturbance categories over about 65 years has several probable causes: Fire suppression allowed ingrowth to stands with the most severe fire (e.g., those on xeric and subxeric sites, Harrod et al. 2000), and loss of American chestnut and Fraser fir due to forest pests resulted in increased sapling and small tree densities. Hemlock, one of the species that likely increased following loss of the chestnut, had not yet experienced decline when the 1990s data were collected. Thus, hemlock may have contributed to increased density in the 1990s plots, although it was undergoing largescale mortality in all size classes in 2017. Loss of dominant

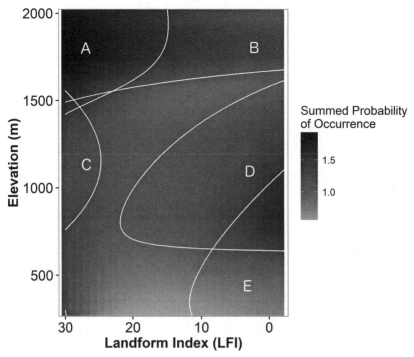

FIGURE 4-2. Distribution of five indirect human disturbances that cause mortality of large canopy trees in Great Smoky Mountains National Park as predicted from species distribution models for affected species based on elevation and landform index, an index developed for mountain landscapes that incorporates slope shape, slope position, and topographic shading (McNab 1993). A = *Fagus grandifolia* (a proxy for beech bark disease, which began in the 1990s), B = *Abies fraseri* (a proxy for balsam woolly adelgid, which began in the 1960s), C = *Tsuga canadensis* (a proxy for hemlock woolly adelgid, which began in the 2000s), D = *Castanea dentata* (a proxy for chestnut blight, which began in the 1930s), and E = *Pinus rigida* (a proxy for historic fire regime following suppression that began in the 1930s). The gray scale indicates the intensity of disturbance. (Overlapping intensities are summed, although the disturbances may not actually be overlapped on individual plots). White lines indicate the 50 percent probability levels for each of the disturbances. The models are constructed from species occurrence data from the 1930s data set described in the text. Other indirect human disturbances are not mapped, although they also affect large areas (e.g., pollutant deposition, which increases from low to high elevations, and near ground ozone exposure). See text for further discussion.

TABLE 4-I. The density and basal area of old-growth forests in Great Smoky Mountains National Park in the 1930s and 1990s (extracted from Tuttle and White 2016). Undisturbed plots are in forests that were neither logged nor subject to settlement; diffuse disturbance plots had some direct human impact, such as grazing, fire, and selective logging. The 1930s mark the beginnings of chestnut blight impacts and the fire suppression period. In the 1990s, hemlock woolly adelgid had not yet caused heavy mortality. The datasets have been matched so that they cover similar environments. See Tuttle and White (2016) for details of methods and analyses. BA = basal area, CV = coefficient of variation, QMD = quadratic mean stem diameter (a measure that is weighted towards larger trees).

| | N | BA $m^2/ha$ | CV | Density Stems/ha | CV | QMD cm |
|---|---|---|---|---|---|---|
| | | **Undisturbed plots** | | | | |
| 1930s | 132 | 39.3 | .69 | 338.2 | .98 | 40.1 |
| 1990s | 167 | 37.8 | .50 | 451.3 | .66 | 34.5 |
| | | **Diffuse disturbance plots** | | | | |
| 1930s | 250 | 21.2 | .86 | 253.4 | .51 | 31.5 |
| 1990s | 188 | 32.4 | .45 | 475.8 | .46 | 31.3 |

American chestnut and Fraser fir trees likely also contributed to the decrease in basal area and quadratic mean diameter in undisturbed plots between the 1930s and 1990s. Although these same factors (fire suppression and loss of American chestnut and Fraser fir as dominant trees) likely contributed to the increase in density in diffuse disturbance plots, much lower 1930s basal area indicates the effects of prepark human disturbances in these plots, as well. Looking to the future, the ongoing mortality of hemlock and beech, and the continued response to earlier disturbances, means that changes in the structure in undisturbed plots will continue in this landscape.

## A Framework for Old Growth in the Southern Appalachians

Given the history and ongoing natural and human disturbances in the southern Appalachians, what is the best framework for evaluating, delineating, and conserving old-growth forests? We answer this question by organizing old-growth criteria on three axes (figure 4-3; table 4-2). Two of these axes rep-

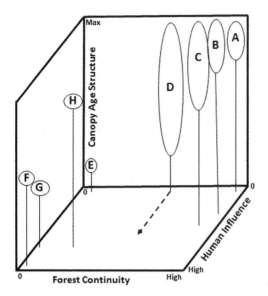

FIGURE 4-3. A three-dimensional framework for old-growth forests, incorporating canopy age structure, human influence, and forest continuity (see also table 4-2), with examples from Great Smoky Mountains National Park. Canopy age structure is an index relative to the maximum values expected for particular environments and the scale appropriate to natural disturbance and regeneration patterns. (Note that "age structure" is used rather than maximum tree age for the vertical axis because forest canopies usually contain a mix of tree ages; table 4-2.) Forest continuity is measured by the numbers of generations of trees during which the site has been a forest without direct, land-clearing disturbances by humans. Human influence ranges from none to direct, intentional, land clearing (logging and settlement), with nonland-clearing disturbances (invasive pests, fire regime change, atmospheric exposure to pollutants, and climate change) between these extremes. Together, canopy age structure and forest continuity produce late successional characteristics such as patterns of coarse woody debris and standing snags. The vertical length of the ovals represents variation in age structure at the landscape scale. Examples are from Great Smoky Mountains National Park: A. cove forest on sites originally without chestnut or hemlock and dominated by small scale gap dynamics; B. cove forests on sites originally with chestnut and/or hemlock as dominants; C. spruce-fir forests affected by the balsam woolly adelgid (forests with lower diversity and, thus, occupying more space on the vertical axis than cove forests); D. xeric pine-oak forests with stand-initiating natural fire regimes (the dashed line shows displacement for human set fires); E. debris-avalanche sites with open bedrock and primary succession; F. forests clear-cut during unregulated phase of mechanized logging pre-1930; G. abandoned farmed sites, settlement era; H. abandoned Native American fields. Note that ovals for logged and farmed stands are smaller because recovery began in a narrow window of time when the park was established. The large gap between forests of higher (A–D) and lower continuity (F–H) is due to the exclusion of logging and settlement after 1934 in Great Smoky Mountains National Park.

resent the most frequently used old-growth criteria: age structure of the tree canopy and "naturalness," that is, lack of disturbance by humans. We add a third axis: old growth defined by forest continuity through time (Norden et al. 2014), regardless of the current age structure of the tree canopy. In a report on old growth in Great Smoky Mountains National Park, Johnson (1995) used the phrase "ecosystem age" to represent this concept. Forests with long continuity, through multiple generations of trees without direct human disturbance, have also been called "primary forest" (Ervin 2016; chapter 7). The three axes are seemingly self-evident descriptors of old-growth forest, yet they are not always simultaneously true of a particular forest and all of them present problems even when considered individually (table 4-2; Leverett 1996). It is not surprising, then, that papers often recommend flexibility in applying old-growth criteria (e.g., Region 8 Old-Growth Planning Team 1997). After discussing these three criteria (figure 4-3; table 4-2), we add a fourth dimension at the landscape scale.

Many old-growth characteristics result from the interaction of these axes. For instance, mixed age structures, the presence of standing snags, and large logs and coarse woody debris in all stages of decay are the result not only of the agedness of the canopy (axis 1), but also the number of previous generations of trees (axis 3). Late successional composition and structure are the result of multiple generations of forest trees (axis 3) but imply both aged trees (axis 1) and low human influence (axis 2). Tree age criteria (axis 1) are often set to a value that dates tracts to presettlement, preindustrial, or prelogging times, thus also identifying forests with moderate to long continuity as forest ecosystems (axis 3). However, some forests that originated on Native American old fields would qualify based on tree age alone (axis 1), even though they had direct human disturbance (axis 2). In the late 1990s, for instance, the age criterion in the southern Appalachians is often set to greater than or equal to 150 years for trees, but this criterion depends on species composition and is sometimes set as low as 80 years for some forest types (Region 8 Old-Growth Planning Team 1997).

Figure 4-3 suggests two alternatives for designating old-growth forests in our landscape: a narrow definition and an inclusive definition. Under the narrow definition, in which all three axes are at maximum values, no area of old growth remains in the southern Appalachians today because 1) indirect, diffuse human disturbances like forests pests and climate change mean that no areas are pristine (axis 2), and 2) natural disturbances result in stands that lack old trees (axis 1). The narrow view is not tenable as the basis of policy and would cause land managers to overlook the value of primary forests with long continuity as forests. Including continuity (axis 3) as a factor emphasizes the value of

TABLE 4-2. Criteria for old growth forest (expanded from Leverett 1996; see also Pyle 1985; Yost et al. 1994; Johnson 1995). The criteria are organized by three axes (see also figure 4-3). Axis 2 (human influence) and axis 3 (forest continuity) are sometimes implicitly combined when old growth is referred to as virgin, primeval, original, primary, ancient, or presettlement forest.

### Canopy age structure (axis 1)

*Tree-age criteria: Tree age relative to maximum longevity*

In all-aged forests, only a percentage of trees are near maximum age, leading to: >majority of canopy at least ½ maximum age; median age >½ maximum lifespan (Cogbill 1996); mean age >150 years (Cogbill 1996); oldest trees >200 years (Runkle 1996); trees old enough for owl nesting sites, 524 years (Norden et al. 2014); >150 years of age or older than settlement; older than rotational age for a managed stand (Tyrell 1996).

*Tree-size criteria: Tree size relative to maximum size on a given site type*

In all-aged forests, only a percentage of trees are near maximum size, leading to such criteria as: large trees over 75 centimeters DBH, >7 per hectare (Runkle 1996).

*Challenges with age and size*

Trees can be difficult to age because of rotting or hollow cores; maximum tree age and size vary along gradients and trees on unfavorable sites might not be large; density and size and age structure varies with scale (Busing and White 1993); trees near maximum size and life spans occur on long-abandoned Native American fields; old and large trees were left by loggers if they were noncommercial species or had poor form.

### Successional status (axes 1, 2, and 3; also referred to as "climax" structure/composition)

*Successional criteria: Composition*

Steady state composition and structure; late successional composition; understory plants with long generation time, low reproductive allocation, limited dispersal (Norden et al. 2014); presence of preferred timber species (Paulson et al. 2016).

*Successional criteria: Structure*

Shade-tolerant canopy and reproduction, all-aged or uneven-aged structure; low light level at herb layer, 0.3–2 percent of light (Greenberg et al. 1997); trees with evidence of competition in shady environments (no large lower branches or branch scars, undivided trunks that only branch in or just below the canopy); pit-and-mound topography; soil with thick organic layer and macropores (Greenberg et al. 1997); large snags and coarse woody debris, logs in all stages of decay ( e.g., >10 snags per hectare at least 10 centimeters DBH; Runkle 1996), >19 logs per hectare at least 30 centimeters DBH (Runkle 1996); 12 +6 decadent trees per hectare (Greenberg et al. 1997); tree crowns rounded or flattened; emergent tree crowns damaged by wind and lightning.

*continued on next page*

TABLE 4-2. *continued*

### Successional status (axes 1, 2, and 3; also referred to as "climax" structure/composition)

*Successional criteria: Dynamics*

Continuous reproduction; uneven aged; no net growth, equilibrial, mature or over-mature; gaps or turnover rate 0.6–2 percent per year (Runkle 1996); tree rings show several long growth suppression periods (Di Filippo et al. 2017).

*Challenges with successional criteria*

Large trees create large gaps, so light-demanding successional species can be retained in old growth; forest structure exhibits scale dependence and this varies by species and disturbance type (Busing and White 1993); natural disturbances can reset communities to earlier successional states; coarse woody debris is removed by some natural disturbances (flood scour) and redistributed to flood-plains downstream; disturbance regimes vary with site, community, and time (e.g., climate variation); on extreme sites, such as thin-soiled, steep, talus slopes, early successional trees can reach old ages and be in stable populations.

### Human influence (axis 2), (naturalness/pristineness)

*Human-influence criteria*

No or minimal human impact; affected only by natural conditions; forests never logged or farmed; developed only after natural disturbances (Bradshaw et al. 2015).

*Challenges with human-influence criteria*

Historical data may be inaccurate or absent; diffuse indirect human disturbances are universal; fire is hard to attribute to human or natural influence.

### Continuity (axis 3)

*Continuity criteria*

Primary forest never logged regardless of current tree age, > 1,000 years as continuous forest (Bradshaw et al. 2015); ancient woodland defined as sites with trees since 1600, 1700, or 1775 AD (Peterken 1996), regardless of human use; stand age > time since settlement.

*Challenges of continuity criteria*

Forest history is often hard to document for specific sites; highly selective logging may be difficult to map on the landscape; some chestnut trees were felled to reduce the risk of fire-initiating lightning strikes, with logs left in place, but this can be confused with selective logging; forests are hard to document prior to 1500–1800, depending on location (Norden et al. 2014).

forests that were never directly disturbed by logging or agriculture, even if they have—and will continue to have—natural disturbances and diffuse human impacts that prevent trees from attaining old age. Areas with little disturbance and high continuity, as well as old trees, can be mapped to represent the narrow definition of "old growth"—"old growth" in the more traditional sense.

Continuity has been emphasized as an important descriptor for old growth, particularly in Europe (Norden et al. 2014). Recently, Veldman et al. (2015) applied the concept to "grassy biomes," including woodlands, savannas, and grasslands, in Africa. They pointed to the role of grazing animals and fire, along with the difficulty of applying Clementsian successional concepts to define late-successional or climax status, as leading to a lack of recognition, appreciation, and conservation of old habitats that did not necessarily have old trees yet were old and continuous belowground. In this book, Peet et al. (chapter 3) convincingly characterize fire-maintained pine savannas of the southeast US as meriting old-growth status. These examples question the narrow approach of restricting old growth to self-reproducing, mixed-aged forests with late successional composition and structure. Continuity gives us the opportunity to place older tree canopies in the context of their successional development, thus incorporating the heterogeneity and absence of direct human impact that can be important for biodiversity (Flensted et al. 2016).

We propose adding a fourth set of criteria for assessing potential old-growth value: temporal and spatial scale. Although even isolated old trees have ecological value (Lindenmayer 2016), the successional dynamics of old-growth forests suggest that area and spatial context are critical to old growth function and resilience. Pickett and Thompson (1978) introduced the concept of "minimum dynamic area" for an ecosystem, defined as the area required to encompass patches of all successional ages, which are thus large enough to support ecosystem regeneration based on the pattern of natural disturbances (see chapter 8). As natural disturbances range from smallscale gap dynamics to largerscale disturbances in old-growth forests, the minimum dynamic area would need to be, following Shugart's (1984) estimates, a minimum of 50 times the scale of these patches. Larger-scale heterogeneity is needed to incorporate the biodiversity in all phases of disturbance and succession across the range of patch types and sizes.

Temporal scales must also be considered to capture functional values of old-growth forests. Turner et al. (1993) used two axes to graphically model dynamic stability: the ratio of disturbed area to landscape area, the concept behind Shugart's 50-to-1 rule, and the ratio of disturbance interval to recovery time, which represents the capacity of the system to recover between disturbances. As the continuity axis of old growth implies variation in the frequency and size of natural disturbance patches, the density and population structure of tree species would be expected to reflect species' responses to the historical range of variation of natural disturbance (Keane et al. 2009). In southern Appalachian mesic cove forests, Busing and White (1993) found that the coefficient of variation for population structure was scale dependent and that more light-demanding species, like tulip tree, had higher coefficients of variation across scales than shade-tolerant species, suggesting that the density of these species was a function of the history and intensity of wind disturbance. The fourth dimension proposed here, then, which includes both spatial and temporal components, emphasizes crucial functional aspects of old-growth tracts that depend on their size and heterogeneity, in addition to their age, naturalness, and continuity.

## Conclusion

Valuable old-growth forests remain in the southern Appalachians, but an inventory of these forests needs to be completed using a consistent, regionwide approach. This is particularly needed for national forest lands to ensure conservation of tracts beyond those currently protected. An inventory should be based on a careful review and centralized archiving of logging and settlement records (completed or underway in the work of Quentin Bass and others), followed by field evaluation, as long-recommended by Messick (pers. comm.). LiDAR is a valuable new tool that should be used to map forest structure across the landscape as an adjunct to further historical research and field evaluation (Ervin 2016). Adding the idea of forest continuity to more traditional evaluation criteria is especially important in helping to identify areas for protection. Forests with continuity but lacking dominance by old trees and/or lacking late-successional structural and compositional status, especially those on sites with small maximum tree sizes (e.g., xeric sites), have been termed *cryptic old growth* by White (1995). There may be substantial acreage of such forests, and completing an inventory is a high priority (Messick pers. comm.). Field work should also assess how soils in ecosystems with continuity differ in physical and biological properties from those with varying degrees of past logging and settlement (chapter 10). Finally, an assessment of biodiversity of forests of long continuity, whether or not they are presently characterized by old trees, is needed, including insects, fungi, and other taxa that characterize hollow trees, snags, decaying logs, and undisturbed soils (chapter 11).

Larger spatial scales that capture structural and temporal heterogeneity contribute to old-growth functions, such as resilience and retention of biodiversity. The focus should be not just on the big tree stands or the older-successional states, but the complete mosaic of successional states that is possible at larger spatial scales to create tracts of minimum dynamic area and landscape resilience (Shugart 1984; Turner et al. 1993). For this reason, efforts should be made to designate old-growth landscapes where the value of old growth is improved by including second-growth lands on logged sites that have a high potential for recovery as old growth. As we argue here for the oldest patches, we posit that earlier successional patches are also needed and cannot be duplicated by highly modified and intensely managed successional patches. We believe that we can not only complete the identification and conservation of current old growth but we can also build larger landscapes that can themselves become the old-growth forests of the future.

## Acknowledgments

We are grateful to Robert Messick for sharing his considerable knowledge on southern Appalachian old growth and for the dedication he has shown over many years in advocating for its documentation and protection. We also thank Andrew Barton, one of the editors of this volume, for providing valuable review comments.

## References

Ayers, H. B., and W. W. Ashe. 1905. "The Southern Appalachian forests." Professional Paper #37, Series H, Forestry; 12. US Geological Survey.

Bartram, W. 1791. *Travels of William Bartram*. M. Van Doren, ed. New York: Dover Publications.

Bradshaw, R. H. W., C. S. Jones, S. J. Edwards, and G. E. Hannon. 2015. "Forest continuity and conservation value in Western Europe." *The Holocene 25*: 194–202.

Busing, R. T. 2005. "Tree mortality, canopy turnover, and woody detritus in old cove forests of the southern Appalachians." *Ecology 86*: 73–84.

Busing, R. T., and P. S. White. 1993. "Effects of area on oldgrowth forest attributes: implications for the equilibrium landscape concept." *Landscape Ecology 8*: 119–126.

Butler, S. M., A. S. White, K. J. Elliott, and R. S. Seymour. 2014. "Disturbance history and stand dynamics in secondary and old-growth forests of the Southern Appalachian Mountains, USA." *The Journal of the Torrey Botanical Society 141*: 189–204.

Cogbill, C. V. 1996. "Black growth and fiddlebutts: the nature of old-growth red spruce." In *Eastern Old-Growth Forests: Prospects for Rediscovery and Recovery*, edited by M. B. Davis, 113–125. Washington, DC: Island Press.

Cohen, D., B. Dellinger, R. Klein, and B. Buchanan. 2007. "Patterns in lightning-caused fires at Great Smoky Mountains National Park." *Fire Ecology Special Issue 3*: 68–82.

Davis, D. E. 2000. *Where There are Mountains: an Environmental History of the Southern Appalachians*. Athens, GA: University of Georgia Press.

Davis, M. B., ed. 1996. "Extent and location." In *Eastern Old-Growth Forests: Prospects for Rediscovery and Recovery*, edited by M. B. Davis, 18–34. Washington, DC: Island Press.

Delcourt, H. R. 1987. "The impact of prehistoric agriculture and land occupation on natural vegetation." *Trends in Ecology and Evolution 2*: 39–44.

Di Filippo, A., F. Biondi, G. Piovesan, and E. Ziaco. 2017. "Tree ring-based metrics for assessing old-growth forest naturalness." *Journal of Applied Ecology 54*: 737–749.

Ervin, J. 2016. "Describing forest structure in Southern Blue Ridge cove forests: a LiDAR-based analysis." Master's Thesis. Burlington, VT: University of Vermont.

Fakhraei, H., C. T. Driscoll, J. R. Renfro, M. A. Kulp, T. F. Blett, P. F. Brewer, and J. S. Schwartz. 2016. "Critical loads and exceedances for nitrogen and sulfur atmospheric deposition in Great Smoky Mountains National Park, United States." *Ecosphere 7*(10): e01466. doi: 10.1002/ecs2.1466.

Flensted, K. K., H. H. Bruun, R. Ejrnaes, A. Eskildsen, P. F. Thomsen, and J. Heilmann-Clausen. 2016. "Red-listed species and forest continuity – A multi-taxon approach to conservation in temperate forests." *Forest Ecology and Management 378*: 144–159.

Greenberg, C. H., D. E. McLeod, and D. L. Loftis. 1997. "An old-growth definition for western and mixed mesophytic forests." General Technical Report SRS-16. Southern Research Station. USDA Forest Service.

Greenberg, K., F. R. Thompson, and B. Collins. 2011. *Sustaining Young Forest Communities – Ecology and Management of Early Successional Habitat in the US Central Hardwood Region*. Berlin: Springer-Verlag.

Harmon, M., S. P. Bratton, and P. S White. 1983. "Disturbance and vegetation response in relation to environmental gradients in the Great Smoky Mountains." *Vegetatio 55*: 129–139.

Harrod, J. C., M. E. Harmon, and P. S. White. 2000. "Post-fire succession and twentieth century reduction in fire frequency on xeric southern Appalachian sites." *Journal of Vegetation Science 11*: 465–472.

Jenkins, M. A., S. Jose, and P. S. White. 2007. "Impacts of a forest fungal disease on forest community composition and structure and the resulting effects on foliar calcium cycling." *Ecological Applications 17*: 869–881.

Johnson, K. 1995. "Eastern hemlock delineations in Great Smoky Mountains National Park (1993-1995)." Unpublished report. Great Smoky Mountains National Park, Resource Management and Science Division. USDI National Park Service.

Kaylor, S. D., M. L. Hughes, and J. A. Franklin. 2017. "Recovery trends and predictions of Fraser fir (*Abies fraseri*) dynamics in the Southern Appalachian Mountains." *Canadian Journal of Forest Research 47*: 125–133.

Keane, R. E., P. F. Hessburg, P. B. Landres, and F. J. Swanson. 2009. "The use of historical range and variability (HRV) in landscape management." *Forest Ecology and Management 258*: 1025–1037.

Kinkaid, J. A., and A. J. Parker. 2008. "Structural characteristics and canopy dynamics of *Tsuga canadensis* in forests of the southern Appalachian Mountains, USA." *Plant Ecology 199*: 265–280.

Lafon, C. W., A. T. Naito, H. D. Grissino-Mayer, S. P. Horn, and T. A. Waldrop. 2017. "Fire history of the Appalachian region: review and synthesis." General Technical Report SRS-219. Southern Research Station. USDA Forest Service.

Leverett, R. 1996. "Definitions and history." In *Eastern Old-growth Forests: Prospects for Rediscovery and Recovery*, edited by M. B. Davis, 3–17. Washington, DC: Island Press.

Lindenmayer, D. B. 2016. "Conserving large old trees as small natural features." *Biological Conservation*. ddx.doi.org/10.1016/j.biocon.2016.11.012.

MacKenzie, M. D., and P. S. White. 1998. "The vegetation of Great Smoky Mountains National Park: 1935–1938." *Castanea 63*: 323–336.

Madden, M., R. Welch, T. Jordan, P. Jackson, R. Seavey, and J. Seavey. 2004. "Digital vegetation maps for the Great Smoky Mountains National Park." Final report. Center for Remote Sensing and Mapping Science. Athens: University of Georgia.

McNab, W. H. 1993. "A topographic index to quantify the effect of mesoscale landform on site productivity." *Canadian Journal of Forest Research 23*: 1100–1107.

Messick, R. E. 2004. "High quality reconnaissance and verification in old-growth forests of the Blue Ridge Province (cataloged mainly on National Forest lands)." Funded by 11 organizations including SAFC.

Norden, B., A. Dahlberg, T. E. Brandrud, O. Fritz, R. Ejrnaes, and O. Ovaskainen. 2014. "Effects of ecological continuity on species richness and composition in forests and woodlands: a review." *Ecoscience 21*: 34–45.

Paulson, A. K., S. Sanders, J. A. Kirschbaum, and D. M. Waller. 2016. "Post-settlement ecological changes in the forests of the Great Lakes National Parks." *Ecosphere 7*: e01490. doi: 10.1002/ecs2.1490.

Pederson, N., J. M. Dyer, R. W. McEwan, A. E. Hessl, C. J. Mock, D. A. Orwig, H. E. Rieder, and B. I. Cook. 2014. "The legacy of episodic climatic events in shaping temperate, broadleaf forests." *Ecological Monographs 84*: 599–620.

Peterken, G. F. 1996. *Natural Woodland: Ecology and Conservation in Northern Temperate Regions*. Cambridge, England: Cambridge University Press.

Pickett, S. T. A., and J. N. Thompson. 1978. "Patch dynamics and the design of nature reserves." *Biological Conservation 13*: 27–37.

Pyle, C. 1985. "Vegetation disturbance of the Great Smoky Mountains National Park: an analysis of archival maps and records." Research/Resource Management Report SER-77. Southeast Region. USDI National Park Service.

Region 8 Old-Growth Team. 1997. "Guidance for conserving and restoring old-growth forest communities in the National Forests in the Southern Region." USDA Forest Service.

Runkle, J. R. 1985. "Disturbance regimes in temperate forests." In *The Ecology of Natural Disturbance and Patch Dynamics*, edited by S. T. A. Pickett and P. S. White, 17–33. Orlando, FL: Academic Press.

Runkle, J. R. 1996. "Central mesophytic forests." In *Eastern Old-Growth Forests: Prospects for Rediscovery and Recovery*, edited by M. B. Davis, 161–177. Washington, DC: Island Press.

Shugart, H. H. 1984. *A Theory of Forest Dynamics*. Berlin: Springer-Verlag.

Sullivan, T. J. 2017. *Air Pollution and Its Impacts on US National Parks*. Boca Raton, FL: CRC Press.

Turner, G. M., W. H. Romme, R. H. Gardner, R. V. O'Neill, and T. K. Kratz. 1993. "A revised concept of landscape equilibrium: disturbance and stability on scaled landscapes." *Landscape Ecology 8*: 213–227.

Tuttle, J. P., and P. S. White. 2016. "Structural and compositional change in Great Smoky Mountains National Park since protection, 1930s–2000s." In *Natural Disturbances and Historic Range of Variation*, edited by C. H. Greenberg, and B. S. Collins, 263–294. Berlin: Springer International Publishing.

Tyrrell, L. E. 1996. "National forests in the Eastern Region: land allocation and planning for old growth." In *Eastern Old-growth Forests: Prospects for Rediscovery and Recovery*, edited by M. B. Davis, 245–273. Washington, DC: Island Press.

Veldman, J. W., E. Buisson, G. Durigan, G. W. Fernandez, S. Le Stradic, G. Mahy, D. Negreiros, et al. 2015. "Toward an old-growth concept for grasslands, savannahs, and woodlands." *Frontiers of Ecology and the Environment 13*: 154–162. doi:10.1890/140270.

White, P. S. 1995. "Conserving biodiversity: lessons from the Smokies." *FORUM for Applied Research and Public Policy 10(2)*: 114–120.

White, P. S., B. Collins, and G. R. Wein. 2011. "Natural disturbances and early successional habitats." In *Sustaining Young Forest Communities – Ecology and Management of Early Successional Habitat in the US Central Hardwood Region*, edited by K. Greenberg, F. R. Thompson, and B. Collins, 27–40. Berlin: Springer-Verlag.

Whittaker, R. H. 1956. "Vegetation of the Great Smoky Mountains." *Ecological Monographs 26*: 1–80.

Yost, E. C., K. E. Johnson, and W. F. Blozan. 1994. "Old-growth project: stand delineation and disturbance rating, Great Smoky Mountains National Park." Technical Report. Southeast Region. USDI National Park Service.

# Chapter 5

# Topography and Vegetation Patterns in an Old-Growth Appalachian Forest: Lucy Braun, You Were Right!

*Julia I. Chapman and Ryan W. McEwan*

The biologically diverse Appalachian forests of eastern North America are an especially interesting and important example of the complex relationship between physiographic factors (e.g., elevation), disturbance processes, and long-term shifts in forest composition. Due to the widespread and often intense land-use practices of Euro-Americans, particularly circa 1880 to 1930, past human activity is an important component of the pattern and process we observe in forests of eastern North American today. Only a few small parcels of forest remain where dynamics have been driven mainly by nonanthropogenic phenomena, and these old-growth forests provide a crucial window to the past and an important baseline for the present and future. Understanding the historical and contemporary drivers of long-term dynamics has become an increasingly important goal in ecology as anthropogenically driven declines in biodiversity, including extinctions, and undesirable shifts in community composition threaten the performance of ecosystems as well as the benefits derived from them by humans (Pimm et al. 2014).

In the brilliant series of papers that preceded her defining book, *Deciduous Forests of Eastern North America* (1950), E. Lucy Braun assessed the relationship between topographic features (e.g., elevation, aspect) and the distributions and associations of various plant species to describe how forest communities assemble according to landscape features. In her manuscript, "An Ecological Transect of Black Mountain, Kentucky," Braun (1940) describes the forces that delineate species community patterns, noting (p. 239–240): "The individualistic concept suggested by Gleason (1926) might be applicable here, where communities with rather definite visible expression are the result of sorting under influences which have been in continuous operation for long periods of time." Two of the most commonly identified influences on species sorting that may operate for long periods in forests are environmental filtering and competitive interactions. These two drivers are not mutually exclusive, though, and their relative importance is hypothesized to vary with environmental context. This phenomenon has been termed the "stress-dominance hypothesis" (SDH) and encompasses two of the primary strategies for plant survival under a low-intensity disturbance regime outlined by Grime (1977; see also Coyle et al. 2014; Weiher and Keddy 1995). In harsh environments, species' survival is dependent upon adaptations that allow persistence despite resource limitations. In these "filtered" communities, selection for a stress-tolerant strategy results in a convergence of physical characteristics with the community being comprised of species that have similar traits for coping with a common limiting resource (Weiher and Keddy 1995; Cornwell and Ackerly 2009). For example, only a subset of possible species in an Appalachian watershed may be successful growing on south-facing upper slopes with shallow, rocky, and acidic soils. The strong filtering effect exerted by such an environment results in a local floristic assemblage characterized by a common set of traits for coping with low water and nutrient availability. In contrast, in resource-rich environments, species are not selected for their ability to tolerate abiotic stress but rather their ability to compete for resources. In this case, strong competition is hypothesized to result in a set of species possessing a diverse range of traits for acquiring the essentials of life. This selection for a variety of unique trait syndromes allows the exploitation of a variety of different niches, which helps to minimize competition between species (Hardin 1960; Weiher et al. 1998; Chesson 2000).

While the importance of the local environment to community assembly was clearly evident to Braun in her detailed descriptions of various community types and their associated habitats, the ability to identify the underlying mechanisms (environmental filtering, competition) using statistical methods would not be possible for several more decades. In fact, Braun herself recognized that "detailed statistical studies" could have enhanced her work (p. 417, Braun 1942). Unfortunately, much of the old-growth forest in eastern North American has been lost, and many of the sites that Braun characterized have since been destroyed or disrupted by human activity, such as logging and land clearing for agriculture. Old-growth forests that do remain serve as crucial model systems for understanding baseline forest dynamics under minimal human influence. Analyzing long-term trends of these sites with modern analytical techniques can provide important insights on natural ecosystem processes that would otherwise be obscured by anthropogenic disturbance.

## Beyond Linneaus: Trait-Based Approaches to Quantifying Diversity

Understanding how communities assemble has often led scientists to analyze diversity. The standard approach has focused on quantifying the number and abundance of different species in an area based on taxonomic classifications. An important limitation of this approach is that the unique characteristics or traits of the species present in a community are ignored. Imagine two communities having an equal number of trees: one community contains three different oak species while the other consists of a pine, a maple, and a magnolia. Even though species richness is the same, the two communities represent very different collections of life histories, adaptations, and survival strategies. The oaks, being closely related within the same genus *Quercus*, have more in common in terms of their physical characteristics than the pine, maple, and magnolia, which are more distantly related across three different genera (*Pinus*, *Acer*, and *Magnolia*, respectively). Understanding functional differences among species can provide different insights into how communities were formed and how they operate.

This alternative view of diversity has led to the development of new metrics that go beyond taxonomic identities by either directly quantifying physical traits (functional) or using evolutionary relatedness to approximate trait similarity among species (phylogenetic). Functional diversity indices use direct measurements of growth-form characteristics (e.g., leaf size, height) and physiological processes (e.g., water use, photosynthetic rate) to understand within-community variation. Phylogenetic diversity indices use cladograms (phylogenetic trees) to measure the evolutionary distance among species in a community to approximate trait similarity based on the principle of "phylogenetic niche conservatism"—the assumption that closely related species are more similar in morphology and physiology than distantly related species and, thus, have similar adaptations (Wiens et al. 2010). Following this approach, ecological communities, which are made up of species with higher levels of phylogenetic and functional diversity, represent a wide array of strategies and suggest environmental conditions where resources are plentiful and that competition may be the most important factor determining species composition. In contrast, communities with lower phylogenetic and functional diversity indicate harsh environments are "filtering" the community by selecting for a narrow set of survival strategies (Webb et al. 2002; Spasojevic and Suding 2012).

Functional and phylogenetic diversity indices provide a new means for understanding how abiotic environmental gradients influence plant community composition. Support for the stress-dominance hypothesis (SDH) has been found in tropical forest communities where many studies have shown that trait convergence is more prevalent in stressful conditions, indicating that topography and soil fertility can act as a strong environmental filter (Kraft et al. 2008; Culmsee and Leuschner 2013; Lasky et al. 2013; Fortunel et al. 2014). Similar studies have been conducted in temperate forests of North America and Asia, but they are fewer, and the results seem to show less consistent support for SDH (Coyle et al. 2014; Sabatini et al. 2014; Spasojevic et al. 2014; Kitagawa et al. 2015). We are unaware of any studies explicitly testing SDH at local scales in old-growth Appalachian forests. The Appalachians offer an interesting opportunity for testing this theory because the region is considered a biodiversity hot spot for North America (Ricketts et al. 1999), and the importance of local topographic variation in promoting biological diversity has long been recognized in this geographic area (Braun 1940; Whittaker 1956; Boerner 2006). In addition, there has been a widespread, ongoing shift in tree species composition across eastern US forests where oak species (*Quercus* spp.) are exhibiting reduced regeneration in some areas, while mesophytic species such as maple (*Acer* spp.) are thriving (Abrams 1998; Nowacki and Abrams 2008). The contrast in life histories and traits between these two groups is a point of interest, because a transition from oaks to maples could have important implications for wildlife populations, ecosystem function, and forest management (Arthur et al. 2012; Alexander and Arthur 2010). Understanding the role of topographic variation in structuring communities based on the life histories (i.e., functional traits) of species rather than just the number of species present can provide greater insight on this oak-to-maple shift and the degree of uniformity to which it is occurring across eastern North American forests. We wondered whether there are some habitat types where this successional dynamic is occurring more slowly because strong environmental filtering is maintaining a particular assemblage of species despite the changes in climate, disturbance regime, and human impacts that are the proposed drivers of the oak-to-maple shift (McEwan et al. 2011; Pederson et al. 2014).

## The Lilley Cornett Woods Forest Dynamics Project

Big Everidge Hollow (BEH) is a 52-hectare watershed containing a stand of old-growth forest located within the Lilley Cornett Woods Appalachian Research Station in eastern Kentucky (Martin 1975). The research history of this site is rich, including a long-term dataset on overstory trees (woody stems greater than or equal to 2.5 centimeters in diameter at breast height [DBH]). A set of 80 permanent sampling plots (0.04 hectares each) was established in 1979 by Robert Muller throughout the watershed, arrayed across north-, east-, and south-facing slopes at low, mid, and high elevations. After the first overstory sampling that same year, Muller (1982) classified each plot into one of three community types based on the dominant overstory tree species in the plot: 32 chestnut oak, 31 beech, and 17 mixed mesophytic. Since 1979, the overstory has been resampled at approximately decadal intervals, in 1989, 1999, and 2010. The local environment of each plot has also been characterized for topography (slope, aspect, and elevation) and a number of soil parameters (McEwan et al. 2017).

For this analysis, tree data were divided into two strata, overstory (greater than or equal to 25 centimeters DBH) and midstory (2.5 to 24.99 centimeters DBH), and we assessed changes of these strata through time from 1979 to 2010. To measure phylogenetic diversity, we used mean pairwise distance (MPD), which represents the mean evolutionary distance among the species present in a community. It is calculated from the branch lengths separating species in a cladogram (diagram representing evolutionary relationships among species). To measure functional diversity, we used functional dispersion ($F_{Dis}$; Laliberté and Legendre 2010), which is a trait-based analog of the Shannon diversity index that uses functional trait characteristics of each species present, such as wood density, maximum height, and seed mass, to name a few, to calculate diversity instead of the taxonomic identities of species. In this manner, plots containing tree species with a wide range of values for functional traits of interest would have higher functional diversity than plots where the functional trait values of the species were all roughly equivalent. Lastly, weighted mean values for each functional trait were calculated for each plot (community-weighted mean, CWM) to represent the dominant trait values (Garnier et al. 2004). Mean pairwise distance, functional dispersion, and the trait community-weighted means are all abundance-weighted metrics, so basal area was used as the abundance measure for each species within each plot. More detailed information about these calculations and statistical analyses is given at the end of the chapter.

## The Influence of Topography on Patterns of Forest Diversity

Previous analyses of soil conditions within our old-growth central Appalachian forest site revealed a gradient of soil fertility across the highly varied topography and demonstrated that tree communities separate along this gradient (Muller 1982; McEwan et al. 2005). South-facing higher elevation portions of our site tend to have low soil moisture and nutrients (McEwan et al. 2005; McEwan and Muller 2006), and our new analyses revealed that these more water stressed environments were associated with species possessing a narrower range of trait values. These findings align with other studies that have shown functional trait convergence and phylogenetic clustering to be associated with more stressful environments (Kraft et al. 2008; Culmsee and Leuschner 2013; Coyle et al. 2014; Spasojevic et al. 2014). Aspect and slope were less important than elevation in our study, but where relationships existed, functional diversity was higher on steeper and more north-facing slopes.

Our study of local patterns of phylogenetic and functional diversity showed that these measures were strongly linked with elevation and, less strongly, with aspect. Recalling the stress-dominance hypothesis mentioned earlier, our results suggest that there is a stress gradient within our study site where, at one end of the gradient, stressful conditions (dry, south-facing areas at high elevation) act as an environmental filter as plant communities develop and result in assemblages of tree species that are (a) closely related (low phylogenetic diversity), (b) have similar life history characteristics (low functional diversity), and (c) possess physical characteristics that are adaptations for survival in moisture stressed conditions (based on trait CWMs, explained below). In other areas of our site with more abundant resources (lower elevation, north-facing), tree communities tend to (a) be more distantly related (high phylogenetic diversity; figure 5-1), (b) have more divergent life history characteristics (high functional diversity; figure 5-2), and (c) possess physical characteristics associated with a competitive strategy, such as fast growth. The trait community-weighted means associated with stressful habitats indicated a conservative, drought-tolerant strategy: On average, the tree community there is dominated by species with high wood density (slow growth), low specific leaf area and leaf nitrogen content (smaller but thicker leaves with low photosynthetic rates), and low maximum height (Stahl et al. 2013; table 5-1). In the more resource-abundant habitats, the tree communities were dominated by species with low wood density (fast growth), high specific leaf area and leaf nitrogen content (broader but thinner leaves with high photosynthetic rates), and high maximum height (ability to compete for light).

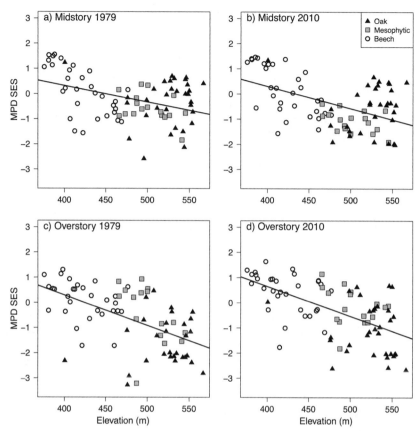

FIGURE 5-1. Phylogenetic diversity (standardized effect sizes of mean pairwise distance, MPD) in relation to the elevation gradient through time (1979–2010) in Big Everidge Hollow. Symbols designate the dominant overstory community type of each plot: chestnut oak (dark triangle), beech (open circle), or mixed mesophytic (gray square). The midstory includes stems 2.5–24.99 centimeters DBH, and the overstory includes stems greater than or equal to 25 centimeters DBH.

## Temporal Trends of Phylogenetic and Functional Diversity

We were surprised to find that the relationships between functional diversity, phylogenetic diversity, and topography generally became stronger over the 30-year period (table 5-1). Originally, we hypothesized that the relationships between tree functional and phylogenetic diversity and the environment would weaken as a result of a long-term shift in species composition at our site where juvenile maples are abundant and

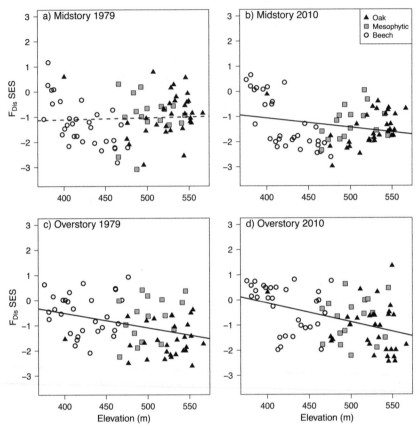

FIGURE 5-2. Functional diversity (standardized effect sizes of functional dispersion, $F_{Dis}$) in relation to the elevation gradient through time (1979–2010) in Big Everidge Hollow. Symbols designate the dominant community type of each plot: chestnut oak (dark triangle), beech (open circle), or mixed mesophytic (gray square). The midstory includes stems 2.5–24.99 centimeters DBH, and the overstory includes stems greater than or equal to 25 centimeters DBH.

are increasingly dominant across the watershed (Chapman and McEwan 2016). Hypothetically, this pattern should be driving all communities toward more similar phylogenetic and functional diversity values as mesophytic species, such as maple, become more dominant throughout the watershed and vegetation becomes more homogeneous. We expected that this homogenization would make the filtering effect of the local environment less detectable. Instead, we found that trait-environment relationships were largely consistent through time, mirroring the find-

TABLE 5-1. Multiple linear regression model results testing for relationships between topography and phylogenetic diversity (mean pairwise distance, MPD), functional diversity (functional dispersion, $F_{Dis}$), or community-weighted means (CWM) of traits within the midstory (2.5–24.99 centimeters DBH) and overstory ($\geq 25$ centimeters DBH) strata. MPD, $F_{Dis}$, and CWMs are weighted by basal area. Significant explanatory variables are topographic variables (slope, aspect, elevation) that were statistically significant predictors of functional or phylogenetic diversity (N.S. indicates that none of the topographic variables were significant predictors in the model). Model $R^2$ values represent the strength of the associations between the explanatory and response variables where values closer to 1 indicate a stronger relationship. The midstory includes stems 2.5–24.99 centimeters DBH, and the overstory includes stems $\geq 25$ centimeters DBH.

| Response | Significant Explanatory Variables (Coefficient) | Model $R^2$ | Model P |
|---|---|---|---|
| **Midstory 1979** | | | |
| MPD | Elevation (-0.007) | 0.153 | <0.001 |
| $F_{Dis}$ | Aspect (0.257) | 0.037 | 0.049 |
| Wood Density | Elevation (0.0003) | 0.261 | <0.001 |
| Specific Leaf Area | N.S. | -0.003 | 0.433 |
| Leaf Nitrogen | Elevation (-0.0001) | 0.048 | 0.029 |
| Maximum Height | Aspect (3.20), Elevation (-0.116) | 0.372 | <0.001 |
| Seed Mass | N.S. | -0.024 | 0.763 |
| **Midstory 2010** | | | |
| MPD | Aspect (-0.277), Elevation (-0.009) | 0.281 | <0.001 |
| $F_{Dis}$ | Elevation (-0.004), Slope (0.015) | 0.107 | 0.005 |
| Wood Density | Elevation (0.0003) | 0.185 | <0.001 |
| Specific Leaf Area | Elevation (0.179) | 0.103 | 0.002 |
| Leaf Nitrogen | N.S. | -0.027 | 0.823 |
| Maximum Height | Aspect (4.157), Elevation (-0.106), Slope (0.181) | 0.381 | <0.001 |
| Seed Mass | N.S. | 0.003 | 0.367 |
| **Overstory 1979** | | | |
| MPD | Elevation (-0.012) | 0.300 | < 0.001 |
| $F_{Dis}$ | Elevation (-0.006), Slope (0.017) | 0.178 | < 0.001 |
| Wood Density | Aspect (-0.017), Elevation (0.0003) | 0.149 | 0.001 |

*continued on next page*

TABLE 5-1. *continued*

| Response | Significant Explanatory Variables (Coefficient) | Model R² | Model P |
|---|---|---|---|
| | **Overstory 1979** | | |
| Specific Leaf Area | Aspect (21.4), Elevation (-0.332) | 0.328 | < 0.001 |
| Leaf Nitrogen | Aspect (0.016) | 0.102 | 0.002 |
| Maximum Height | Elevation (-0.067) | 0.156 | < 0.001 |
| Seed Mass | Aspect (-0.270), Elevation (0.007) | 0.453 | < 0.001 |
| | **Overstory 2010** | | |
| MPD | Elevation (-0.012) | 0.343 | <0.001 |
| $F_{Dis}$ | Elevation (-0.008) | 0.175 | <0.001 |
| Wood Density | Elevation (0.0004) | 0.199 | 0.001 |
| Specific Leaf Area | Elevation (-0.311) | 0.162 | <0.001 |
| Leaf Nitrogen | N.S. | 0.007 | 0.323 |
| Maximum Height | Elevation (-0.089) | 0.270 | <0.001 |
| Seed Mass | Aspect (-0.179), Elevation (0.007) | 0.462 | <0.001 |

ings of Amatangelo et al. (2014) that functional traits in Wisconsin tree communities remained strongly linked to soil environmental gradients over a period of 50 years, despite losses in species richness and changes in forest composition.

From 1979 to 2010, the midstory and overstory experienced different shifts in phylogenetic and functional diversity. Plot-level tree communities in the midstory became more closely related (phylogenetic diversity decreased) and traits became more similar (functional diversity decreased), while communities in the overstory became less closely related (phylogenetic diversity increased) and traits became more divergent (functional diversity increased; table 5-2). This may represent the long-term shift noted by McEwan and Muller (2006) where the midstory (but not the overstory) frequency of oaks decreased markedly—a loss that could create a more compositionally homogenous midstory throughout the site. In contrast, as maples transition from the midstory into the overstory, they join the existing and somewhat stable previously-dominant overstory oaks, effectively increasing the phylogenetic and functional diversity of the dominant trees.

The temporal trends seen in the community-weighted trait means also

TABLE 5-2. Paired t-test results comparing levels of phylogenetic diversity (mean pairwise distance, MPD), functional diversity (functional dispersion, $F_{Dis}$), and community-weighted means of traits, CWM) between 1979 and 2010 across Big Everidge Hollow. *P*-values less than 0.05 indicate a statistically significant change in phylogenetic or functional diversity or trait mean through time from 1979 to 2010. The direction of this change (increase or decrease) is indicated by the *t*-values, where a negative value indicates a decrease in diversity or trait mean. The midstory includes stems 2.5–24.99 centimeters DBH, and the overstory includes stems ≥25 centimeters DBH.

| | Midstory | | Overstory | |
|---|---|---|---|---|
| | t | P | t | P |
| MPD | -2.29 | 0.025 | 3.45 | 0.001 |
| $F_{Dis}$ | -3.28 | 0.002 | 2.31 | 0.024 |
| Wood Density | -3.74 | <0.001 | -1.63 | 0.107 |
| Specific Leaf Area | -2.91 | 0.005 | 0.22 | 0.830 |
| Maximum Height | 6.82 | <0.001 | 1.44 | 0.154 |
| Leaf Nitrogen | -3.86 | <0.001 | -1.39 | 0.168 |
| Seed Mass | -3.21 | 0.002 | -1.62 | 0.109 |

support this notion of a dynamic midstory and stable overstory. There were significant, sitewide trait shifts in the midstory of increased dominance by mesophytic species. In particular, lower wood density and greater maximum height reflect increased recruitment of fast-growing, shade-tolerant species (table 5-2). The significant decrease in midstory seed mass reflected the loss of oak (*Quercus*) and hickory (*Carya*) species (heavy seeds) and an increase in species with lighter seeds, such as sugar maple (*Acer saccharum*), red maple (*A. rubrum*), American beech (*Fagus grandifolia*), and eastern hemlock (*Tsuga canadensis*). Decreasing leaf nitrogen content and specific leaf area (i.e., thinner leaves) in the midstory appear to be driven by large increases in the abundance of *A. saccharum* and *T. canadensis* within the beech community, influencing leaf nitrogen and SLA, and *A. rubrum* within the oak community, influencing leaf nitrogen. Overstory traits did not change significantly during this period (table 5-2), which may reflect the relative long-term stability of overstory dominants across the watershed (chestnut oak in higher elevation, mixed-mesophytic in the midslope, and beech in the lower elevation).

# Conclusion

The assembly of ecological communities in any ecosystem at any point in time represents multiple interacting drivers—both biotic and abiotic—historical and contemporary (McEwan et al. 2011). Competitive interactions and environmental filtering are two of the most commonly implicated influences that may "sort" species over "long periods of time" (see E. Lucy Braun quote above). The recent rise in popularity of phylogenetic and functional trait analyses is allowing ecologists to analyze the relative importance of these factors across a variety of ecosystems. Our study of an old-growth Appalachian forest revealed that topography is an important driver of functional and phylogenetic diversity patterns. We found that tree communities in more stressful environments (xeric, high elevation) tended to be more functionally and phylogenetically clustered, suggesting a strong environmental filtering effect as predicted by the Stress-Dominance Hypothesis and supporting E. Lucy Braun's vision of species sorting in regional mixed mesophytic forests (Braun 1940). These patterns were temporally robust, especially within the overstory stratum, which suggests that the oak-to-maple successional dynamic is occurring slowly enough that either (a) 30 years is not a long enough timeframe to see a disruption of community-environment relationships at our site, or (b) perhaps there are some strong, localized environmental filtering effects (like those on high-elevation, south-facing slopes) that are counteracting this species transition in specific areas.

We found shifts in dominant functional traits over the course of the study that may reflect the ongoing, widespread oak-to-maple compositional shift described both at our site (Chapman and McEwan 2016) and in many eastern United States forests (Abrams 1998; Lorimer 1984; Shotola et al. 1992; McEwan et al. 2011). Considering that the contributions of individual species to ecosystem functions are linked to their morphological and physiological traits, there may be serious ecological implications for these shifting trait means. Maple species, which are becoming increasingly dominant, are known to influence nitrogen cycling in forests (Lovett and Mitchell 2004; Alexander and Arthur 2010). Shifts in dominance toward species with lower wood density or higher specific leaf area could have implications for carbon storage dynamics and decomposition rates (Alexander and Arthur 2010). Oak and hickory species produce large seeds that are an important food source for a variety of mammals, and if forests continue to shift towards increased dominance of smaller-seeded mesophytic species, there may be impacts on food availability for animal species that currently depend on oak and hickory seeds as part of their diet.

## Acknowledgements

This is contribution No. 49 of Lilley Cornett Woods Appalachian Ecological Research Station, Eastern Kentucky University. Much thanks goes to project initiator Robert N. Muller and all those who assisted in data collection over the years. This work has been supported in part by the University of Dayton Office for Graduate Academic Affairs through the Graduate Student Summer Fellowship Program. The study has been supported by the TRY initiative on plant traits (http://www.try-db.org). The TRY initiative and database is hosted, developed, and maintained by J. Kattge and G. Bönisch (Max Planck Institute for Biogeochemistry, Jena, Germany). TRY is currently supported by DIVERSITAS/Future Earth and the German Centre for Integrative Biodiversity Research (iDiv) Halle-Jena-Leipzig.

## Details of the Data Analysis Methods

The site phylogeny of tree species was generated using Phylomatic (version 3; based on stored tree R20120829) and Phylocom (version 4.2; branch lengths added using "wikstrom.ages") software (Webb et al. 2008). Standardized effect sizes of mean pairwise distance (MPD; Webb et al. 2002) were calculated using standardized effect size of MPD in the R package 'picante' (Kembel et al. 2010) with null models generated by randomly shuffling taxa labels across the tips of the phylogeny for 999 runs and 10,000 iterations per run. Trait data used to calculate functional dispersion ($F_{Dis}$) and community-weighted means were obtained from GLOPNET (Wright et al. 2004), the TRY Plant Trait Database (Kattge et al. 2011), Harmon et al. (2008), Chave et al. (2009), the Kew Royal Botanic Gardens Seed Information Database (Royal Botanic Gardens Kew 2017), and the USDA Plants Database (USDA 2017). Leaf nitrogen content and seed mass trait data were log-transformed to attain normality, and all traits were standardized to a mean of zero and unit variance prior to functional richness and dispersion calculations. Standardized effect sizes of $F_{Dis}$ were calculated using null models generated by shuffling the names of traits on the trait matrix (999 runs; Mason et al. 2013; Swenson 2014). Community-weighted means of each trait were calculated using untransformed SLA, maximum height, and wood density trait values, and log-transformed leaf nitrogen and seed mass values.

Multiple linear regressions (backward selection) were used to test for relationships between phylogenetic and functional diversity measures and topographic variables (slope, aspect, elevation) within each combination of forest stratum (midstory and overstory) and time point (1979 and 2010). Paired t-tests were used to test for significant changes in diversity measures over time for the watershed as a whole (all 80 plots) and within each of the three community types (oak, beech, mixed mesophytic).

# References

Abrams, M. D. 1998. "The red maple paradox: What explains the widespread expansion of red maple in eastern forests?" *BioScience 48*: 355–364.

Alexander, H. D., and M. A. Arthur. 2010. "Implications of a predicted shift from upland oaks to red maple on forest hydrology and nutrient availability." *Canadian Journal of Forest Research 40*: 716–726.

Amatangelo, K. L., S. E. Johnson, D. A. Rogers, and D. M. Waller. 2014. "Trait-environment relationships remain strong despite 50 years of trait compositional change in temperate forests." *Ecology 95*: 1780–1791.

Arthur, M. A., H. D. Alexander, D. C. Dey, C. J. Schweitzer, and D. L. Loftis. 2012. "Refining the oak-fire hypothesis for management of oak-dominated forests of the eastern United States." *Journal of Forestry 110*: 257–266.

Boerner, R. E. J. 2006. "Unraveling the Gordian Knot: Interactions among vegetation, topography, and soil properties in the central and southern Appalachians." *The Journal of the Torrey Botanical Society 133*: 321–361.

Braun, E. L. 1940. "An ecological transect of Black Mountain, Kentucky." *Ecological Monographs 10*: 193–241.

Braun, E. L. 1942. "Forests of the Cumberland Mountains." *Ecological Monographs 12*: 413–447.

Braun, E. L. 1950. *Deciduous Forests of Eastern North America*. Philadelphia, PA: Blakiston.

Chapman, J. I., and R. W. McEwan. 2016. "Thirty years of compositional change in an old-growth temperate forest: The role of topographic gradients in oak-maple dynamics." *PloS ONE 11*: e0160238.

Chave, J., D. Coomes, S. Jansen, S. L. Lewis, N. G. Swenson, and A. E. Zanne. 2009. "Towards a worldwide wood economics spectrum." *Ecology Letters 12*: 351–366.

Chesson, P. 2000. "Mechanisms of maintenance of species diversity." *Annual Review of Ecology and Systematics 31*: 343–366.

Cornwell, W. K., and D. D. Ackerly. 2009. "Community assembly and shifts in plant trait distributions across an environmental gradient in coastal California." *Ecological Monographs 79*: 109–126.

Coyle, J. R., F. W. Halliday, B. E. Lopez, K. A. Palmquist, P. A. Wilfahrt, and A. H. Hurlbert. 2014. "Using trait and phylogenetic diversity to evaluate the generality of the stress-dominance hypothesis in eastern North American tree communities." *Ecography 37*: 814–826.

Culmsee, H., and C. Leuschner. 2013. "Consistent patterns of elevational change in tree taxonomic and phylogenetic diversity across Malesian mountain forests." *Journal of Biogeography 40*: 1997–2010.

Fortunel, C., C. E. T. Paine, P. V. A. Fine, N. J. B. Kraft, and C. Baraloto. 2014. "Environmental factors predict community functional composition in Amazonian forests." *Journal of Ecology 102*: 145–155.

Garnier, E., J. Cortez, G. Billès, M. L. Navas, C. Roumet, M. Debussche, G. Laurent, et al. 2004. "Plant functional markers capture ecosystem properties during secondary succession." *Ecology 85*: 2630–2637.

Gleason, H. A. 1926. "The individualistic concept of the plant association." *Bulletin of the Torrey Botanical Club 53*: 7–26.

Grime, J. P. 1977. "Evidence for the existence of three primary strategies in plants and its relevance to ecological and evolutionary theory." *The American Naturalist 111*: 1169–1194.

Hardin, G. 1960. "The competitive exclusion principle." *Science 131*: 1292–1297.

Harmon, M. E., C. W. Woodall, B. Fasth, and J. Sexton. 2008. "Woody detritus density and density reduction factors for tree species in the United States: A synthesis." General Technical Report NRS-29. Northern Research Station. Newtown Square, PA: USDA Forest Service.

Kattge, J., S. Diaz, S. Lavorel, I. C. Prentice, P. Leadley, G. Bonisch, E. Garnier, et al. 2011. "TRY - a global database of plant traits." *Global Change Biology 17*: 2905–2935.

Kembel, S. W., P. D. Cowan, M. R. Helmus, W. K. Cornwell, H. Morlon, D. D. Ackerly, S. P. Blomberg, and C. O. Webb. 2010. "Picante: R tools for integrating phylogenies and ecology." *Bioinformatics 26*: 1463–1464.

Kitagawa, R., M. Mimura, A. S. Mori, and A. Sakai. 2015. "Topographic patterns in the phylogenetic structure of temperate forests on steep mountainous terrain." *AoB Plants 7*: plv134.

Kraft, N. J. B., R. Valencia, and D. D. Ackerly. 2008. "Functional traits and niche-based tree community assembly in an Amazonian forest." *Science 322*: 580–582.

Laliberté, E., and P. Legendre. 2010. "A distance-based framework for measuring functional diversity from multiple traits." *Ecology 91*: 299–305.

Lasky, J. R., I. F. Sun, S. H. Su, Z. S. Chen, and T. H. Keitt. 2013. "Trait-mediated effects of environmental filtering on tree community dynamics." *Journal of Ecology 101*: 722–733.

Lorimer, C. G. 1984. "Development of the red maple understory in northeastern oak forests." *Forest Science 30*: 3–22.

Lovett, G. M., and M. J. Mitchell. 2004. "Sugar maple and nitrogen cycling in the forests of eastern North America." *Frontiers in Ecology and the Environment 2*: 81–88.

Martin, W. H. 1975. "The Lilley Cornett Woods: A stable mixed mesophytic forest in Kentucky." *Botanical Gazette 136*: 171–183.

Mason, N. W. H., F. De Bello, D. Mouillot, S. Pavoine, and S. Dray. 2013. "A guide for using functional diversity indices to reveal changes in assembly processes along ecological gradients." *Journal of Vegetation Science 24*: 794–806.

McEwan, R. W., J. I. Chapman, and R. N. Muller. 2017. "Old-growth deciduous forest dynamics data archive, Lilley Cornett Woods." https://doi.org/10.26890/gfg2uszb8v.

McEwan, R. W., J. M. Dyer, and N. Pederson. 2011. "Multiple interacting ecosystem drivers: Toward an encompassing hypothesis of oak forest dynamics across eastern North America." *Ecography 34*: 244–256.

McEwan, R. W., and R. N. Muller. 2006. "Spatial and temporal dynamics in canopy dominance of an old-growth central Appalachian forest." *Canadian Journal of Forest Research-Revue Canadienne De Recherche Forestiere 36*: 1536–1550.

McEwan, R. W., R. N. Muller, and B. C. McCarthy. 2005. "Vegetation-environment relationships among woody species in four canopy-layers in an old-growth mixed mesophytic forest." *Castanea 70*: 32–46.

Muller, R. N. 1982. "Vegetation patterns in the mixed mesophytic forest of eastern Kentucky." *Ecology 63*: 1901–1917.

Nowacki, G. J., and M. D. Abrams. 2008. "The demise of fire and "mesophication" of forests in the eastern United States." *Bioscience 58*: 123–138.

Pederson, N., J. M. Dyer, R. W. McEwan, A. E. Hessl, C. J. Mock, D. A. Orwig, H. E. Rieder, and B. I. Cook. 2014. "The legacy of episodic climatic events in shaping temperate, broadleaf forests." *Ecological Monographs 84*: 599–620.

Pimm, S. L., C. N. Jenkins, R. Abell, T. M. Brooks, J. L. Gittleman, L. N. Joppa, P. H. Raven, et al. 2014. "The biodiversity of species and their rates of extinction, distribution, and protection." *Science 344*: 1246752.

Ricketts, T. H. T., E. Dinerstein, D. M. D. Olson, and C. Loucks. 1999. "Who's where in North America? Patterns of species richness and the utility of indicator taxa for conservation." *BioScience 49*: 369–381.

Royal Botanic Gardens Kew. 2017. "Seed Information Database (SID)." Version 7.1. http://data.kew.org/sid.

Sabatini, F. M., J. I. Burton, R. M. Scheller, K. L. Amatangelo, and D. J. Mladenoff. 2014. "Functional diversity of ground-layer plant communities in old-growth and managed northern hardwood forests." *Applied Vegetation Science 17*: 398–407.

Shotola, S. J., G. T. Weaver, P. A. Robertson, and W. C. Ashby. 1992. "Sugar maple invasion of an old-growth oak-hickory forest in southwestern Illinois." *American Midland Naturalist 127*: 125–138.

Spasojevic, M. J., J. B. Grace, S. Harrison, and E. I. Damschen. 2014. "Functional diversity supports the physiological tolerance hypothesis for plant species richness along climatic gradients." *Journal of Ecology 102*: 447–455.

Spasojevic, M. J., and K. N. Suding. 2012. "Inferring community assembly mechanisms from functional diversity patterns: The importance of multiple assembly processes." *Journal of Ecology 100*: 652–661.

Stahl, U., J. Kattge, B. Reu, W. Voigt, K. Ogle, J. Dickie, and C. Wirth. 2013. "Whole-plant trait spectra of North American woody plant species reflect fundamental ecological strategies." *Ecosphere 4*: 128.

Swenson, N. G. 2014. *Functional and Phylogenetic Ecology in R.* New York: Springer.

USDA, NRCS. 2017. "The PLANTS Database." http://plants.usda.gov. Greensboro, NC: National Plant Data Team.

Webb, C. O., D. D. Ackerly, and S. W. Kembel. 2008. "Phylocom: software for the analysis of phylogenetic community structure and trait evolution." *Bioinformatics 24*: 2098–2100.

Webb, C. O., D. D. Ackerly, M. A. McPeek, and M. J. Donoghue. 2002. "Phylogenies and community ecology." *Annual Review of Ecology and Systematics 33*: 475–505.

Weiher, E., G. D,. P. Clarke, and P. A. Keddy. 1998. "Community assembly rules, morphological dispersion, and the coexistence of plant species." *Oikos 81*: 309–322.

Weiher, E., and P. A. Keddy. 1995. "Assembly rules, null models, and trait dispersion, new questions from old patterns." *Oikos 74*: 159–164.

Whittaker, R. H. 1956. "Vegetation of the Great Smoky Mountains." *Ecological Monographs 26*: 1–80.

Wiens, J. J., D. D. Ackerly, A. P. Allen, B. L. Anacker, L. B. Buckley, H. V. Cornell, E. I. Damschen, et al. 2010. "Niche conservatism as an emerging principle in ecology and conservation biology." *Ecology Letters 13*: 1310–1324.

Wright, I. J., P. B. Reich, M. Westoby, D. D. Ackerly, Z. Baruch, F. Bongers, J. Cavender-Bares, et al. 2004. "The worldwide leaf economics spectrum." *Nature 428*: 821–827.

# Chapter 6

# Old-Growth Disturbance Dynamics and Associated Ecological Silviculture for Forests in Northeastern North America

*Anthony W. D'Amato, Patricia Raymond, and Shawn Fraver*

Our understanding of natural disturbance dynamics of old-growth forest ecosystems in eastern North America has expanded significantly over the past several decades. This expansion has largely stemmed from an active period of discovery of remnant old-growth stands across the region beginning in the late 1980s (Davis 1996; Tyrrell et al. 1998) and parallel development of methodological advances for applying dendroecological techniques to reconstruct disturbance regimes from these areas (summarized in Frelich 2002). Our aim in this chapter is to summarize the disturbance patterns observed in old-growth forests and the resulting developmental dynamics, particularly for common old-growth forest types in northeastern North America. The translation of this information into the development of ecological silviculture systems is also presented with further elaboration in chapters 8 and 13.

Disturbance has long been recognized as a central driver of compositional and structural conditions in old-growth forests, including early descriptions of these systems in the northeastern United States (Cline and Spurr 1942). Various definitions have been proposed for what constitutes a natural disturbance. Following White and Pickett (1985), we define these as *any relatively discrete event in time that disrupts ecosystem, community, or population structure and changes resources, substrate availability, or the physical environment.* We have allowed ourselves some flexibility regarding the "relatively discrete" criteria by including pathogens and drought in our discussions below, both of which may cause more prolonged periods of tree mortality when compared to truly discrete events such as wind storms or wildfires.

## Historical Disturbance Agents in Forests of
## Northeastern North America

Our knowledge of natural disturbances arises from historical accounts, from land survey records conducted prior to extensive European settlement, from paleoecological studies of pollen and charcoal in sediments, and from dendrochronological studies of extant old-growth forests. Dendrochronological data in particular is best obtained from remaining old-growth forests, given their lack of direct human alteration, which provides insights typically unavailable from managed forests. Although dendrochronological methods provide reliable estimates for the timing and extent of past disturbances, they often fall short in identifying the agent of disturbance. For this we may rely on circumstantial historical or physical evidence. This section provides an overview of the major disturbance agents in our region (plate 2). The forest structure and composition, resulting from natural disturbance, provide the basis for ecological forestry prescriptions addressed later in this chapter.

### Wind Storms

In the presettlement forests of the region, particularly in New England, wind may have been the prevalent natural disturbance agent. That is, it may have caused more tree mortality than any other natural agent. Windstorms occur in a wide range of types, including thunderstorm downbursts, hurricanes, derechos, tornados, and nor'easters (cyclonic winter storms originating in the north Atlantic Ocean), and can occur at any time of the year. Storms also span a wide range of spatial extents and severities, from individual tree deaths (forming canopy gaps) to stand-replacing events covering extensive areas (table 6-1). However, the spatial extent of most windstorms is difficult to delineate, as the affected area is often an interwoven and patchy complex of varying disturbance severities.

Less intense windstorms occurring over localized areas create single or multiple treefall gaps, a topic extremely well studied in this region (e.g., Frelich and Lorimer 1991). Gaps form frequently in many of the regional forest types, often accounting for circa 0.5 to 2.0 percent canopy loss per year across a range of forest types (Runkle 1985). These gaps are typically small, even when they include several trees. A review by Seymour et al. (2002) reports mean gap sizes of 24 to 126 square meters.

At the other extreme, windstorms can cause severe damage—more than 60 percent canopy removal—over extensive areas throughout this re-

TABLE 6-1. Disturbances and their severity in northeastern forests. Light, moderate, and severe, causing < 25 percent, 25–75 percent, and > 75 percent canopy mortality in a given location, based on Kern et al. (2017). Scale ranges from individual tree, to stand (4–20 hectares), to landscapes.

| | | Scale | | |
|---|---|---|---|---|
| Agent | Severity | Landscape | Stand | Tree |
| **Wind** | light to moderate | + | + | + |
| **Insects** | moderate to severe | + | + | + |
| Spruce budworm (*C. fumiferana*) | | + | + | + |
| Hemlock looper (*L. fiscellaria*) | | | + | + |
| Emerald ash borer (*A. planipennis*) | | | + | + |
| Hemlock woolly adelgid (*A. tsugae*) | | | + | + |
| Forest tent caterpillar (*M. disstria*) | | + | + | + |
| Gypsy moth (*L. dispar*) | | + | + | + |
| **Pathogens** | light to moderate | | | |
| White pine blister rust (*C. ribicola*) | | | + | + |
| Beech bark disease (*C. fagi*) | | | + | + |
| *Armillaria* | | | + | + |
| **Fire** | light to severe | + | + | |
| **Ice** | light to moderate | | + | + |
| **Drought** | light | | + | + |
| **Background mortality** | light | | | + |

gion, as evidenced by recent storms. A 1977 storm in Wisconsin severely affected 24,000 hectares, a 1995 storm in New York severely affected 15,400 hectares, and a 1999 storm in Minnesota severely affected 57,000 hectares (Jenkins 1995; Schulte and Mladenoff 2005). Early land survey records estimate return intervals for such storms at 450 to 10,500 years (Schulte and Mladenoff 2005), 541 years (Zhang et al. 1999), 980 to 3,190 years (Seischab and Orwig 1991), 1,150 years (Lorimer 1977), 1,210 years (Canham and Loucks (1984), and 1,220 years (Whitney 1986), all much longer than the life span of dominant trees in the region.

In between the two extremes mentioned above are moderate-severity windstorms, which have been overlooked until somewhat recently. In their review of forest disturbances in New England, Seymour et al. (2002) found no studies explicitly addressing moderate-severity events. More recently, Stueve et al. (2011) suggest that the aggregated impact of moderate-severity windstorms may equal that of the less-frequent, higher-severity events. Worrall et al. (2005) point out that moderate-severity disturbances (windstorms and other agents) form part of a *nested bicycle*, with a pattern of frequent, smallscale canopy gaps nested within episodic pulses of moderate-severity disturbance. These pulses have a profound and lasting effect on forest development, susceptibility to future disturbance, insect and pathogen dynamics, and tree species composition (Worrall et al. 2005; Hanson and Lorimer 2007; Fraver et al. 2009). Virtually all regional reconstructed disturbance histories based on dendrochronological data show these episodic pulses, which occur every few decades (e.g., Frelich and Lorimer 1991; Ziegler 2002; Fraver et al. 2009; D'Amato and Orwig 2008; chapters 4 and 7).

## *Insects*

The number of insects that cause tree mortality and forest disturbance across this region is far too large to address in this overview. Instead, we focus on two of the more prominent native insects—the spruce budworm and the forest tent caterpillar—both of which exhibit periodic outbreaks affecting large areas. The eastern spruce budworm (*Choristoneura fumiferana*) is an insect that defoliates both balsam fir (its primary host) and spruce species. Affected trees may die after several years of defoliation, causing widespread mortality (plate 2). The total area affected by major outbreaks can exceed 50 million hectares (MacLean 1984), making it the foremost insect pest in our region. Dendrochronological records, some of which extend back to circa 1700, reveal outbreaks occurring on a roughly 40-year cycle (Boulanger and Arseneault 2004; Fraver et al. 2007), with spatial synchrony of recent outbreaks approaching 2,000 kilometers (Williams and Liebhold 2000). However, in the western portion of the budworm's range (western Ontario and Minnesota), outbreaks appear to be patchier and less synchronized (Robert et al. 2012).

The forest tent caterpillar (*Malacosoma disstria*), another native defoliating insect, exhibits periodic outbreaks in hardwood forests of the region, targeting *Populus* species (primary host) but also *Acer* and *Fraxinus* species. Several consecutive years of defoliation can cause widespread mortality (Churchill et al. 1964; Roland 1993)—particularly when coinciding with droughts (Worrall et al. 2013)—with caterpillar outbreaks historically showing periodicities of 9 to 13 years (Cooke and Lorenzetti 2006).

In recent years, two additional insects, both nonnative, are causing extensive tree mortality in our region (chapters 4 and 12). The hemlock woolly adelgid (*Adelges tsugae*) is a sapfeeder that targets eastern hemlock exclusively. It became established in the southeastern United States in the 1950s and has since migrated to New England (McClure 1990), where it is causing widespread mortality, and, hence, forest-type conversion, in managed and old-growth forests of our region (Orwig et al. 2012). Finally, the emerald ash borer (*Agrilus planipennis*), a phloemfeeder, targets our native ash species. Since its first detection on this continent in 2002, it has caused widespread mortality throughout our region.

## Pathogens

As with insects, pathogens as disturbance agents are too numerous to cover in this overview; instead, we focus on two, both nonnative, that have dramatically altered forest composition and structure (chapters 4 and 12). First, the American chestnut bight, caused by the fungus *Cryphonectria parasitica*, was introduced to North America in 1904 and quickly spread throughout the chestnut's range. In the decades that followed, the formerly prominent chestnut was virtually eliminated from the canopy, being replaced by oaks, hickories, or hemlock (McCormick and Platt 1980). Second, beech bark disease, which was introduced to Nova Scotia in the 1890s (Ehrlich 1934), continues to spread throughout our region, killing mature beech (Cale et al. 2017). The disease results from an infection by several species of the fungus *Neonectria*, which enter the cambium via wounds produced by the scale insect *Cryptococcus fagi* (Houston 1975). The impact of this disease is rather unusual. Because of the beech's ability to reproduce by root suckers following death of the central stem, beech stems may be more numerous now than before the infestation, even though large trees (greater than about 31 centimeters in diameter) are uncommon because of disease-related mortality (Houston 1975).

## Wildfires

The role that wildfire played in the presettlement forest varied markedly across this region, from the fire-dependent pine systems of the northern Lake States (Michigan, Minnesota, Wisconsin, and southern Ontario) to the "asbestos" hardwood forests of New England (Bormann and Likens 1979).

Considering pine forests of northern Minnesota, Heinselman (1996) estimated a fire rotation period (time required for the fire regime to burn an area
equivalent to the area in question) of 122 years, with high severity and le-
thal crown fires dominating. Such a fire regime created a patchwork of forest
stands in various stages of recovery postfire. Using the location of fire scars
and tree ages, Heinselman (1996) estimated that the largest fire burned circa
112,000 hectares during the drought of 1864. Considering the more mesic
forests of the northeastern region, fire return intervals (mean time between
disturbances at a given location) have been estimated between circa 800 to
9,000 years (reviewed by Seymour et al. 2002) and circa 700 to 93,000 years
(Schulte and Mladenoff 2005). Fire sizes reconstructed by Lorimer (1977)
spanned from 10,000 to 80,000 hectares across a range of forest types in
northern Maine. Fires occur much more frequently in temperate-boreal transition forests immediately to the north of our region, with cycles averaging
150 to 300 years in the boreal forest (Bergeron and Brisson 1990; Bergeron
et al. 2004; Boucher et al. 2011; chapter 8). Finally, the use of fire by Native
Americans in this region remains speculative: Day (1953) claims anthropogenic fire use was pervasive, while Russell (1983) views it as infrequent and
local in extent. Regardless, given the rarity of dry lightning in much of this
region, many historic fires likely had human ignition sources.

## Ice Storms

Ice storms are more common in this region than any forested region of
the globe (Lafon 2004), with severe storms occurring every year in some
part of the region. Little is known about the extent of ice storms in the
presettlement forests; however, major ice storms of the last century were
extensive enough to affect several US states (Irland 2000). The well-docu-
mented 1998 ice storm damaged circa 10 million forested hectares in the
northeastern United States and eastern Canada, although the damage was
quite patchy within the affected area. Contemporary evidence suggests that
ice storms cause widespread branch and stem failure, yet they rarely result
in true stand-replacing events (Irland 2000). Hooper et al. (2001) provide
an example of ice-storm damage in an extant old-growth hardwood forest
in Quebec, reporting that the storm removed 7 to 10 percent of the total
aboveground biomass. Thus, ice storms likely contribute to the pulses of
intermediate-severity disturbance often reported in dendrochronological
studies of old-growth forests in the Northeast (Ziegler 2002; D'Amato and
Orwig 2008; Fraver et al. 2009).

PLATE I. Map of ecological regions of North America. Level I–III ecoregions within the purview of this book (eastern North America). For each three-digit number (x.x.x), the first refers to Level I, the coarsest division; the second refers to Level II; and the final refers to Level III, the finest division. The identity of ecoregions at Levels I and II are as follows.

LEVELS 3.0 and 4.0 – BOREAL FORESTS: 3.1 – Alaska Boreal Interior; 3.2 – Taiga Cordillera; 3.3 – Taiga Plain; 3.4 – Taiga Shield; 4.1 – Hudson Plain.

LEVEL 5.0 – NORTHERN FORESTS: 5.1 – Softwood Shield; 5.2 – Mixed Wood Shield; 5.3 – Atlantic Highlands; 5.4 – Boreal Plain.

LEVEL 8.0 – EASTERN TEMPERATE FORESTS: 8.1 – Mixed Wood Plains; 8.2 – Central US Plains; 8.3 – Southeastern US Plains; 8.4 – Ozark, Ouchita-Appalachian Forests; 8.5 – Mississippi Alluvial and Southeast US Coastal Plains.

LEVEL 15.0 – TROPICAL WET FORESTS: 15.4 – Everglades.

Further details are available at Commission on Environmental Cooperation, 2006. *Ecological Regions of North America, Levels I-III*. Commission for Environmental Cooperation, Montreal, Canada. https://www.epa.gov/eco-research/ecoregions-north-america.

PLATE 2. Main disturbances in northeastern North America: (a) windstorm in a subboreal aspen mixed-wood forest in northern Minnesota (Photo credit: S. Fraver); (b) spruce budworm damage in Quebec's subboreal forest (Photo credit: M. Brousseau); (c) background mortality, causing canopy gaps in temperate mixed woods (Photo credit: P. Raymond);

(d) wildfire in subboreal jack pine forest in northern Minnesota (Photo credit: S. Fraver); (e) ice storm damage in northern hardwood forest, New Hampshire (Photo credit: A. W. D'Amato).

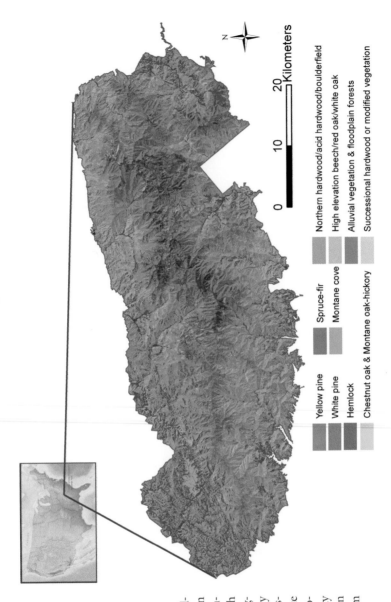

PLATE 3. Topographic shading and major vegetation types of Great Smoky Mountains National Park, North Carolina and Tennessee. Vegetation map was created by the Center for Remote Sensing and Mapping Science (now the Center for Geospatial Research), University of Georgia, Athens (Madden et al. 2004). Reprinted from Tuttle and White 2016.

Yellow pine    Spruce-fir    Northern hardwood/acid hardwood/boulderfield

White pine    Montane cove    High elevation beech/red oak/white oak

Hemlock    Alluvial vegetation & floodplain forests

Chestnut oak & Montane oak-hickory    Successional hardwood or modified vegetation

N

0    10    20 Kilometers

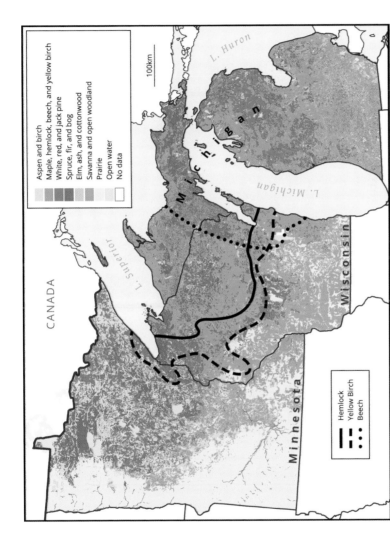

PLATE 4. Generalized vegetation of the Lake States, compiled from the US Public Land Survey data (1800s), and western range limits of major mesic species.

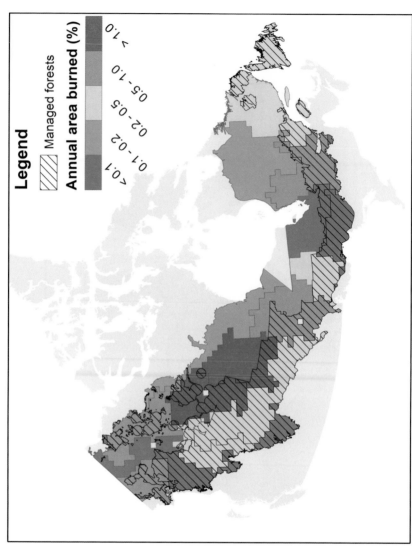

PLATE 5. Zonation of fire regimes in boreal Canada, based on historical average annual area burned and fire frequency, with mean annual area burned (in percent) given for each zone. Hashed area represents managed (commercial) forests. Based on and modified from Boulanger et al. (2014).

PLATE 6. Examples of epiphyte growth forms characterizing communities on large- and small-diameter trees. (a) *Ulota crispa* (cushion) with *Platygyrium repens* (a smooth mat with stems tightly appressed to bark surface), sprawling between the cushions. The cushion and smooth-mat growth forms dominate small-diameter trees and are adaptations to xeric conditions. The outer portions of cushions insulate stems in the interior from desiccation, and the tightly appressed stems of the smooth mat protect foliage from desiccation by keeping it within the boundary layer of the substrate. *Neckera pennata* (b) and *Leucodon brachypus* (c), which tend to occur on only larger-diameter trees, are examples of thick mats and wefts with erect, ascending or pendant branches. These growth forms are limited to mesophytic habitats, since foliage is held away from the substrate and beyond the boundary layer, and, therefore, are subject to greater desiccation rates. (d) Foliose and green-algal hair lichens inhabit the upper bole of an Adirondack sugar maple. Photo credits: G. G. McGee and H. Root.

PLATE 7. Hillside of old-growth hemlock trees killed by the hemlock woolly adelgid in the Caldwell Fork region near Cataloochee, North Carolina.

## *Droughts*

The role of drought as a disturbance agent in our presettlement forest remains poorly understood. Pederson et al. (2014) present evidence that drought occurring from 1772 to 1775 fostered tree establishment (following the death of overstory trees) and subsequently reduced competition (evident as growth spurts in surviving trees) at a subcontinental scale, including the Northeast. This finding presents a new perspective on disturbance in the Northeast, where drought was not thought to cause extensive tree mortality, given the generally humid and mesic conditions. This finding is also supported by site-level work suggesting that even humid regions may indeed be vulnerable to drought (Parshall 1995), with canopy mortality at times ultimately caused by pathogens, such as *Armillaria* spp. (Cline and Spurr 1942).

## Ecological Silviculture Approaches Based on Natural Dynamics

The disturbance regimes affecting the primary forest types in northeastern North America vary considerably across the region reflecting variability in prevailing weather patterns, biophysical settings, climate, and host population structures. The following sections build on several decades of research to summarize the natural disturbance dynamics for three predominant forest types in the region: northern hardwoods, northern pine forests, and temperate mixedwoods. For each forest type, we also briefly introduce ecological silviculture systems derived from these natural dynamics to illustrate approaches to restoring old-growth compositional and structural conditions (see chapter 13) to potentially offset the dramatic, historic loss of old-growth forests observed in this region and the resultant reductions in biodiversity and provisions of ecological services (Davis 1996; Wirth 2009). The primary information guiding the development of these silvicultural systems is an understanding of the frequency and severity of natural disturbances and the resultant structural and compositional conditions (Spurr and Cline 1942; Seymour and Hunter 1992; Franklin et al. 2007; D'Amato et al. 2017).

Given the general rarity of truly stand-replacing disturbances across northeastern old-growth forest types, the ecological silvicultural systems described below broadly rely on multicohort approaches that restore and maintain two-aged and uneven-aged structures (Seymour and Hunter 1999)—a clear departure from the even-aged conditions maintained in these areas through traditional forest management practices. Moreover, forest management has historically tended to homogenize stand structure, particularly regarding attributes associated with old-growth stages, such as large trees and standing and downed woody debris (Goodburn and Lorimer 1998). Similarly, the numbers and sizes of gaps are also more variable and irregular in old growth than in managed stands (Hanson and Lorimer 2007). Thus, each ecological silvicultural system described below includes provisions for varying harvest intensity and spatial patterns to emulate these natural patterns. These provisions assume that generating within-stand heterogeneity will stem the loss in species diversity that can occur from decades of traditional management (Hanson and Lorimer 2007). These systems also include provisions for restoring key old-growth structural attributes, including permanent retention of large living and dead tree legacies, which are critical for promoting biodiversity (Franklin et al. 2007; Keeton 2006; Bauhus et al. 2009). Chapter 13 will return to these concepts, providing more detail on the development of ecological silviculture approaches based on desired old-growth structural outcomes.

## Northern Hardwoods

The greatest amount of information on natural disturbance dynamics for old-growth forests in northeastern North America is from northern hardwood forests. These forests are generally defined by sugar maple; however, American beech, eastern hemlock, yellow birch, and American basswood can also make up significant components, depending on site conditions and geography. A distinguishing feature of these forests is the rarity of stand-replacing events, with rotation periods for stand-replacing windstorms and fires often exceeding 1,000 to 3,000 years (Canham and Loucks 1984; Frelich and Lorimer 1991; Lorimer and White 2003; Talon et al. 2005). The rarity of stand replacement, coupled with a general disturbance regime characterized by frequent, light, and moderate disturbances (Frelich and Lorimer 1991; Fraver et al. 2009) favor the development of old-growth stands with an uneven-aged structure (Seymour et al. 2002; table 6-2). Windthrow causes high within-stand heterogeneity and a variety of opening sizes, ranging from 10 to 5,000 square meters (Hanson and Lorimer 2007). Gap dynamics from background mortality trigger frequent (less than every 200 years) gap-phase replacement that favors the establishment of late-successional, shade-tolerant species, such as sugar maple, American beech, and eastern hemlock in the smaller gaps, and favors establishment

TABLE 6-2. Primary forest types in northeastern North America covered in this chapter and their disturbance agents.

| Type group | Forest Types | Disturbance agents | |
|---|---|---|---|
| | | Main | Secondary |
| Northern hardwoods | beech-maple-basswood northern hardwoods-hemlock | wind | senescence, ice, snow, pathogens, insects, fire |
| Temperate mixedwoods | spruce-fir | spruce budworm | wind, senescence |
| | yellow birch-spruce-fir | spruce budworm | fire, wind, senescence |
| Northern pines | red pine eastern white pine | fire | wind, drought, root disease, senescence |

of moderately shade-tolerant species, such as yellow birch, American bass-wood, and ashes in the larger gaps (Foster 1988; Dahir and Lorimer 1996; Webster and Lorimer 2005). Under the more continental climate of Quebec and Ontario, local fire events can regenerate patches of eastern white pines in a matrix of shade-tolerant hardwood species (Carleton et al. 1996; Fahey and Lorimer 2014). Nevertheless, moderate-severity disturbances caused by wind seem to play a central role in the regeneration process of the less shade-tolerant species in these northern hardwood ecosystems (Foster 1988; Hanson and Lorimer 2007).

The earliest silvicultural guidelines for managing northern hardwood forests emphasized uneven-aged approaches, particularly single-tree selection, to generate "balanced" (equal canopy area allocated to each age class), uneven-aged structures. This approach was based on natural size structures observed in the old-growth shade-tolerant forests of Europe and eastern North America (Meyer 1952). It resulted in the unfortunate misapplication of size structures observed at broad spatial scales (reverse J-shaped distributions) to management guides designed for stand-scale application. An important ecological consequence of this management approach has been the general loss of midtolerant species from these forests because such species cannot become established in the small gaps resulting from single-tree removal, which often lack suitable substrates for establishment (Webster and Lorimer 2005). The repeated creation of only small gaps runs counter to the wide range of gap sizes created by natural disturbances in these forests, as described above. To rectify these issues, ecological approaches have emphasized emulating small- to medium-canopy openings to perpetuate the diversity of shade-tolerant classes; however, the effectiveness of this strategy has varied across the geography of northern hardwood forests given the range of site conditions (Kern et al. 2017). In the Lake States, relatively small openings (200 to 1,000 square meters) foster recruitment of yellow birch on poorer quality sites where competition from sugar maple advance regeneration does not limit recruitment (Webster and Lorimer 2005). Historically, such competition-free microsites were created by downed woody debris and tip-up mounds that elevated germinants above advance regeneration (chapter 11). In northeastern North America, dense thickets of American beech saplings have developed as a result of beech bark disease and historic selective cutting, creating the need for treatments that simultaneously remove beech saplings and create larger canopy openings to favor midtolerant species (Leak and Filip 1977). The application of hybrid selection (i.e., combination of single-tree and group selection) and irregular shelterwood methods (i.e., two- and multi-aged systems) that emulate the heterogeneity of canopy disturbances in northern hardwoods has become the primary approach to ecological silviculture in these forests (Raymond et al. 2009; Bédard et al. 2014). In concert with these regeneration methods has been an emphasis on silviculture practices aimed at promoting old-growth structures, including creating tip-up mounds and living and dead legacy trees, to restore these attributes to the landscape (Keeton 2006; Bauhus et al. 2009; D'Amato et al. 2015).

## Temperate Mixedwoods

Temperate mixedwoods constitute the ecotone between the temperate and the boreal vegetation zones in northeastern North America, where both temperate and boreal tree species grow in mixtures across latitudinal or elevational gradients (Frelich et al. 2012). Mixed species were common in the presettlement landscapes of the region and were usually dominated by either conifer or hardwood species, such as red and white spruce, balsam fir, yellow birch, and maple species, depending on site conditions and disturbance history (Lorimer 1977). Largescale catastrophic disturbances were rare, and stand dynamics were historically regulated by a mix of frequent light-severity and episodic moderate-severity disturbances, such as the spruce budworm outbreaks, which historically occurred every 30 to 40 years (Barrette and Bélanger 2007; Fraver et al. 2007). The importance of stand-replacing fires varied across the region. These fires were less frequent in the Acadian region along the Atlantic coast (e.g., approximately every 800 years; Lorimer 1977) than further northwest on the continent (e.g., 200 to 400 years; Boucher et al. 2011). Consequently, old-growth stands of the Acadian region are often dominated by shade-tolerant conifers (e.g., red spruce, northern white cedar, and eastern hemlock) (Fraver et al. 2009). These long-lived conifers tend to increase in proportion over the course of stand development (Barrette and Bélanger 2007) by accessing the canopy via multiple release events caused by low severity disturbances (Fraver and White 2005; Kneeshaw and Prévost 2007). As with northern hardwood forests, moderate disturbances by windstorms are important for maintaining mid-tolerant species (e.g., eastern white pine, northern red oak, and yellow birch) (Lorimer 1977). Under the more continental climate of Quebec and Ontario, yellow birch, eastern white pine, northern white cedar, and white spruce are more abundant. With increased fire frequency, old-growth stands are composed of yellow birch, maples, and conifers on mesic sites (Hébert 2003; Messier et al. 2005). Nonetheless, spruce budworm strongly influences stand dynamics in the mixedwood forest (Baskerville 1975), and the cyclic nature of outbreaks generate stands with balanced or irregular uneven-aged (multicohort) structures (Fortin et al. 2003; Barrette et al. 2007). Gap dynamics are also important, especially in yellow birch-conifer stands, where the gap fraction can reach 30 percent (Messier et al. 2005; Kneeshaw and Prévost 2007).

Natural dynamics of the temperate mixedwood forest have been poorly understood for a long time. More often than not, maladapted practices using heavy harvesting (e.g., clear-cutting and diameter-limit cutting) triggered heavy brush competition that hindered conifer regeneration (Archambault et al. 1998; Hébert 2003). As a result, most managed mixedwood forests of northeastern America experienced a decline in the number and abundance of conifers species, and in particular of traditional companion species (e.g., red and white spruce, eastern hemlock, eastern white pine, and northern white cedar) compared to preindustrial times (Boucher et al. 2006; Dupuis 2011; Danneyrolles et al. 2017). Improved understanding of natural disturbance dynamics and recent trends in ecological forestry systems point to silvicultural systems inspired by small- and mesoscale disturbances (Seymour et al. 2002; Raymond et al. 2009). As with northern hardwood systems (above), single-tree selection was attempted in gap-driven mixedwood systems but failed to maintain balanced uneven-aged structures (Seymour and Kenefic 1998). Studies addressing the large end of the gap-size range have shown that patch-selection using large gaps

(300 to 1,400 square meters) with systematic gap placement was too rigid to accommodate removals of balsam fir before natural mortality occurred. Consequently, patch-selection failed to regenerate uncommon companion species, such as red spruce, in late-successional stands (Prévost et al. 2010; Raymond et al. 2016). Thus, it appears that silvicultural systems offering more flexibility in gap size and placement, similar to those systems found under a natural disturbance regime, could be more efficient in promoting tree regeneration, while also favoring species diversity in managed mixed-wood forests (Kneeshaw and Prévost 2007; Raymond et al. 2016). A recent study comparing patterns of selection cutting confirmed that intermediate-size gaps (100 to 300 square meters) could regenerate a mixture of species, including yellow birch, red spruce, and balsam fir (Prévost and Charrette 2015). This finding implies that hybrid selection cutting methods could be a sound option for late-successional mixedwood stands where gap dynamics prevail (Prévost and Charrette 2015). In stands harboring multicohort or irregular stand structures, a continuous-cover irregular shelterwood system could simulate the effects of spruce budworm outbreaks and windstorms while maintaining irregular stand structures (Raymond et al. 2009; Raymond and Bédard 2017). In mature, even-aged stands, an expanding-gap irregular shelterwood system could be used to increase heterogeneity and sustain diversity in second-growth stands (Arseneault et al. 2011). As with other approaches, silviculture for old-growth attributes should include retention practices (e.g. Raymond and Bédard 2017) aimed at protecting large living trees of uncommon species, large snags, downed woody debris, and cavity trees (chapter 13).

## Northern Pines

The northern pine forests of the region represent a broad range of forest types and corresponding disturbance regimes; we address three such systems—jack, white, and red pine—to demonstrate this range. Historically, jack pine systems within the Lake States region formed nearly pure stands and were subject to a fire regime characterized by high-intensity, stand-replacing crown fires with mean return intervals of 50 to 75 years (Heinselman 1996). This species' serotinous cone habit, coupled with its shade-intolerance, typically led to stand self-replacement of single-cohort stands; however, more complex age structures often characterize these forests at jack pine's range margin where open-coned varieties predominate (Gill et al. 2015). Nevertheless, given the limited longevity of jack pines and prevailing disturbance regimes, these forests rarely achieved advanced ages or stages of development that could be considered as old growth, according to age-based criteria (chapters 1 and 4).

The disturbance regimes related to white pine are much more difficult to categorize because this species usually occurred in various admixtures with both conifers and hardwoods in our presettlement forests (Abrams 2001). Pure stands of old-growth white pines were a minor component of forests across the Northeast (Lorimer 1977; Ziegler 2010); however, these forests were more abundant at the northern and western range of the species. This greater abundance of pure stands, particularly in the Great Lakes and St. Lawrence Valley regions, reflected higher disturbance frequencies, especially on poorer sites with outwash or shallow soils over bedrock (Abrams 2001). Given its intermediate shade tolerance, white pine can exploit a variety of disturbance types, ranging from moderate-sized canopy gaps to largescale events, resulting from fire or wind storms (Abrams 2001); however, the resulting age structures were typically multiaged (Guyette and Dey 1995; Fahey and Lorimer 2014).

Given that red pine (*Pinus resinosa*) forms nearly pure stands and achieves advanced stages of development that could be considered as old growth, it lends itself well to the development of silviculture approaches based on natural dynamics, the focus of this chapter. The greatest extent of red pine forests is in the Lake States region, particularly in areas of extensive glacial outwashes or shallow soils over bedrock. The historical fire regime for this species is described as mixed severity, with surface fires occurring on average every 5 to 50 years and crown fires occurring every 150 to 250 years (Heinselman 1996), resulting in single-cohort stands. However, this characterization may not apply universally. Detailed dendrochronological work in old-growth red pine stands, combined with spatial data, suggests that surface fires in places advance into the canopy, thus killing overstory trees in patches. New cohorts establish in these patches postfire, resulting in two- to three-cohort stands (Fraver and Palik 2012). In addition, moderate-severity windstorms likewise create canopy openings that may foster red pine regeneration (Bergeron and Gagnon 1987). Under this scenario, red pine stands may persist without stand-replacing fire (Harvey 1922; Bergeron and Brisson 1990; Fraver and Palik 2012) with white pine becoming an increasingly important canopy species over time. This scenario also creates heterogeneous stand structures and compositional conditions (Fraver and Palik 2012).

Despite the complexity of natural developmental pathways for old-growth red pine forests, this species has been primarily managed using even-aged plantations throughout much of its range (Gilmore and Palik 2006). This traditional management approach reflects several economic and ecological factors, including the commercial importance of the species, its intolerance of shade, and the relatively infrequent seed crops produced, which generate challenges for using silvicultural systems based on natural reproduction. Nonetheless, there have been several advances in ecological silvicultural approaches for these forests over the past two decades that build on an understanding of natural dynamics to restore historic patterns in structural complexity and composition (Gilmore and Palik 2006). Given that the mixed severity fire, wind, and disease dynamics naturally influencing these forests generated irregular, uneven-aged structures (Fraver and Palik 2012), the prevailing approach to ecological silviculture has been to use variable retention harvests or variable density thinning (i.e., extended and continuous cover irregular shelterwoods) to increase heterogeneity in spatial conditions, recruit new cohorts, and maintain mature tree legacies (Roberts et al. 2016). These approaches have been particularly effective in restoring native diversity and recruiting new cohorts of red and white pine when treatments such as prescribed fire, mechanical scarification, or competition control have been applied to generate the understory and forest floor conditions historically maintained by frequent fire (D'Amato et al. 2012; Roberts et al. 2016). Shelterwood systems combined with scarification have also proven to be effective in regenerating pure even-aged white pine stands (Burgess and Wetzel 2002; Boucher et al. 2007). Thinnings and irregular shelterwood variants could be used to increase structural diversity over time (Bebber et al. 2004) and restore the irregular stand structures typical of old-growth stands.

## Conclusion

Our understanding of the natural dynamics of forests in northeastern North America has grown considerably over the last several decades due to a multitude of studies documenting disturbance patterns and processes in old-growth forest systems. This work has underscored the general rarity of stand-replacing disturbances in this region, with frequent gap and intermediatescale disturbances generating diverse forest composition and complex structural conditions at local and landscape scales. Of note, this natural disturbance regime is in stark contrast to those characterizing many of the forest types that have historically guided our understanding of forest developmental processes and ecosystem dynamics (e.g., Oliver 1980; Spies et al. 1988). This contrast highlights the need for revisiting these concepts and models in the context of northeastern forests. This is particularly important when developing ecological silvicultural systems for this region, as systems developed from regions where stand replacement prevails may overemphasize structural retention during regeneration harvests and extended rotation ages to emulate longer development periods (Franklin et al. 2007); however, such concepts do not directly translate to the historically multiaged, compositionally diverse forests of the Northeast. Instead, the use of selection and irregular shelterwood systems that focus not only on creating structural heterogeneity through various harvest opening sizes and structural retention but also on the resource needs of less-tolerant yet historically important canopy species, is critical for restoring old-growth forest conditions in northeastern North America.

# References

Abrams, M. D. 2001. "Eastern white pine versatility in the pre-settlement forest." *Bioscience* 51: 967–979.

Archambault, L., J. Morissette, and M. Bernier-Cardou, 1998. "Forest succession over a 20-year period following clearcutting in balsam fir-yellow birch ecosystems of eastern Québec, Canada." *Forest Ecology and Management 102*: 61–74.

Arseneault, J. E., M. R. Saunders, R. S. Seymour, and R. G. Wagner. 2011. "First decadal response to treatment in a disturbance-based silviculture experiment in Maine." *Forest Ecology and Management 262*: 402–412.

Barrette, M., and L. Bélanger. 2007. "Reconstitution historique du paysage préindustriel de la région écologique des hautes collines du Bas-Saint-Maurice." *Canadian Journal of Forest Research 37*: 1147–1160.

Baskerville, G. L. 1975. "Spruce budworm: super silviculturist." *The Forestry Chronicle 51*: 138–140.

Bauhus, J., K. Puettmann, and C. Messier. 2009. "Silviculture for old-growth attributes." *Forest Ecology and Management 258*: 525–537.

Bebber, D. P., S. C. Thomas, W. G. Cole, and D. Balsillie. 2004. "Diameter increment in mature eastern white pine *Pinus strobus* L. following partial harvest of old-growth stands in Ontario, Canada." *Trees 18*: 29–34.

Bédard, S., F. Guillemette, P. Raymond, S. Tremblay, C. Larouche, and J. DeBlois. 2014. "Rehabilitation of northern hardwood stands using multicohort silvicultural scenarios in Québec." *Journal of Forestry 112*: 276–286.

Bergeron, Y., and J. Brisson. 1990. "Fire regime in red pine stands at the northern limit of the species' range." *Ecology 71*: 1352–1364.

Bergeron, Y., and D. Gagnon. 1987. "Age structure of red pine (*Pinus resinosa* Ait.) at its northern limit in Quebec." *Canadian Journal of Forest Research 17*: 129–137.

Bergeron, Y., S. Gauthier, M. Flannigan, and V. Kafka. 2004. "Fire regimes at the transition between mixedwood and coniferous boreal forest in northwestern Quebec." *Ecology 85*: 1916–1932.

Bormann, F. H., and G. E. Likens. 1979. "Catastrophic disturbance and the steady state in northern hardwood forests: A new look at the role of disturbance in the development of forest ecosystems suggests important implications for land-use policies." *American Scientist 67*: 660–669.

Boucher, J-F., P. Y. Bernier, H. A. Margolis, and A. D. Munson. 2007. "Growth and physiological response of eastern white pine seedlings to partial cutting and site preparation." *Forest Ecology and Management 240*: 151–164.

Boucher, Y., D. Arseneault, and L. Sirois. 2006. "Logging-induced change (1930-2002) of a preindustrial landscape at the northern range limit of northern hardwoods, eastern Canada." *Canadian Journal of Forest Research 36*: 505–517.

Boucher, Y., M. Bouchard, P. Grondin, and P. Tardif. 2011. "Le registre des états de référence : intégration des connaissances sur la structure, la composition et la dynamique des paysages forestiers naturels du Québec méridional. " Mémoire recherche forestière no. 161. Direction recherche forestière. Gouvernement du Québec.

Boulanger, Y., and D. Arseneault. 2004. "Spruce budworm outbreaks in eastern Quebec over the last 450 years." *Canadian Journal of Forest Research 34*: 1035–1043.

Burgess, D., and S. Wetzel. 2002. "Recruitment and early growth of eastern white pine (*Pinus strobus*) regeneration after partial cutting and site preparation." *Forestry 75*: 419–423.

Cale, J. A., M. T. Garrison-Johnston, S. A. Teale, and J. D. Castello. 2017. "Beech bark disease in North America: Over a century of research revisited." *Forest Ecology and Management 394*: 86–103.

Canham, C. D., and O. L. Loucks. 1984. "Catastrophic windthrow in the pre-settlement forests of Wisconsin." *Ecology 65*: 803–809.

Carleton, T. J., P. F. Maycock, R. Arnup, and A. M. Gordon. 1996. "*In situ* regeneration of *Pinus strobus* and *P. resinosa* in the Great Lakes forest communities of Canada." *Journal of Vegetation Science 7*: 431–444.

Churchill, G. B., H. H. John, D. P. Duncan, and A. C. Hodson. 1964. "Long-term effects of defoliation of aspen by the forest tent caterpillar." *Ecology 45*: 630–633.

Cline, A. C., and S. H. Spurr. 1942. "The virgin upland forest of central New England: A study of old growth stands in the Pisgah mountain section of southwestern New Hampshire." *Harvard Forest Bulletin 21*: 58.

Cooke, B. J., and F. Lorenzetti. 2006. "The dynamics of forest tent caterpillar outbreaks in Quebec, Canada." *Forest Ecology and Management 226*: 110–121.

D'Amato, A. W., P. F. Catanzaro, and L. S. Fletcher. 2015. "Early regeneration and structural responses to patch selection and structural retention in second-growth northern hardwoods." *Forest Science 61*: 183–189.

D'Amato, A. W., and D. A. Orwig. 2008. "Stand and landscape-level disturbance dynamics in old-growth forests in Western Massachusetts." *Ecological Monographs 78*: 507–522.

D'Amato, A. W., B. J. Palik, J. F. Franklin, and D. R. Foster. 2017. "Exploring the origins of ecological forestry in North America." *Journal of Forestry 115*: 126–127.

D'Amato, A. W., J. Segari, and D. Gilmore. 2012. "Influence of site preparation on natural regeneration and understory plant communities within red pine shelterwood systems." *Northern Journal of Applied Forestry 29*: 60–66.

Dahir, S. E., and C. G. Lorimer. 1996. "Variation in canopy gap formation among developmental stages of northern hardwood stands." *Canadian Journal of Forestry Research 26*: 1875–1892.

Danneyrolles, V., S. Dupuis, D. Arseneault, R. Terrail, M. Leroyer, A. de Römer, G. Fortin, Y. Boucher, and J-C. Ruel. 2017. "Eastern white cedar long-term dynamics in eastern Canada: Implications for restoration in the context of ecosystem-based management." *Forest Ecology and Management 400*: 502–510.

Davis, M. B., ed. 1996. *Eastern Old-Growth Forests: Prospects for Rediscovery and Recovery*. Washington, DC: Island Press.

Day, G. M. 1953. "The Indian as an ecological factor in the northeastern forest." *Ecology 34*: 329–346.

Dupuis, S., D. Arseneault, and L. Sirois. 2011. "Change from pre-settlement to present-day forest composition reconstructed from early land survey records in eastern Québec, Canada." *Journal of Vegetation Science 22*: 564–575.

Ehrlich, J. 1934. "The beech bark disease: a *Nectria* disease of *Fagus*, following *Cryptococcus fagi* (Baer.)." *Canadian Journal of Research 10*: 593–692.

Fahey, R. T., and C. G. Lorimer. 2014. "Persistence of pine species in late-successional forests: evidence from habitat-related variation in stand age structure." *Journal of Vegetation Science 25*: 584–600.

Fortin, M., J. Bégin, and L. Bélanger. 2003. "Évolution de la structure diamétrale et de la composition des peuplements mixtes de sapin baumier et d'épinette rouge de la forêt primitive après une coupe à diamètre limite sur l'Aire d'observation de la rivière Ouareau." *Canadian Journal of Forest Research 33*: 691–704.

Foster, D. R. 1988. "Disturbance history, community organization and vegetation dynamics of the old-growth Pisgah Forest, south-western New Hampshire, USA." *Journal of Ecology 76*: 105–134.

Franklin, J. F., R. J. Mitchell, and B. Palik. 2007. "Natural disturbance and stand development principles for ecological forestry." General Technical Report NRS-19, USDA Forest Service.

Fraver, S., and B. J. Palik. 2012. "Stand and cohort structures of old-growth *Pinus resinosa*-dominated forests of northern Minnesota, USA." *Journal of Vegetation Science 23*: 249–259.

Fraver, S., R. S. Seymour, J. H. Speer, and A. S. White. 2007. "Dendrochronological reconstruction of spruce budworm outbreaks in northern Maine." *Canadian Journal of Forest Research 37*: 523–529.

Fraver, S., and A. S. White. 2005. "Disturbance dynamics of old-growth *Picea rubens* forests of northern Maine." *Journal of Vegetation Science 16*: 597–610.

Fraver, S., A. S. White, and R. S. Seymour. 2009. "Natural disturbance in an old-growth landscape in northern Maine, USA." *Journal of Ecology 97*: 289–298.

Frelich, L. E. 2002. *Forest Dynamics and Disturbance Regimes*. Cambridge, UK: Cambridge University Press.

Frelich, L. E., and C. G. Lorimer. 1991. "Natural disturbance regimes in hemlock–hardwood forests of the upper Great Lakes region." *Ecological Monographs 61*: 145–164.

Frelich, L. E., R. O. Peterson, M. Dovčiak, P. B. Reich, J. A. Vucetich, and N. Eisenhauer. 2012. "Trophic cascades, invasive species and body-size hierarchies interactively modulate climate change responses of ecotonal temperate–boreal forest." *Philosophical Transactions of the Royal Society of London B: Biological Sciences 367*: 2955–2961.

Gill, K. G., A. W. D'Amato, and S. Fraver. 2015. "Multiple developmental pathways for range-margin *Pinus banksiana* forests." *Canadian Journal of Forest Research 46*: 200–214.

Gilmore, D. W., and B. Palik. 2006. "A revised manager's handbook for red pine in the North Central region." General Technical Report NC-264. USDA Forest Service.

Goodburn, J. M., and C. G. Lorimer. 1998. "Cavity trees and coarse woody debris in old-growth and managed northern hardwood forests in Wisconsin and Michigan." *Canadian Journal of Forest Research 28*: 427–438.

Guyette, R. P., and D. C. Dey. 1995. "Age, size and regeneration of old-growth white pine at dividing Lake Nature Reserve Algonquin Park, Ontario." Forest Research Report, no. 131. Ontario Forest Research Institute. Ontario Ministry of Natural Resources.

Hanson, J. J., and C. G. Lorimer. 2007. "Forest structure and light regimes following moderate wind storms: Implications for multi-cohort management." *Ecological Applications 17*: 1325–1340.

Harvey, L. H. 1922. "Yellow-white pine formation at Little Manistee, Michigan." *Botanical Gazette 73*: 26–43.

Hébert, R. 2003. "Are clearcuts appropriate for the mixed forest of Québec?" *The Forestry Chronicle 79*: 664–671.

Heinselman, M. L. 1996. *The Boundary Waters Wilderness Ecosystem*. Minneapolis, MN: University of Minnesota Press.

Hooper, M. C., K. Arii, and M. J. Lechowicz. 2001. "Impact of a major ice storm on an old-growth hardwood forest." *Canadian Journal of Botany 79*: 70–75.

Houston, D. R. 1975. "Beech bark disease: the aftermath forests are structured for a new outbreak." *Journal of Forestry 73*: 660–663.

Irland, L. C. 2000. "Ice storms and forest impacts." *Science of the Total Environment 262*: 231–242.

Jenkins, J. 1995. "Notes on the Adirondack blowdown of July 15th, 1995." New York: NY Wildlife Conservation Society.

Keeton, W. S. 2006. "Managing for late-successional/old-growth characteristics in northern hardwood-conifer forests." *Forest Ecology and Management 235*: 129–142.

Kern, C. C., J. I. Burton, P. Raymond, A. W. D'Amato, W. S. Keeton, A. A. Royo, M. B. Walters, C. R. Webster, and J. L. Willis. 2017. "Challenges facing gap-based silviculture and possible solutions for mesic northern forests in North America." *Forestry 90*: 4–17.

Kneeshaw, D. D., and M. Prévost. 2007. "Natural canopy gap disturbances and their role in maintaining mixed-species forests of central Quebec, Canada." *Canadian Journal of Forest Research 37*: 1534–1544.

Lafon, C. W. 2004. "Ice-storm disturbance and long-term forest dynamics in the Adirondack Mountains." *Journal of Vegetation Science 15*: 267–276.

Leak, W. B., and S. M. Filip. 1977. "Thirty-eight years of group selection in New England northern hardwoods." *Journal of Forestry 75*: 641–643.

Lorimer, C. G. 1977. "The pre-settlement forest and natural disturbance cycle of northeastern Maine." *Ecology 58*: 139–148.

Lorimer, C. G., and A. S. White. 2003. "Scale and frequency of natural disturbances in the northeastern US: implications for early successional forest habitats and regional age distributions." *Forest Ecology and Management 185*: 41–64.

MacLean, D. A. 1984. "Effects of spruce budworm outbreaks on the productivity and stability of balsam fir forests." *The Forestry Chronicle 60*: 273–279.

McClure, M. S. 1990. "Role of wind, birds, deer, and humans in the dispersal of hemlock woolly adelgid (*Homoptera: Adelgidae*)." *Environmental Entomology 19*: 36–43.

McCormick, J. F., and R. B. Platt. 1980. "Recovery of an Appalachian forest following the chestnut blight." *American Midland Naturalist 104*: 264–273.

Messier, J., D. Kneeshaw, M. Bouchard, and A. de Römer. 2005. "A comparison of gap characteristics in mixedwood old-growth forests in eastern and western Quebec." *Canadian Journal of Forest Research 35*: 2510–2514.

Meyer, H. A. 1952. "Structure, growth, and drain in balanced uneven-aged forests." *Journal of Forestry 50*: 85–92.

Oliver, C. D. 1980. "Forest development in North America following major disturbances." *Forest Ecology and Management 3*: 153–168.

Orwig, D. A., J. R. Thompson, N. A. Povak, M. Manner, D. Niebyl, and D. R. Foster. 2012. "A foundation tree at the precipice: *Tsuga canadensis* health after the arrival of *Adelges tsugae* in central New England." *Ecosphere 3*: 1–16.

Parshall, T. 1995. "Canopy mortality and stand-scale change in a northern hemlock–hardwood forest." *Canadian Journal of Forest Research 25*: 1466–1478.

Pederson, N., J. M. Dyer, R. W. McEwan, A. E. Hessl, C. J. Mock, D. A. Orwig, H. E. Rieder, and B. I. Cook. 2014. "The legacy of episodic climatic events in shaping temperate, broadleaf forests." *Ecological Monographs 84*: 599–620.

Prévost, M., and L. Charette. 2015. "Selection cutting in a yellow birch-conifer stand, in Quebec, Canada: Comparing the single-tree and two hybrid methods using different sizes of canopy opening." *Forest Ecology and Management 357*: 195–205.

Prévost, M., P. Raymond, and J-M. Lussier. 2010. "Regeneration dynamics after patch cutting and scarification in yellow birch – conifer stands." *Canadian Journal of Forest Research 40*: 357–369.

Raymond, P., and S. Bédard. 2017. "The irregular shelterwood system as an alternative to clearcutting to achieve compositional and structural objectives in temperate mixedwood stands." *Forest Ecology and Management 398*: 91–100.

Raymond, P., S. Bédard, V. Roy, C. Larouche, and S. Tremblay. 2009. "The irregular shelterwood system: review, classification, and potential application to forests affected by partial disturbances." *Journal of Forestry 107*: 405–413.

Raymond, P., M. Prévost, and H. Power. 2016. "Patch cutting in temperate mixedwood stands: what happens in the between-patch matrix?" *Forest Science 62*: 227–236.

Robert, L. E., D. Kneeshaw, and B. R. Sturtevant. 2012. "Effects of forest management legacies on spruce budworm (*Choristoneura fumiferana*) outbreaks." *Canadian Journal of Forest Research 42*: 463–475.

Roberts, M. W., A. W. D'Amato, C. C. Kern, and B. J. Palik. 2016. "Long-term impacts of variable retention harvesting on ground-layer plant communities in *Pinus resinosa* forests." *Journal of Applied Ecology 53*: 1106–1116.

Roland, J. 1993. "Large-scale forest fragmentation increases the duration of tent caterpillar outbreak." *Oecologia 93*: 25–30.

Runkle, J. R. 1985. "Disturbance regimes in temperate forests." In *The Ecology of Natural Disturbance and Patch Dynamics*, edited by S. T. A. Pickett, and P. S. White, 17–33. San Diego, CA: Academic Press.

Russell, E. W. B. 1983. "Indian-set fires in the forests of the northeastern United States." *Ecology 64*: 78–88.

Schulte, L. A., and D. J. Mladenoff. 2005. "Severe wind and fire regimes in northern forests: historical variability at the regional scale." *Ecology 86*: 431–445.

Seischab, F. K., and D. Orwig. 1991. "Catastrophic disturbances in the pre-settlement forests of western New York." *Bulletin of the Torrey Botanical Club 118*: 117–122.

Seymour, R. S., and M. L. Hunter, Jr. 1992. "New forestry in eastern spruce-fir forests: principles and applications to Maine." College of Forest Resources. Orono, ME: University of Maine.

Seymour, R. S., and M. L. Hunter, Jr. 1999. "Principles of ecological forestry." In *Maintaining Biodiversity in Forest Ecosystems*, edited by M. L. Hunter, Jr, 22–64. Cambridge, UK: Cambridge University Press.

Seymour, R. S., and L. S. Kenefic. 1998. "Balance and sustainability in multiaged stands: A northern conifer case study." *Journal of Forestry 96*: 12–17.

Seymour, R. S., A. S. White, and P. G. deMaynadier. 2002. "Natural disturbance regimes in northeastern North America – evaluating silvicultural systems using natural scales and frequencies." *Forest Ecology and Management 155*: 357–367.

Spies, T. A., J. F. Franklin, and T. B. Thomas. 1988. "Coarse woody debris in Douglas-fir forests of western Oregon and Washington." *Ecology 69*: 1689–1702.

Spurr, S. H., and A. C. Cline. 1942. "Ecological forestry in central New England." *Journal of Forestry 40*: 418–420.

Stueve, K. M., C. H. H. Perry, M. D. Nelson, S. P. Healey, A. D. Hill, G. G. Moisen, W. B. Cohen, D. D. Gormanson, and C. Huang. 2011. "Ecological importance of intermediate windstorms rivals large, infrequent disturbances in the northern Great Lakes." *Ecosphere 2*: 1–21.

Talon, B., S. Payette, L. Filion, and A. Delwaide. 2005. "Reconstruction of the long-term fire history of an old-growth deciduous forest in Southern Québec, Canada, from charred wood in mineral soils." *Quaternary Research 64*: 36–43.

Tyrrell, L. E., G. J. Nowacki, D. S. Buckley, E. A. Nauertz, J. N. Niese, J. L. Rollinger, and J. C. Zasada. 1998. "Information about old growth for selected forest type groups in the eastern United States." General Technical Report NC-197, USDA Forest Service.

Webster, C. R., and C. G. Lorimer. 2005. "Minimum opening size for canopy recruitment of midtolerant tree species: a retrospective approach." *Ecological Applications 15*: 1245–1262.

Wirth, C. 2009. Old-growth forests: function, fate and value–a synthesis. In *Old-Growth Forests*, edited by C. Wirth, G. Gleixner, and M. Heiman, 465–491. Berlin: Springer.

White, P. S., and S. T. A. Pickett. 1985. "Natural disturbance and patch dynamics: an introduction." In *The Ecology of Natural Disturbance and Patch Dynamics*, edited by S. T. A. Pickett, and P. S. White, 3–13. San Diego, CA: Academic Press.

Whitney, G. G. 1986. "Relation of Michigan's pre-settlement pine forests to substrate and disturbance history." *Ecology 67*: 1548–1559.

Williams, D. W., and A. M. Liebhold. 2000. "Spatial synchrony of spruce budworm outbreaks in eastern North America." *Ecology 81*: 2753–2766.

Worrall, J. J., T. D. Lee, and T. C. Harrington. 2005. "Forest dynamics and agents that initiate and expand canopy gaps in *Picea–Abies* forests of Crawford Notch, New Hampshire, USA." *Journal of Ecology 93*: 178–190.

Worrall, J. J., G. E. Rehfeldt, A. Hamann, E. H. Hogg, S. B. Marchetti, M. Michaelian, and L. K. Gray. 2013. "Recent declines of *Populus tremuloides* in North America linked to climate." *Forest Ecology and Management 299*: 35–51.

Zhang, Q., K. S. Pregitzer, and D. D. Reed. 1999. "Catastrophic disturbance in the pre-settlement forests of the Upper Peninsula of Michigan." *Canadian Journal of Forest Research 29*: 106–114.

Ziegler, S. S. 2002. "Disturbance regimes of hemlock-dominated old-growth forests in northern New York, U.S.A." *Canadian Journal of Forest Research 32*: 2106–2115.

Ziegler, S. S. 2010. "The past and future of white pine forests in the Great Lakes region." *Geography Compass 4*: 1179–1202. https://doi.org/10.1111/j.1749-8198.2010.00369.x.

# Chapter 7

# Historical Patterns and Contemporary Processes in Northern Lake States Old-Growth Landscapes

*David J. Mladenoff and Jodi A. Forrester*

Understanding the value of remaining old-growth forest requires that we evaluate it in the spatial context in which it occurs and in the alterations of that context over the last 150 years. We can see this visually at the regional scale. The northern Lake States forests of Minnesota, Wisconsin, and Michigan, are at the northwestern edge of the north temperate forest of eastern North America (plate 4). This generalized map of forests before widespread Euro-American settlement shows a legacy of responses to physical environment and disturbance, with the upper Midwest ecotone or tension zone generally separating the more northern evergreen-deciduous forested region from the more southwestern oak savanna and prairie. Western limits of beech, hemlock, and yellow birch as major dominants are indicated by black lines. The edges of the beech and hemlock range were particularly abrupt. Yellow birch was in scattered stands and locally abundant as it declined in abundance into northeastern Minnesota. Pines are generalized here but usually occurred over large areas as one or two dominant species.

Significant research efforts have focused on old-growth in the northern Lake States since the review of old growth status in the eastern United States by Mary Byrd Davis (1996). Frelich and Reich (1996) estimated that about 1.1 percent or less than one million acres of an original 80.6 million of original primary forest (unlogged) remains in the three northern Lake States. Some of this original primary forest would have been early successional, recovering from large disturbances and, thus, not considered old growth by some definitions (see chapters 1, 3, 4, 11, and 13). The largest portion of the remnant old growth is in northern lowland forests with low economic value and less likely to be logged. Most of the old growth occurs in northeastern Minnesota and western upper Michigan in a few large reserves, with the largest trees being red and white pine, and the southern boreal forest in the Boundary Waters Canoe Area Wilderness (BWCAW) in the Superior National Forest, Minnesota (figure 7-1; Heinselman 1996). Many very small remnant old-growth patches, often tens of hectares or less in size, are scattered across public and private ownership.

It seems likely that Davis (1996) erred significantly in describing the size and amount of old growth on Isle Royale National Park (figure 7-1). Davis (1996, 24) quotes J. Oelfke as describing Isle Royale Park as 231,395 hectares—nearly four times its actual size—with an estimate of 23,000 to 34,800 hectares or 10 to15 percent of the Park as old growth. This does not seem to be correct. Previant et al. (2013), in the first detailed forest inventory of the park, noted that there is negligible forest over 120 years old.

The boundaries of the major reserves have not changed since the thorough review by Frelich (1995; figure 7-1). However, natural disturbances continue to operate. Large areas of the BWCAW old growth have been reduced by major windstorms and fire (L. B. Johnson, USGS, pers. comm.; Rich et al. 2007). At the same time, fire suppression over the last century has moved much boreal forest in the BWCAW into an old-growth category, something that may not have occurred in the boreal forest type in the past (Frelich 1995). It is also worth noting that even in managed, nonreserve forest lands, some areas are gradually aging to the old-growth threshold. These areas are intrinsically interesting, because they are not primary forest and have often undergone severe human disturbance in the past 100 to 150 years.

Remnant old-growth forests in the Lake States differ from those of pre–Euro-American settlement for at least three reasons. First, the remaining large old-growth forests in reserves are not necessarily representative of the region but are, in various ways, historical artifacts in the sense that they were not cut because of their unusual locations or ownership. Second, because such a very small percentage of old growth of any forest type remains, relatively little of the variation within an ecosystem type is captured in existing reserves. Third, fire disturbance regimes and other broadscale factors, such as deer-browsing levels, have drastically changed in both frequency and severity in the last 75 years compared to what they were in the 1800s (VanDeelen et al. 1996; chapter 12)

Many of the tree species in remnant old growth in the Lake States are the same as those in the northeastern United States and southeast-

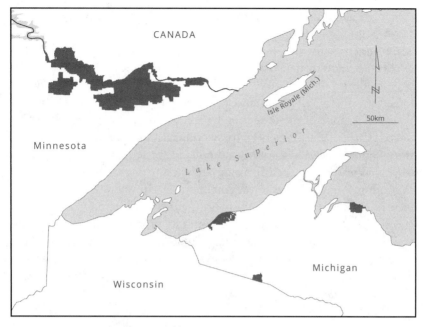

FIGURE 7-1. Current major old-growth reserves in the northern Lake States. Shown are approximate administrative or ownership boundaries and do not indicate the area of old growth contained therein. From west to east: Boundary Waters Canoe Area Wilderness, Superior National Forest (Minnesota); Porcupine Mountains State Park (Michigan); Sylvania Wilderness, Ottawa National Forest and the private Huron Mountain Club (Michigan). See text for extent of old-growth forest in these units.

ern Canada. There are, however, significant differences in environment, dominant species, and the combination of ecological factors that control forest composition, structure, and processes. For example, the Northeast contains greater topographic relief, higher elevations, and greater precipitation. Also, major species, such as American beech (*Fagus grandifolia*), yellow birch (*Betula alleghaniensis*), and eastern hemlock (*Tsuga canadensis*), reach the edge of their range within northern Wisconsin in the northern Lake States, while species such as red spruce (*Picea rubens*), common in the Northeast, are lacking in the Lake States (Curtis 1959). Groundlayer communities respond and change due to these gradients from the Lake States to New England (Rogers 1980). In this chapter, we will concentrate on research by our lab and others in the major forest type of the northern Lake States: northern hardwoods-hemlock.

## Lake States Northern Hardwood-Hemlock Forests

The northern hardwood-hemlock forest types, typically sugar maple (*Acer saccharum*), eastern hemlock, and yellow birch, historically occupied the largest portion of the region (Frelich and Reich 1996), over one-third of mostly northern Wisconsin and Michigan. These shade-tolerant species occupied a range of mesic sites and were not as monolithic as often portrayed (Curtis 1959). American basswood (*Tilia americana*) was important throughout, especially toward southern and western Wisconsin. In lower Michigan, eastern upper Michigan, and along Lake Michigan in eastern Wisconsin, American beech was common. Beech, yellow birch, and hemlock reach western limits in Wisconsin, either their actual physical limit (beech, hemlock) or their functional limit as an important dominant tree (yellow birch; figure 7-1). Sugar maple quickly becomes less important in Minnesota. Eastern white pine (*Pinus strobus*) was a frequent, if minor, species in northern hardwoods-hemlock, as were a number of other tree species (Curtis 1959; Schulte et al. 2002; Fahey et al. 2012).

Wind was the dominant disturbance in these mesic forests, with significant fire being rare, although evidence shows it did occur (Fahey et al. 2012). Given the predominantly mesic soils and lack of flammable understory or other significant fuel, fire would have required unusual circumstances, such as drought following windthrow, which could lead to very low-intensity fires in the litter and duff of the forest floor (Frelich and Reich 1996). Severe crown fires could have occurred if fire reached a dense hemlock canopy.

Unlike fire-dominated systems, remnant old-growth northern hardwood-hemlock forests did not suffer drastic changes in disturbance regime in the last 100 years because of the predominance of wind as a disturbance. Of course, other profound disturbances, such as deer browsing, disease and pests, and others have increased, shifting their relative contribution and significance to the overall disturbance regime (chapter 12).

Today, significant northern hardwood-hemlock old-growth forests occur in reserves in the Porcupine Mountains in the western Upper Peninsula of Michigan, which include about 14,000 hectares, and in the nearby Sylvania Wilderness, with about 6,000 hectares. Small stands also exist in Wisconsin and Michigan, ranging from tens to a few hundreds of acres. Up to 2,000 hectares occur in the private Huron Mountain Club reserve on the shore of Lake Superior (figure 7-1).

The Porcupine Mountains are unusual because they have the most varied topography outside of northeastern Minnesota, with high, rugged bluffs and hills, 240 meters of topographic relief, stream valleys, and clay flats. As a result, the Porcupine Mountains capture a good deal of local site variation, within a strong Lake Superior mesoclimate. Sylvania is a more uniform, rolling area of terminal moraine. Particularly lacking in old-growth representation are the large areas of the central northern Wisconsin till plain. This area was variously dominated by hemlock and sugar maple but also often yellow birch (Schulte et al. 2002), something rarely encountered today (Curtis 1959). This area contained the highest biomass development of any forests in the region (Rhemtulla et al. 2009a). The only significant reserve in this area, a 400-hectare old-growth northern hardwood-hemlock forest, was completely blown down in a 1977 derecho that affected much of northern Wisconsin (Dunn et al. 1983; Canham and Loucks 1984).

## Historical Ecological Context of Northern Hardwood-Hemlock Forests

The US Public Land Survey (PLS) greatly expanded our understanding of the regional forest composition of the 1800s (Whitney 1994). The PLS was conducted primarily from Ohio westward around 1800 to the early 1900s (Stewart 1935). The major value of the PLS is the systematic procedure by which land surveyors recorded tree species or genus, diameter, and location from a survey point. They also mapped major disturbances. When mapped on a broad scale, the data provide the earliest and most complete picture of the landscape before major conversion by logging and agricultural settlement. Use of the data comes with caveats, which have been thoroughly explored (Liu et al. 2011; Cogbill et al. 2018). This remarkable data source has allowed a deeper understanding of what our remaining old growth represents as well as an understanding of the processes and dynamics of old and natural forest systems from the landscape to stand scales (Schulte et al. 2005a, 2005b).

The mapping of the presettlement forest in Wisconsin has been a major effort in our lab and by others for parts of Minnesota (Friedman et al. 2001) and Michigan (Albert et al. 2008). This mapping is particularly valuable within this region of the eastern United States for two main reasons. First, most of the pre–Euro-American settlement forest (probably 70 to 80 percent) was in fact old growth, depending on the age definition of old growth (Lorimer and Frelich 1994; chapters 1 and 4). Second, because so little old growth remains, historical pattern reconstruction is invaluable to infer processes in these extant stands.

The PLS began in Michigan in 1826, in Wisconsin in 1832, and in Minnesota in 1847, each time starting at the southern boundary of the state and proceeding north, following quickly on the heels of Indian treaty cessions. The survey was completed in the 1860s in northern Wisconsin, somewhat earlier in Upper Michigan, and about 1900 in northern Minnesota (Stewart 1935).

One could argue about how well this snapshot in time captures the nature of the regional landscape before Euro-American settlement, but pollen studies have supported the major results, showing that the patterns are broadly representative of the previous several thousand years (Cole et al. 1998). About 3,000 years ago, all of the tree species had migrated into the region following deglaciation and had achieved quasi-equilibrium (Davis 1996). At finer scales within the region, there could be oscillations among dominance of pine versus oak in the northwestern Wisconsin sand plain (Hotchkiss et al. 2007). But that ecosystem was still a dry, fire-controlled, pine and oak landscape throughout. Within the mesic regions, dominance might vary among hemlock, sugar maple, and yellow birch, with white pine less common across the landscape, but these changes were within the bounds of a late successional, mesic north temperate plant community (Curtis 1959). Climate is not static on the broad temporal scale of millennia, however, and there is an ecologically significant climate gradient across the region. Fire- and wind-disturbance gradients also exist across the region and vary subregionally with the substrate (Schulte and Mladenoff 2005) and prehistorical and historical human use. These disturbance gradients also vary through time. Again, pollen records have not shown drastic changes in ecosystem species composition or general distribution for the past few thousand years, until Euro-American settlement.

Therefore, our baseline is the only comprehensive spatial data we have, derived quantitatively from the PLS (figure 7-1), spanning the 1800s. These broadscale patterns are, in effect, another model, and, as such, they reduce high complexity to an understandable level. Nevertheless, the resolution and extent of the PLS data give us a remarkable view of the landscape at the critical time before Euro-American settlement and industrialization changed the landscape more in just 100 years than it had changed in thousands of years prior to that (Cole et al. 1998). Given the relatively low level of complete, stand-replacing disturbance on these mesic landscapes, over 70 percent of the landscape was uneven aged, old and mature stands, with a median tree age of 200 to 250 years (Frelich and Lorimer 1991) in northern hardwood-hemlock forests.

## Reconstructing Forests from the Sylvania Wilderness to the State of Wisconsin

Researchers began developing a more sophisticated understanding of broad landscape dynamics by focusing on existing landscapes of Sylvania and using the PLS records to reconstruct past regional patterns of forest and disturbance regimes. The old-growth northern hardwood-hemlock landscape was mapped, delineating the different forest compositional patches, which were often larger than what was typically considered a forest stand. These were analyzed for their patch size (area), class structure, distribution, shape, and spatial relationship among patches (Pastor and Broschart 1990). This landscape was then compared with an adjacent landscape on the same substrate but of postlogging origin (Mladenoff et al. 1993). Findings showed strong differences: The Sylvania old-growth landscape had a broader range of patch sizes, with a few but much larger patches among more abundant smaller patches. The postlogging landscape was limited to smaller patches (for example, less than 200 hectares). In the old-growth landscapes, larger patches were over 1,000 hectares. Also, patch shapes were more complex on the old landscape, resulting in more interdigitated boundaries with adjacent patches. In this way, both forest interior habitat and landscape complexity were maximized (Mladenoff et al. 1993). Subsequently, we tried to apply this knowledge to how an adjacent landscape with small old-growth remnants within young second growth might be managed to restore the connectivity and patch structure found on the Sylvania old-growth landscape (Mladenoff et al. 1994).

At the same time in Sylvania, paleoecological research was showing that landscape and stand processes operated to maintain both hemlock and hardwood forest patches in the same location for the 2,000 years since hemlock migrated back to the region (Frelich and Reich 1996). This suggested that the broad mosaic of patches has been maintained by strong positive feedbacks with ecosystem processes. Hemlock reached its current, furthest extension west about 1,000 years ago (plate 4; figure 7-1; Davis et al. 1986).

This work in Sylvania informed reconstruction of the regional historical forests, focusing on all of Wisconsin using the PLS data. Beginning with the small subset of the Wisconsin PLS data that was computerized, we tested methods of using the PLS data in classification and mapping. We compared PLS survey data of the Sylvania landscape, where the primary forest remains (Manies and Mladenoff 2000), and quantitatively assessed variability in the surveyor data (Manies et al. 2001). Eventually, data for the entire state were computerized and the dataset and methods were updated (summarized in Liu et al. 2011), allowing the mapping work to be greatly improved. Recent comprehensive analyses have further evaluated forest density calculation methods (Cogbill et al. 2018).

Following the massive effort to computerize all the data recorded in the PLS notebooks, we conducted a statistical analysis and classification of the entire northern half of the state of Wisconsin (Schulte et al. 2002), producing the largest statistically based map of a historic landscape of which we are aware. With this, we began to truly better understand the pattern of the broadscale forest landscape, which historically was that of a largely old and mature forest region. We found that the historical mosaic of forest patches in Wisconsin was influenced primarily by broadscale patterns in soils, with largely pine systems on sands and northern hardwoods-hemlock on more mesic soils (Schulte et al. 2002). However, the mapped historical data showed dramatically that these two general forest types were not monolithic and uniform. Rather they varied widely in species composition in relation to finer soil differences, effects of bordering types, and likely disturbance interactions (Schulte et al. 2002). Such a large spatial dataset allowed us to begin exploring the drivers of broadscale spatial variability in forest pattern, age, and composition (Bolliger et al. 2004; Bolliger and Mladenoff 2005).

Another interesting finding was how disturbance shaped the composition and forest pattern of the region across a range of scales. Our compositional maps were derived primarily from analysis of the survey corner witness tree data. However, additional details could be mined from the line notes where surveyors entered and exited disturbed areas. Using a combination of the line notes and witness tree data, we reconstructed the statewide historic disturbance regimes from fire and windthrow data recorded by the land surveyors (Schulte et al. 2005a; Schulte and Mladenoff 2005). The analysis validated estimates of catastrophic wind and fire disturbances, suggested from tree ring studies, and highlighted the spatial complexity of wind and fire disturbance. We discovered that disturbances were spatially clustered at several scales, operating within substrate patterns, which were nested within the climate pattern. Feedbacks with existing vegetation also increase the spatial complexity. Wind operates across forest types, thus interacting with fire where it occurs, with more clustering on sandy outwash plains. Thus, there is a dynamic where wind has broadscale, positive feedback with fire (creating fuel), but fire has a negative feedback with wind, keeping forests in younger age classes that are less wind susceptible (Schulte et al. 2005a; Schulte and Mladenoff 2005).

While the surveyor notes provided the details for intense disturbance events, we detected limitations of the database for lower-intensity events. The range of long-lived, shade-tolerant species important in mesic old growth suggests that lower-intensity disturbances occurred that were not detectable with the PLS dataset. But we could detect severe-but-partial canopy removal, largely by wind (Schulte et al. 2005a). Evidence for this was in data of Frelich and Lorimer (1991) and Dahir and Lorimer (1996). Hanson and Lorimer (2007) more fully explored turnover and mortality rates, explaining that significant wind events removing 30 to 60 percent of canopy trees occur in northern hardwoods at frequencies of several hundred years, which would be significant to stand composition and structure in these systems where complete catastrophic blowdowns occur only every several thousand years. Such intermediate disturbances may have longer-term consequences than small gaps or the rare complete canopy removal (see chapters 4 and 6). These dynamics likely maintained midtolerants such as yellow birch at the extremely high basal area that occurred in parts of northern Wisconsin, as well as lower levels of some others, such as white pine, white ash (*Fraxinus americana*), and American basswood. This would also release the ubiquitous understory cohorts of the dominant, shade-tolerant sugar maple. In this way, a diverse suite of canopy species often occurred with the leading dominants—hemlock and sugar maple (Frelich and Lorimer 1991; Dahir and Lorimer 1996; Schulte et al. 2002, 2005a; Hanson and Lorimer 2007).

Later, linking this approach with the tree-size data from the PLS more explicitly revealed the extent of former old growth on the landscape by calculating high biomass forest on the regional scale (Rhemtulla et al. 2009a). As well as revealing patterns of old forest, the combined data showed how much greater potential for biomass accumulation and carbon storage was untapped today (see chapter 14). Using the historical PLS data, 1930s and recent inventories showed the magnitude of change on the landscape from historic times and the low level of recovery of both structure and composition (Rhemtulla 2009b). Looking at the statewide landscape, Rhemtulla et al. (2009a) found potential for aboveground live biomass that exceeded field studies today. This shows that reserve systems are not representative of the full range of site conditions (e.g., productivity) across which old growth occurred historically (Rhemtulla et al. 2009a). Our broadscale spatial research provided new context to regional old-growth patterns and helped to fill in our understanding of the high variability of old-growth composition due to environment and disturbance.

## Empirical Investigations of Regional Drivers

These studies of regional patterns complemented extensive and often intensive studies of the processes inherent to the Lake States old-growth northern hardwood-hemlock forests. In the 1980s, the importance of small treefall gaps had been clarified by Runkle and others in mesic Appalachian forests (Runkle 1982) and to a large extent this paradigm became dominant in thinking about mesic forests in eastern North America.

Our understanding of the range, frequency, and variable severities of wind disturbances had been poor in the region. Small gap dynamics were known to be important because of the average long periods without large disturbances (greater than 2,000 years for more than 60 percent canopy removal, based on field data from tree rings; Frelich and Lorimer 1991). Larger wind disturbances were known to occur as well, based on historic data (Stearns 1949; Canham and Loucks 1984) and striking evidence from the large derecho system that moved across northern Wisconsin in 1976 (Dunn et al. 1983). Some evidence of fire existed, but significant fire is extremely rare in this system because of the generally moist site conditions and the lack of understory fuel due to high shade (Frelich and Lorimer 1991).

Process studies also examined effects of treefall openings in the region. In an early process study, Mladenoff (1987) found that canopy dominance (hardwood versus hemlock) was important for total nitrogen mineralization but that small treefall gaps were a finer-scale control: Nitrification declined with greater hemlock abundance but increased in gaps. Further study clarified that gap size and location within the gap influenced nitrogen dynamics. Gap edges were shown to be hot spots of higher microbial biomass nitrogen, and higher nitrogen mineralization (Sharenbroch and Bockheim 2007). Soil-surface carbon dioxide flux also followed these microscale spatial patterns (Schatz et al. 2012). Larger gaps (up to 590 square meters) were found to have reduced microbial biomass though greater ammonium concentration, suggesting a microbial constraint on nitrogen cycling (Schliemann and Bockheim 2014).

A number of studies have compared composition and structural characteristics of old-growth northern hardwood forests and young stands; Tyrrell et al. (1998) summarizes many of the available metrics in a voluminous, invaluable compilation of work across the eastern United States (see chapter 11 for an updated review). Here we highlight studies that have reported biomass or net primary productivity (table 7-1). Tyrrell and Crow (1994a; 1994b) studied 25 old-growth stands with maximum tree ages of

177 to 374 years. They found that forests of these types reached a suite of old-growth structural features at 275 to 300 years. Based on log-decay rates and tree mortality, stands older than 350 years appeared to be at an equilibrium of production and loss (Tyrrell and Crow 1994a). Subsequently, Lorimer et al. (2001) explained that in old-growth stands with trees of 200 to 350 years old, average mortality occurred at 216 years for sugar maple and 301 years for hemlock. This suggests that these forests do not reach true old-growth, when canopy mortality becomes significant, until 180 to 250 years, and downed wood and other structural characteristics approach a quasi-equilibrium, that is, a relatively stable, though variable range. Simulations from a model derived from long-term population monitoring in primary forests reinforced that, in this region, a quasi-equilibrium biomass is reached after about 275 years, and showed biomass no longer increasing after this age (Halpin and Lorimer 2016a, 2016b; see chapter 14). These works, in addition to the results from Tyrrell and Crow, have important implications for assumptions about old-growth processes and habitat, and belie the 120-year criterion for many systems, which derives from forest management and economic rotation ages (Tyrrell et al. 1998).

Comparative studies have highlighted how soil type, topography, and disturbance influence the regional variation in spatial patterns and composition. Surprisingly, no consistent differences in soil morphological properties, including soil horizons, presence of mor or mull humus (compacted layer of humus), and fragipan (the subsurface layer restricting water flow and root penetration) development existed between hemlock and northern hardwood stands, occurring on a range of sites and soils. Nutrient availability was not measured, but there were significant differences in exchangeable calcium, magnesium, and potassium sum of bases between hemlock and northern hardwoods, with levels higher under hemlocks, and greater calcium in aboveground biomass of hardwoods and in the forest floor and soil under hemlocks (Bockheim 1997), suggesting a mechanism for calcium depletion in sugar maple stands.

In old-growth stands in Sylvania where maple was dominant, Campbell and Gower (2000) measured higher net nitrogen mineralization than in hemlock old-growth stands but with high finescale variation across the forest cover types. Net primary productivity was much higher in the hemlock stands as opposed to the hardwood stands (Campbell and Gower 2000). Tang et al. (2008) found that hemlock old-growth stands respire more carbon from stems but less from leaves than the mixed hardwood-hemlock stands. Goodburn and Lorimer (1998; table 7-1) had quantified coarse woody debris stocks, noting that hardwoods-hemlock had nearly

TABLE 7-1. Biomass and net primary productivity estimated from northern hardwood study areas in the Lake States. The first two columns show the variability in either the composition and/or developmental stage. DWD: downed woody debris; CWD: coarse woody debris.

| Forest type | Age category | Variable | Biomass (Mg/ha) | NPP (Mg/ha/yr) | Citation |
|---|---|---|---|---|---|
| Hardwood | Old growth | Standing and DWD (>10 cm diam) | 36.4 | | Goodburn and Lorimer 1998 |
| Hardwood-hemlock | Old growth | Standing and DWD (>10 cm diam) | 40.1 | | Goodburn and Lorimer 1998 |
| Hardwood-hemlock | Old growth | DWD | 11.4–110.3 | | Tyrrell and Crow 1994a |
| Hemlock | Old growth | DWD | 1.8–28.3 | | Tyrrell and Crow 1994a |
| Hemlock | Old growth | Aboveground live | 189 | 4.8 | Campbell and Gower 2000 |
| | | CWD | 22 | | |
| | | Litterfall | 3.4 | | |
| | | Forest floor | 29 | | |
| Hardwood | Old growth | Aboveground live | 330 | 8.7 | |
| | | CWD | 9 | | |
| | | Litterfall | 4.4 | | |
| | | Forest floor | 22 | | |
| Hardwood-hemlock | Old growth | Aboveground live | 360–450 | | Woods 2014 |
| Hardwood-hemlock | Old growth | Aboveground live | 394.9 | 1.8 | Gries 1995 |
| | | Sapling-immature | | 2.9–2.2 | |
| | | Mature | | 0.6 | |
| Hardwood | Old growth | Live and detrital | 350 | | Fisk et al. 2002 |

*continued on next page*

TABLE 7-1. *continued*

| Forest type | Age category | Variable | Biomass (Mg/ha) | NPP (Mg/ha/yr) | Citation |
|---|---|---|---|---|---|
| Hardwood | Second growth | Live and detrital | 280 | | Fisk et al. 2002 |
| Hardwood-hemlock | Pole to mature | Aboveground live tree | 176–278 | | Halpin and Lorimer 2016 |
| | Early to mid-transition | | 261–331 | | |
| | Late transition | | 266 | | |
| | Steady-state | | 261 | | |

equal proportions of dead wood in standing and down classes and nearly three times the biomass in standing wood as the hardwood stands. These differences in the magnitude and proportions of component respiration and storage result in different carbon budgets between the cover types, despite the stands sharing similar climate and ecological zones.

Other studies were designed to minimize differences in soil type and overstory composition among research stands and focus on comparing successional stage. Fisk et al. (2002) identified patterns of greater detrital nitrogen storage and microbial nitrogen cycling in old-growth than second-growth forests, but nitrogen mineralization and cycling were widely variable and independent of forest age. Desai et al. (2005) found greater carbon stocks with high amounts of coarse woody debris, when comparing old-growth northern hardwoods-hemlock in Sylvania older than 300 years with younger mixed northern hardwoods (70 years). Eddy covariance measurements of carbon dioxide exchange showed that the old-growth landscape was still functioning as a sink, though at a lower rate than the younger forest and with significantly greater ecosystem respiration (chapter 14).

By far the greater number of these studies pointed to a few structural elements of old forest, especially variable size gaps and downed woody debris (Goodburn and Lorimer 1998), as key to both the spatial variability in ecosystem processes and quantity of diverse species habitat (see chapter 6). These features also matter to the habitat of other species (chapter 11), especially the microhabitat structure of large trees and large, old conifers—mostly hemlock and some white pine (e.g., Schulte et al. 2005 for bird species).

As was the case in other areas of the United States, studies in the northern Lake States sought to determine if greater species diversity was a value of old growth. However, in this region, such old-growth habitat value was often shown to be more nuanced. For example, these young postglacial ecosystems have few true old-growth obligate species. What has been shown is that old growth often has greater relative abundance of some species over younger forests (see chapter 11). Historically, eastern white pine was broadly distributed across the mesic hardwood-hemlock forest landscapes though largely absent in the modern landscape (Fahey et al. 2012). White pine was in part maintained in these late successional forests by dispersing from refuge habitats interspersed on the regional landscape, sometimes aided by occasional fire. Forest management, fire protection, and browsing may be limiting pine regeneration in its former range.

Understory plant community composition and functional diversity overall can be characterized as different from younger, postlogging forests, and the spatial distribution of species and communities is also characteristic, suggesting a relationship with ecosystem processes that develop over long-term dynamics in old growth (Miller et al. 2002; Scheller and Mladenoff 2002). Lichen species are those that perhaps show the greatest affinity to forest age and structure, with 10 percent of lichen species being restricted to old forests, and, as with plants, 25 percent having greater relative abundance in old rather than younger forests (Will-Wolf and Nelson 2008; see chapter 11).

For some species, the microhabitat structure of large trees and coarse woody material, coupled with the physiognomy of large structured conifers (hemlock and white pine), are particularly important. For example, using bird habitat suitability models based on recent bird inventory data to assess habitat change since the 1800s for several bird species that use large, old conifers, Schulte et al. (2005) found dramatic loss and degradation of habitat. This is not surprising given that white pine in Wisconsin is estimated to be less than 5 percent of its former basal area, and hemlock is estimated to be less than 0.5 percent. This emphasizes the important habitat values of old growth beyond simple obligate status. These patterns of the value of loss and degradation of old growth were shown across a diverse range of taxa, including salamanders, wood-inhabiting fungi, beetles, and others (Czederpiltz et al. 1999; Bergeson 2001; Latty et al. 2006; chapters 10 and 11).

Regionally, an important background for all of this research in present and recent times is nested within the context of historically very high deer populations and other invasive species (chapters 10, 11, and 12). These all can have significant effects on organic matter and nutrient cycling and on herbaceous plant and tree species composition. These broadscale impacts of the last century also underscore how even our largest current landscapes do not represent the past in some important ways. This effect extends to all the old-growth forest types in the region.

## The Flambeau Experiment

The analysis of broad, regional, historical patterns, and the related fines-cale spatial field studies, drove establishment of a landscape scale manipulative experiment begun in 2006. We began the Flambeau Experiment to follow up on regional comparative and descriptive studies to examine within-stand processes over time and over the variation of a maturing northern hardwood forest landscape. The replicated manipulations include varying those factors shown to be most important in the comparative studies, namely canopy gap size and coarse woody debris quantity. This project is ongoing with more than 10 years of data collection. Initial post-treatment research priorities of the experiment focused on the effects of the canopy openings that were created as a result of variable-sized single- and group-selection harvests. We have described the changes occurring in the gap microclimate, soil carbon dioxide flux (Schatz et al. 2012), net primary productivity (Dyer et al. 2010), plant diversity (Burton et al. 2014), sapling dynamics (Forrester et al. 2014), wood-inhabiting fungal community (Brazee et al. 2014), and soil microbial composition (Lewandowski et al. 2015)—immediately following gap creation. Several studies were launched to assess the role of downed woody debris in carbon dynamics (Forrester et al. 2013, 2015) and will be followed through time as decay proceeds.

These types of long-term, largescale research projects will generate valuable, tangible information for management options for future forest conditions (chapters 6 and 13). Under continuing climate change, we can expect the broader environmental context of the regional forest systems to enhance differences from the past or original forest systems (Scheller and Mladenoff 2008). The northern Lake States is a difficult region for global climate models to predict. The transition to the more arid plains is actually quite abrupt. How temperature and precipitation interact in evapotranspiration will be critical for future forests, but these future precipitation and forest responses are particularly difficult to model.

We will continue to have old-growth forests, but increasing change in those forests and further deviation from any concept of a past natural variability is inevitable. Future research and management will need to respond to this.

## Conclusion

Regional research on old-growth northern hardwood-hemlock forests has dramatically changed our understanding of this highly variable ecosystem. A major theme has been clarifying the high degree of spatial variability within the system. This began with improving our understanding of regional compositional differences and how these patterns are nested with interacting substrate and disturbance gradients. The crucial importance of moderate-severity canopy disturbance has been clarified through field studies. These studies show how forests grow and change, while maintaining a suite of tolerant and midtolerant tree species in a system with infrequent complete canopy removal. Concurrently, research has been able to show that fine-grained spatial factors, including gap sizes and microenvironments, control soil and wood microbial composition and the nutrient availability and carbon transformations that drive the systems.

We have used past survey data to reconstruct the presettlement landscapes of primary forests and approximate the density or carbon storage of the forested lands before any cutover occurred. The map or the larger-scale picture is useful in many capacities. State agencies and academics have used the PLS data as a reference point for forest planning, wildlife planning, and the mapping of potential fire regions, to name a few examples. The results not only provide a visualization of historic landscapes but can and have been used to project how they might look in the future. As we understand what and how processes work in these primary forests, we can apply this knowledge to provide guidance and justification for incorporating elements of old growth, such as restoring structure, diversity, and aspects of habitat into the second-growth forests dominant on today's landscape (chapter 13).

At the same time, many of the results point out where we have limitations in our knowledge. Even existing landscapes are not large enough to contain a full regional landscape pattern of ecosystem responses to the broadscale natural disturbances. A perennial challenge is how to restore such systems and disturbances even at the stand scale. True landscape representation seems impossible, unfortunately, more than ever. Our understanding of the scale required to capture all variability in environment and disturbance has grown significantly. Exotic pests and diseases of forest trees are gradually increasing, and this is likely to continue. It is important that research and management become proactive in understanding effects of these threats and finding ways to maintain the diversity of native tree species.

# References

Albert, D. A., P. J. Comer, and H. Enander. 2008. *Atlas of Early Michigan's Forests, Grasslands, and Wetland*. Lansing, MI: Michigan State University Press.

Bergeson, M. T. 2001. "Red-backed salamanders (*Plethodon cinereus*) in hardwood forests in northeastern Wisconsin and the Upper Peninsula of Michigan." Master's thesis. Madison, WI: University of Wisconsin.

Bockheim, J. G. 1997. "Soils in a hemlock-hardwood ecosystem mosaic in the southern Lake Superior uplands." *Canadian Journal of Forest Research 27*: 1147–1153.

Bolliger, J., and D. J. Mladenoff. 2005. "Quantifying spatial classification uncertainties of the historical Wisconsin landscape (USA)." *Ecography 28*: 141–156.

Bolliger, J., L. A. Schulte, S. N. Burrows, T. A. Sickley, and D. J. Mladenoff. 2004. "Assessing ecological restoration potentials of Wisconsin (USA) using historical landscape reconstructions." *Restoration Ecology 12*: 124–142.

Brazee, N. J., D. L. Lindner, A. W. D'Amato, S. Fraver, J. A. Forrester, and D. J. Mladenoff. 2014. "Disturbance and diversity of wood-inhabiting fungi: effects of canopy gaps and downed woody debris." *Biological Conservation 23*: 2155–2172.

Burton, J. I., D. J. Mladenoff, J. A. Forrester, and M. Clayton. 2014. "Diversity and productivity in northern temperate deciduous forest understories: experimentally testing predictions of the intermediate disturbance hypothesis." *Journal of Ecology 102*: 1634–1648.

Campbell, J. L., and S. T. Gower. 2000. "Detritus production and soil N transformations in old-growth eastern hemlock and sugar maple stands." *Ecosystems 3*: 185–192.

Canham, C. D., and O. L. Loucks. *1984*. "Catastrophic windthrow in the presettlement forests of Wisconsin." *Ecology 65*: 803–809.

Cogbill, C. V., A. L. Thurman, J. W. Williams, J. Hu, D. J. Mladenoff, and S. J. Goring. 2018. "A retrospective on the accuracy and precision of plotless forest density estimators in ecological studies." *Ecosphere*. In press.

Cole, K. L., M. B. Davis, F. Stearns, G. Guntenspergen, and K. Walker. 1998. "Historical landcover changes in the Great Lakes region." In *Perspectives on the Land-use History of North America: A Context for Understanding Our Changing Environment*, edited by T. D. Sisk, 43–50. Biological Science Report USGS/BRD/BSR-1998-0003, Biological Resources Division, US Geological Survey.

Curtis, J. T. 1959. *The Vegetation of Wisconsin: An Ordination of Plant Communities*. Madison, WI: University of Wisconsin Press.

Czederpiltz, D. L. L., G. R. Stanosz, and H. H. Burdsall, Jr. 1999. "Forest management and the diversity of wood-inhabiting fungi." *McIlvainea 14*: 34–45.

Dahir, S. E., and C. G. Lorimer. 1996. "Variation in canopy gap formation among developmental stages of northern hardwood stands." *Canadian Journal of Forest Research 26*: 1875–1892.

Davis, M. B. 1996. *Eastern Old-Growth Forests: Prospects for Rediscovery and Recovery*. Washington, DC: Island Press.

Davis, M. B., K. D. Woods, S. L. Webb, and R. P. Futyma. 1986. "Dispersal versus climate: Expansion of Fagus and Tsuga into the Upper Great Lakes region." *Vegetatio 67*: 93–103.

Desai, A. R., P. V. Bolstad, B. D. Cook, K. J. Davis, and E. V. Carey. 2005. "Comparing net ecosystem exchange of carbon dioxide between an old-growth and mature forest in the upper Midwest, USA." *Agricultural and Forest Meteorology 128*: 33–55.

Dunn, C. P., G. R. Guntenspergen, and J. R. Dorney. 1983. "Catastrophic wind disturbance in an old-growth hemlock-hardwood forest, Wisconsin." *Canadian Journal of Botany 61*: 211–217.

Dyer, J. H., S. T. Gower, J. A. Forrester, C. G. Lorimer, D. J. Mladenoff, and J. I. Burton. 2010. "Effects of selective tree harvests on aboveground biomass and net primary productivity of a second-growth northern hardwood forest." *Canadian Journal of Forest Research 40*: 2360–2369.

Fahey R. T., C. R. Lorimer, and D. J. Mladenoff. 2012. "Habitat heterogeneity and life-history traits influence presettlement distributions of early-successional tree species in a late-successional, hemlock-hardwood landscape." *Landscape Ecology 27*: 999–1013.

Fisk, M. C., D. R. Zak, and T. R. Crow. 2002. "Nitrogen storage and cycling in old- and second-growth northern hardwood forests." *Ecology 83*: 73–87.

Forrester, J. A., C. G. Lorimer, J. H. Dyer, S. T. Gower, and D. J. Mladenoff. 2014. "Response of tree regeneration to experimental gap creation and deer herbivory in north temperate forests." *Forest Ecology and Management 329*: 137–147.

Forrester, J. A., D. J. Mladenoff, A. W. D'Amato, S. Fraver, D. L. Lindner, N. J. Brazee, M. K. Clayton, and S. T. Gower. 2015. "Temporal trends and sources of variation in carbon flux from coarse woody debris in experimental forest canopy openings." *Oecologia 179*: 889–900.

Forrester, J. A., D. J. Mladenoff, and S. T. Gower. 2013. "Experimental manipulation of forest structure: Near-term effects on gap and stand scale C dynamics." *Ecosystems 16*: 1455–1472.

Frelich, L E. 1995. "Old forest in the Lake States today and before European settlement." *Natural Areas Journal 15*: 157–167.

Frelich, L. E., and C. G. Lorimer. 1991. "Natural disturbance regimes in hemlock-hardwood forests of the upper Great Lakes region." *Ecological Monographs 61*: 145–164.

Frelich, L. E., and P. B. Reich. 1996. "Old growth in the Great Lakes Region." In *Eastern Old-Growth Forests: Prospects for Discovery and Recovery*, edited by M. B. Davis, 144–160. Washington, DC: Island Press.

Friedman, S. K., P. B. Reich, and L. E. Frelich. 2001. "Multiple scale composition and spatial distribution patterns of the northeastern Minnesota forest." *Journal of Ecology 89*: 538–554.

Goodburn, J. M., and C. G. Lorimer. 1998. "Cavity trees and coarse woody debris in old-growth and managed northern hardwood forests in Wisconsin and Michigan." *Canadian Journal of Forest Research 28*: 427–438.

Gries, J. F. 1995. "Biomass and net primary production for a northern hardwood stand development sequence in the Upper Peninsula of Michigan." Master's thesis. Madison, WI: University of Wisconsin.

Halpin, C. R., and C. G. Lorimer. 2016a. "Long-term trends in biomass and tree demography in northern hardwoods: An integrated field and simulation study." *Ecological Monographs 86*: 78–93.

Halpin, C. R., and C. G. Lorimer. 2016b. "Trajectories and resilience of stand structure in response to variable disturbance severities in northern hardwoods." *Forest Ecology and Management 365*: 69–82.

Hanson, J. J., and C. G. Lorimer. 2007. "Forest structure and light regimes following moderate wind storms: Implications for multi-cohort management." *Ecological Applications 17*: 1325–1340.

Heinselman, M. 1996. *The Boundary Waters Wilderness Ecosystem*. Minneapolis, MN: University of Minnesota Press.

Hotchkiss, S. C., R. Calcote, and E. A. Lynch. 2007. "Response of vegetation and fire to

Little Ice Age climate change: Regional continuity and landscape heterogeneity." *Landscape Ecology 22*: 25–41.

Latty, E. F., S. M. Werner, D. J. Mladenoff, K. F. Raffa, and T. A. Sickley. 2006. "Response of ground beetle (Carabidae) assemblages to logging history in northern hardwood-hemlock forests." *Forest Ecology and Management 222*: 335–347.

Lewandowski, T. G., J. A. Forrester, D. J. Mladenoff, J. L. Stoffel, S. T. Gower, T. Balser, and A. W. D'Amato. 2015. "Soil microbial community response and recovery following group selection harvest: Temporal patterns from an experimental harvest in a US northern hardwood forest." *Forest Ecology and Management 340*: 82–94.

Liu, F., D. J. Mladenoff, N. S. Keuler, and L. S. Moore. 2011. "Broadscale variability in tree data of the historical Public Land Survey and its consequences for ecological studies." *Ecological Monographs 81*: 259–275.

Lorimer, C. G., S. E. Dahir, and E. V. Nordheim. 2001. "Tree mortality rates and longevity in mature and old-growth hemlock-hardwood forests." *Journal of Ecology 89*: 960–971.

Lorimer, C. G., and L. S. Frelich. 1994. "Natural disturbance regimes in old-growth northern hardwoods." *Journal of Forestry 92*: 33–38.

Manies, K. L., and D. J. Mladenoff. 2000. "Testing methods to produce landscape-scale presettlement vegetation maps from the U.S. public land survey records." *Landscape Ecology 15*: 741–754.

Manies, K. L., D. J. Mladenoff, and E. V. Nordheim. 2001. "Assessing large-scale surveyor variability in the historic forest data of the original US Public Land Survey." *Canadian Journal of Forest Research 31*: 1719–1730.

Miller, T. F., D. J. Mladenoff, and M. K. Clayton. 2002. "Spatial autocorrelation and patterns of understory vegetation and environment in old-growth northern hardwood forests." *Ecological Monographs 72*: 487–503.

Mladenoff, D. J. 1987. "Dynamics of nitrogen mineralization and nitrification in hemlock and hardwood treefall gaps." *Ecology 68*: 1171–1180.

Mladenoff, D. J., M. A. White, T. R. Crow, and J. Pastor. 1994. "Applying principles of landscape design and management to integrate old-growth forest enhancement and commodity use." *Conservation Biology 8*: 752–762.

Mladenoff, D. J., M. A. White, J. Pastor, and T. R. Crow. 1993. "Comparing spatial pattern in unaltered old-growth and disturbed forest landscapes." *Ecological Applications 3*: 294–306.

Pastor, J., and M. Broschart. 1990. "The spatial pattern of a northern conifer-hardwood landscape." *Landscape Ecology 4*: 55–68.

Previant, W. J., L. M. Nagel, S. A. Pugh, and C. W. Woodall. 2013. "Forest Resources of Isle Royale National Park 2010." Resource Bulletin NRS-73. Northern Research Station. Newtown Square, PA: United States Department of Agriculture.

Rhemtulla, J. M., D. J. Mladenoff, and M. K. Clayton. 2009a. "Historical forest baselines reveal potential for continued carbon sequestration." *Proceedings of the National Academy of Sciences 106*: 6082–6087.

Rhemtulla, J. M., D. J. Mladenoff, and M. K. Clayton. 2009b. "Legacies of historical land use on regional forest composition and structure in Wisconsin, USA (mid-1800s to 1930s to 2000s)." *Ecological Applications 19*: 1061–1078.

Rich, R. L., L. E. Frelich, and P. B. Reich. 2007. "Wind-throw mortality in the southern boreal forest: effects of species, diameter and stand age." *Journal of Ecology 95*: 1261–1273.

Rogers, R. S. 1980. "Hemlock stands from Wisconsin to Nova Scotia: Transitions in understory composition along a floristic gradient." *Ecology 61*: 178–193.

Runkle, J. R. 1982. "Patterns of disturbance in some old-growth mesic forests of eastern North America." *Ecology 63*: 1533–1546.

Schatz, J. D., J. A. Forrester, and D. J. Mladenoff. 2012. "Spatial patterns of soil surface C flux in experimental canopy gaps." *Ecosystems 15*: 616–623.

Scheller, R. M., and D. J. Mladenoff. 2002. "Species diversity, composition, and spatial patterning of understory plants in old-growth and managed northern hardwood forests." *Ecological Applications 12*: 1329–1343.

Scheller, R. M., and D. J. Mladenoff. 2008. "Simulated effects of climate change, fragmentation, and inter-specific competition on tree species migration in northern Wisconsin, USA." *Climate Research 36*: 191–202.

Schliemann, S. A., and J. G. Bockheim. 2014. "Influence of gap size on carbon and nitrogen biogeochemical cycling in Northern hardwood forests of the Upper Peninsula, Michigan." *Plant Soil*. doi 10.1007/s11104-013-2005-5.

Schulte, L. A., and D. J. Mladenoff. 2005. "Severe wind and fire regimes in northern forests: Historical variability at the regional scale." *Ecology 86*: 431–445.

Schulte, L. A., D. J. Mladenoff, S. N. Burrows, T. A. Sickley, and E. V. Nordheim. 2005a. "Spatial controls of Pre-Euro-American wind and fire in northern Wisconsin (USA) forest landscapes." *Ecosystems 8*: 73–94.

Schulte, L. A., D. J. Mladenoff, and E. V. Nordheim. 2002. "Quantitative classification of a historic northern Wisconsin (U.S.A.) landscape: Mapping forests at regional scales." *Canadian Journal of Forest Research 32*: 1616–1638.

Schulte, L. A., A. M. Pidgeon, and D. J. Mladenoff. 2005b. "One hundred fifty years of change in forest bird breeding habitat: Estimates of species distributions." *Conservation Biology 19*: 1944–1956.

Sharenbroch, B. C., and J. G. Bockheim. 2007. "Impacts of forest gaps on soil properties and processes in old growth northern hardwood-hemlock forests." *Plant Soil 294*: 219–233.

Stearns, F. W. 1949. "Ninety years change in a northern hardwood forest in Wisconsin." *Ecology 30*: 350–358.

Stewart, L. O. 1935. *Public Land Surveys: History, Instructions, Methods*. Ames, IA: Collegiate Press.

Tang, J., P. V. Bolstad, A. R. Desai, J. G. Martin, B. D. Cook, K. J. David, and E. V. Carey. 2008. "Ecosystem respiration and its components in an old-growth forest in the Great Lakes region of the United States." *Agricultural and Forest Meteorology 148*: 171–185.

Tyrrell L. E., and T. R. Crow. 1994a. "Dynamics of dead wood in old-growth hemlock hardwood forests of northern Wisconsin and northern Michigan. *Canadian Journal of Forest Research 24*: 1672–1683.

Tyrrell L. E., and T. R. Crow. 1994b. "Structural characteristics of old-growth hemlock-hardwood forests in relation to age." *Ecology 75*: 370–386.

Tyrrell, L. E., G. J. Nowacki. T. R. Crow, D. S. Buckley, E. A. Nauertz, J. N. Niese, J. L. Rollinger, and J. C. Zasada. 1998. "Information about old growth for selected forest type groups in the eastern United States." General Technical Report NC-197. North Central Forest Experiment Station. St. Paul, MN: USDA Forest Service.

VanDeelen, T. R., K. S. Pregitzer, and J. B. Haufler. 1996. "A comparison of presettlement and present-day forests in two Northern Michigan deer yards." *The American Midland Naturalist 135*: 181–194.

Whitney, G. G. 1994. *From Coastal Wilderness to Fruited Plain: a History of Environmental Change in Temperate North America, 1500 to the Present*. Cambridge, UK: Cambridge University Press.

Will-Wolf, S., and M. P. Nelsen. 2008. "How have Wisconsin's lichen communities changed?" In *The Vanishing present: Wisconsin's Changing Lands, Waters, and Wildlife*, edited by D. M. Waller, and T. R. Rooney, 127–150. Chicago, IL: University of Chicago Press.

Woods, K. D. 2014. "Multi-decade biomass dynamics in an old-growth hemlock-northern hardwood forest, Michigan, USA". *PeerJ 2*: e598. https://doi.org/10.7717/peerj.598.

# Chapter 8

# Is Management or Conservation of Old Growth Possible in North American Boreal Forests?

*Daniel Kneeshaw, Philip J. Burton, Louis De Grandpré, Sylvie Gauthier, and Yan Boulanger*

Old-growth forests are often perceived as cathedral-like stands of large trees in coastal coniferous forests or eastern hardwoods. In most of the boreal region, however, this is not the case, as tree size is limited by cold climate and short growing seasons, while forest age is limited by frequent large natural disturbances. The boreal forest is a disturbance-driven system. Fire is omnipresent and, where it is less prevalent, other disturbances, such as insects and wind, drive forest dynamics. As such, many stands burn or are subject to other disturbances before they have the chance to attain old-growth status. The idea of old growth as a local equilibrium state (i.e., a forest structure that has reached a stable, self-maintaining state in equilibrium with climate and local abiotic conditions), largely developed for other forests, applies poorly to the boreal (chapters 7, 11, and 14). Even when considered over vast regional scales, the probability that a forest burns is not dependent on its age, as young forests have an equal probability of burning as do older ones. Because of the random nature of fire, forest stands of different ages are not equally represented over the landscape (Cumming et al. 1996). Boreal old growth is a dynamic condition that is manifest by varying forest structure (Kneeshaw and Gauthier 2003; Shorohova et al. 2011), stand size, and longevity across the boreal zone of North America. Despite the constraints on old growth in this biome, old growth plays a key role in the maintenance of biodiversity in the boreal forest (chapter 11).

In this chapter, we first describe the extent of old growth in different regions of the North American boreal forest as it relates to species longevity and prevailing natural disturbance regimes. We also briefly describe how forest management has affected different types of old growth in the recent past. We then present different approaches to maintain this forest type and discuss how conservation, forest management, and forest restoration can be intermixed in order to sustain the multiple old-growth phases found in boreal forests across North America.

## Natural Disturbance Regimes Determine Old Growth Abundance

The prevailing fire cycle and the average longevity of dominant tree species can be used to predict the extent of old-growth forest in a given area of the boreal region (Kneeshaw and Gauthier, 2003; Bergeron et al. 1999). A Weibull or negative exponential function provides an estimate of the theoretical proportion of a landscape in any given age class as a function of time since fire (figure 8-1). Where fire cycles are long (i.e., many decades to centuries), there is sufficient time for the original cohort to break up (as trees die and collapse), for new tree species to establish, and for this community to slowly mature to old growth. This is the last stage in the stand development theory proposed by Oliver and Larson (1990), which was used by Kneeshaw and Burton (1998) to quantify old-growth status and assign stands along a forest development gradient. The median longevity of a tree species can be used as a proxy for when a stand will start breaking up and, thus, enter the old-growth stage (Oliver and Larson 1990; Kneeshaw and Burton 1998). The attainment of old growth can be thought of as the time at which the dynamics of the stand are dominated by single-tree replacement processes (i.e., one tree dying at a time) rather than largescale disturbance. Combining the length of the fire cycle and the age of dominant tree species allows us to predict the age of this old-growth transition and the proportion of the forest that can be expected to be older than that age.

Based on the approach described above, the proportion of old-growth forest has been estimated in many North American boreal regions (plate 5). Analysis reveals significant regional variation in old-growth forest proportion, largely related to the length of the fire cycle, which, in turn, is primarily driven by climate (Kneeshaw and Gauthier 2003; Shorohova et al. 2011; Boulanger et al. 2014). Where conditions are drier, fire cycles are shorter, and old growth is less abundant. Fire recurrence (years between fires) in the North American boreal zone varies from less than 100 years to over 1,000 years (Portier et al. 2016). These variations occur along both north-south and west-east gradients related to precipitation, with shorter cycles in drier continental regions. It has also been shown that these cycles

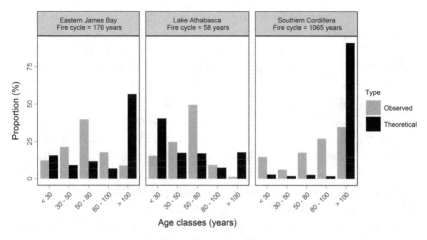

FIGURE 8-1. Examples of the proportion of old-growth forest observed in regions with different fire cycles compared to the distribution for those same regions predicted by the negative exponential function.

change over time: In eastern Canada they have been getting longer over past decades (Portier et al. 2016; Bergeron et al. 2006), again, reflecting the dynamic nature of these forests. Tree longevities are generally shorter in North American boreal forests than in those of Eurasia (Kneeshaw et al. 2011; Shorohova et al. 2011). This may be an evolutionary adaptation to shorter intervals between severe fires in North America compared to Europe (see also Rogers et al. 2015). Even within a continent, the differences can easily be twofold, with species such as trembling aspen (*Populus tremuloides* Michx.), paper birch (*Betula papyrifera* Marsh.), and balsam fir (*Abies balsamea* (L.) Mill.) typically living only 60 to 100 years, whereas the expected lifespan of different spruce (*Picea*) species can exceed 200 years (Burns and Honkala 1990; Nikolov and Helmisaari 1992). This also means that in any given region, the old-growth stage may start earlier or later depending on which species colonize the disturbed sites first.

## Old Growth and Forest Management

Old-growth forests in the boreal region have been called "over-mature" in the forestry literature. In some jurisdictions, forest management agencies use pejorative terms such as "decadent" and "surannée" as synonyms for old growth. Due to a reduced mean annual increment (growth in trunk width) in wood production once stands have passed the "age of culmination," old-growth forests have been targeted by foresters for harvesting so as to make room for more productive younger forests (Kuuluvainen et al. 2009). Empirical evidence confirms that a large reduction of old forests and an increase in young forests has occurred as boreal, hemiboreal, and montane forests are brought under management across North America and Eurasia (Kouki et al. 2001; De Grandpré et al. 2009).

In regions with short fire cycles, young forests are the natural norm and, thus, at the landscape scale, commercial forestry has a smaller effect on truncating the age-class distribution (figure 8-1). Yet, at the same time, the few remaining natural old-growth forests are particularly threatened due to harvesting pressure (Bergeron et al. 2006). In areas with long fire cycles, even-aged forest management has led to a large change in landscape-level forest age-class structure. In these regions, such as northeastern Quebec, historically dominated by old growth, forest management has greatly decreased the percentage of old coniferous forest and increased young mixed or deciduous forest (e.g., Boucher et al. 2015). The proportion of mature and old-growth forest in Quebec alone decreased by almost 10 percent from the mid-1970s to the early 2000s (BFEC 2015). Likewise, if current trends continue in northeastern Alberta, the cumulative effects of forestry and oil and gas development will deplete old-growth forests from historical levels of one-third of the landscape, to practically none within 20 years for conifer-dominated stands (where old growth is defined as more than 140 years old for coniferous stands and more than 65 years old for shade-intolerant broadleaf stands) (Schneider 2001).

As an agent of disturbance that targets mature and older forests, logging has an additive effect on top of the fires that prevail in much of the boreal region. In other words, harvesting combined with fires has resulted in disturbance rates far above what natural ecosystems have undergone in historical times (Bergeron et al. 2017). The lack of old-growth forests within the managed portion of the boreal forest is shown in figure 8-2. Forestry operations can also increase the risk of forest fires, largely because logging slash increases the abundance of fine fuels on the ground and the incidence of human-started fires (Lindenmayer et al. 2009), further compounding overall disturbance levels. In addition, forestry can increase the risk of regeneration failure in burned areas where juvenile stands initiated by timber harvesting are lost to wildfire before attaining sexual maturity and producing seeds (Girard et al. 2008).

In addition to reducing the amount of old growth, logging also reduces the size of large blocks of mature and old forest. Fires burn large areas, but

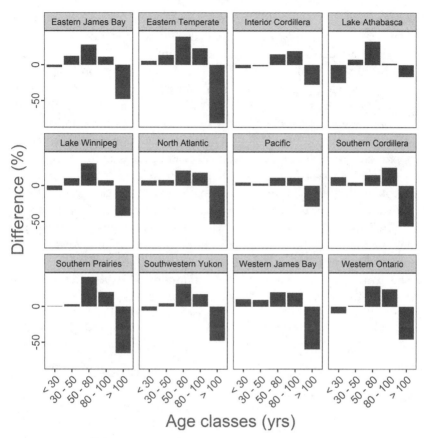

Age classes (yrs)

FIGURE 8-2. Difference between the current (2011) age class structure, as retrieved from Beaudoin et al.'s (2014) forest cover maps, and the one expected (based on the negative exponential function) from the fire cycle observed during the last 56 years (1959–2015) within the managed boreal forest in each of 12 homogeneous fire regime zones. Although the forest cover maps have known limitations (see Beaudoin et al. 2014 for details), this analysis permits a standardized portrait of forest age class structure at the national scale. These departures reflect the effect of logging on the amount of old growth.

where fire frequencies are low, substantial areas of intact older forest also remain between fires. Fire creates a mosaic where young patches of various sizes are randomly distributed within an older landscape matrix. This pattern varies with burn rates such that, when fire cycles are long, the matrix is dominated by old forest, whereas with short fire cycles, old growth naturally occurs as smaller, randomly distributed, remnants (figure 8-3). This is not the case under industrial forest management where logging has historically occurred in a systematic pattern (Dragotescu and Kneeshaw 2012). Logging started in the south, near industrial infrastructure, and progressed northward. The resulting pattern is often described as "unrolling a giant carpet," where logging operations gradually move northward to access the remaining mature stands. Thus, many of the remaining old-growth stands are found in more remote regions or on inaccessible sites within harvested landscapes (Venier et al. 2014). The contrast between the spatial configuration of old growth in managed landscapes and natural landscapes can thus be striking, especially in regions with long fire cycles (Gauthier et al. 2002; Perron et al. 2008), as the matrix of young forest in large areas of old-growth forest typical of natural forests is reversed in managed landscapes.

In Fennoscandia, where forest harvesting has been practiced for several centuries, there has been a large reduction in the area of old growth and in the amount of dead wood typically associated with natural boreal ecosystems. This has threatened populations of dead wood associated boreal species (see, for example, chapter 11), many of which have now been added to The IUCN Red List of Threatened Species (Tikkanen et al. 2006). As suggested by Patry et al. (2017), the negative effect of forest harvesting operations on biodiversity in boreal Fennoscandia is a warning sign for Canadian boreal forests. Their work shows that many functional groups of plants, animals, and fungi that are threatened in Fennoscandia are also present and declining in North America.

## Conservation Challenges

If we accept that it is important to maintain old-growth boreal forests, then we are faced with the question of how to do so. Naturally, old growth is a dynamic entity that commences when stands enter an individual tree-replacement phase and continues to develop until no trees are left from the original postdisturbance cohort, with many structural and compositional attributes potentially changing in the process. Even when no trees from the postdisturbance cohort are left, old growth will continue to change and develop (chapters 6, 7, 10, 11, 13, and 14). Furthermore, as largescale fires are not the only dominant disturbance, different processes may bring forests to an old-growth condition (Shorohova et al. 2011). Surface fires in boreal Eurasia (Rogers et al. 2015) and insect outbreaks and windstorms in eastern North America create complex age-stand structures at both stand and landscape scales (Bouchard et al. 2006). Trees resistant to the distur-

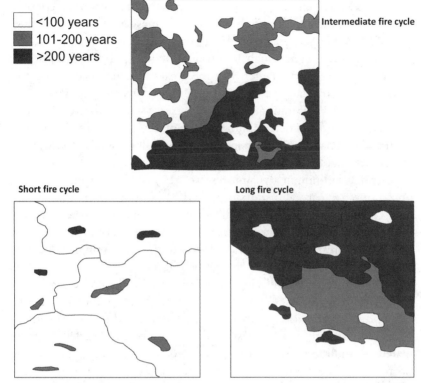

FIGURE 8-3. Diagrammatic portrayal of the proportion and size of old-growth forests in landscapes with different burn rates. With short fire cycles, forest patches older than 100 or 200 years are rare, and young forests dominate the matrix. Whereas with long fire cycles, the situation is reversed, and old forests dominate the landscape.

bances survive individually or in patches. Regeneration is also often patchy among the remnant overstory trees. The forests that remain after such intermediate disturbances typically defy classification as either old growth or as secondary (regenerating) forest. Mortality following these disturbances may in some cases reinitiate a stand (i.e., in even-aged natural monocultures) but will cover a gradient of severity so that, at their lowest levels, they may be difficult to distinguish from background mortality or internal old-growth processes (Shorohova et al. 2011). The challenge in conserving old growth is to be able to allow these different disturbances to occur so that the natural diversity of old-growth forests can still be expressed under a given conservation strategy.

It has been suggested that, for old-growth conservation, a continuous forest area would need to be at least 50 times larger than the average disturbance (Shugart and West 1981) or at least larger than the largest known disturbance, although Johnson and Gutsell (1994) proposed an area at least three times larger than the greatest disturbance. These proposals are part of a larger concept of minimum dynamic area (Pickett and Thompson 1978) or minimum dynamic reserve size (Leroux et al. 2007), which has the goal of determining the size of a conservation area required to include all natural disturbances and the resulting forest-development stages (see chapter 4 for further discussion of this concept). Proponents suggest that a minimum dynamic reserve area would insure internal recolonization after natural disturbances, and, thus, would minimize species loss. However, we question whether these strategies are currently viable in the boreal forest. The great fires of 1919 burned more than 2 million hectares of forest in Alberta and Saskatchewan, and the single largest fire recorded in North America—the Chinchaga fire in 1950—burned over 1.4 million hectares (equivalent to 0.5 percent of the entire productive forest land-base in all of Canada). Similarly, recurrent outbreaks of spruce budworm (*Choristoneura fumiferama*) have defoliated tens of millions of hectares of fir and spruce in eastern Canada every few decades. Approximately 57 million hectares of forest were defoliated during the last outbreak of the twentieth century. Setting aside conservation areas greater than a million hectares would be impossible, as few large natural areas are available due to long-term forest management agreements between the provincial governments and the logging companies. In Canada, only 24 of the existing provincial, territorial, or federal parks are larger than a million hectares and at best only 5 or 6 are dominated by forest, with the rest being in wetlands, tundra, and bare rock.

Instead of creating large reserves that would contain all developmental stages, floating reserves have been proposed (Cumming et al. 1996). In this proposal, the role of old-growth forests compromised by natural disturbances would be filled by (newly protected) mature forests recruiting to the old-growth stage (Cumming et al. 1996). This replacement, however, assumes that there will be future old growth available to replace the old forest stands transformed to younger stages. This limitation was also voiced by Rayfield et al. (2008), who stated, "Our findings emphasize the often-overlooked point that if dynamic conservation planning is to be successful in the long term, the landscape matrix quality surrounding protected areas must be managed in such a way that options remain when it comes to re-planning." Currently such foresight and long-term and largescale planning is an aspirational concept that remains a long way from being made operational. However, as discussed below, it may be possible for this to be made operational by combining it with other approaches, such as extended rotations.

A further problem in using conservation to maintain boreal old growth is that government agencies charged with establishing reserves traditionally have been guided by the goal of achieving representation of all geographical, climatic, and physical regions in conservation plans (Andrew et al. 2014). It could be argued that, under a changing climate, protecting "enduring natural features" should be promoted as a conservation strategy rather than protecting temporally changing features like old forests. However, although a geophysical reserve design is important as a baseline to ensure that all representative ecosystem types have some protection, the lack of emphasis on biological conditions within a region can mean that not all ecological states and thus habitats (e.g., old growth) are well represented. A possible result is that not all species, especially those associated with mature forests and old growth, will be protected in the North American boreal biome (see Chapter 11).

Climate change will impose new challenges for old-growth forest conservation. In the upcoming decades, increased anthropogenic climate forcing is projected to increase the annual area burned throughout much of boreal Canada and Eurasia (Boulanger et al. 2014; Gauthier et al. 2015), therefore decreasing the amount of old-growth forest. In this context, conservation of fire refugia might be of particular importance (Krawchuk et al. 2016). (Fire refugia are zones where fire is known to occur less than expected by regional fire return intervals, including boreal treed wetlands, islands, and lee shores of lakes; see Kasischke et al. 2010; Portier et al. 2016.) These key boreal climate refugia, which are likely to remain relatively resilient to anticipated climate change impacts, may play a critical role in supporting the persistence of boreal biodiversity (Stralberg et al. 2015). Furthermore, maintaining current harvesting rates and practices in regions that will experience sharp increases in fire activity (Gauthier et al. 2015; Bergeron et al. 2017) might further reduce the proportion of old-growth stands, exposing old-growth related species to dramatic negative consequences.

## Solutions to Maintain Boreal Old Growth?

Much of the emphasis in identifying and protecting old-growth forests is concerned with maintaining natural processes and habitats and letting those natural forest dynamics play out in the absence of human intervention (Kneeshaw and Burton 1998). Perhaps we are looking at the conservation of old-growth boreal forests at the wrong scale, focusing on stand development, tree replacement, and gap phase processes that dominate in tropical and temperate forests. In boreal forests, the perspective of looking at old growth from a stand-level perspective might be more appropriately changed to a landscape-level one. Thus, we should think of "gaps" as simply being very large (compared to the small gaps in other forests), with forest regeneration being dominated (depending on climate) by fire and insect outbreaks more than by single-tree mortality and small-gap processes. This concept is explored in table 8-1 where we extrapolate standard old-growth attributes, typically defined at the stand level, to the landscape level. While functional attributes (e.g., a balance of regeneration and mortality) may be more easily transferable than structural attributes (e.g., the abundance of dead wood), the exercise points out the value of thinking of "old-growth landscapes," not just old-growth stands.

Old-growth stands (as classically defined) are still found within old-growth landscapes. The proportion of total area that they cover, as well as their individual size will vary with the fire cycle and ultimately with climate (plate 5 and figure 8-4). Thus, it can be argued that the conservation of old-growth stands, habitats, and processes in the boreal region ultimately depends on the designation and protection (from human exploitation) of large boreal landscapes. Beyond the landscape context, the natural variation in old-growth proportion and configuration in landscapes leads us to argue that we should be thinking at even larger scales. To effectively maintain boreal old growth, we should think regionally, a scale for which we have no management precedent.

As noted above, in parts of the commercial boreal forest, it might already be too late to set aside reserves large enough to accommodate the dynamic nature of old growth. This may then require renegotiating timber agreements with forest products companies and the establishment of interprovincial and international agreements to free up large tracts of land for the long-term maintenance of old growth.

## Designing Reserves for Old-Growth Conservation

As mentioned in earlier sections, landscapes with high burn rates will have small, scattered remnants of old growth located within a matrix of younger forest (figure 8-3). Thus, these small remnants may be important "lifeboats" of biodiversity that are easily threatened by both natural

TABLE 8-1. Extending to the landscape level* some structural and functional attributes (modified from Kneeshaw and Gauthier 2003) that have been used to define old-growth forest stands.

| Structural Attributes | | Functional Attributes | |
| Stand Level | Landscape Level | Stand Level | Landscape Level |
| --- | --- | --- | --- |
| large old trees | has some old stands (variable proportion according to fire cycle) | climax forest | some areas have escaped and show no evidence of historical fire or stand-replacing disturbance |
| gappy canopy | stands show evidence of intermediate disturbances | resources heterogeneously distributed | habitats and ecosystem services heterogeneously distributed |
| structurally complex | diverse patch sizes and shapes | undisturbed by humans (primeval) | undisturbed by humans, primeval |
| multilayered canopy | patches of several different ages and canopy structures | zero net annual growth | standing biomass and productivity reflect climatic potential |
| wide tree spacing | has some stands that meet stand-level old-growth definitions | older than the disturbance cycle | approximately 20–40 percent of stands are older than disturbance return interval |
| large snags | both old and young stands have large numbers of standing dead trees | final stage of development | natural succession and disturbance regimes prevail |
| minimum area | large enough to reflect a variety of times since disturbance / stand ages | secondary recruitment | patches at various stages of stand development; old-growth patch recruitment determined by disturbance regime |
| large logs | both old and young stands have large numbers of fallen logs | steady-state conditions | at equilibrium with natural disturbance regime |
| pit and mounds | physical legacies of disturbance evident in stand structures and landscape patterns | nutrient and carbon retention | watershed retains nutrients; carbon storage in equilibrium with regional disturbance regime |

continued on next page

TABLE 8-1. *continued*

| Structural Attributes | | Functional Attributes | |
|---|---|---|---|
| *Stand Level* | *Landscape Level* | *Stand Level* | *Landscape Level* |
| diverse tree community | high gamma diversity | all-aged structure | stands of various ages since fire / stand-level disturbance |
| patchiness | patchy landscape structure | past commercial maturity | older stands undergoing breakup, paludification. or degradation |
| minimum tree age | minimum stand age since fire or other stand-level disturbance | various states of decay | stands undergoing various stages of stand development, succession, or retrogression |
| uneven-aged stand population structure | stands of all ages and many with complex structures reflecting intermediate disturbances | regeneration occurs through gap dynamics | single-tree mortality and intermediate disturbances dominate more area than stand-level disturbances |
| cohort basal area proportion[†] > 0.5 | uneven-aged and irregular stand structures occupy more area than even-aged stands | mortality is balanced with regeneration | disturbed stands are regenerating |

*Functional attributes are not paired with structural attributes in the same row, but those at the landscape level are paired with those at the stand level for structure and function, respectively.
[†]As proposed and defined by Kneeshaw and Gauthier (2003).

and anthropogenic disturbances (Franklin et al. 2000). Strict conservation of these sites may be needed, while planning for their eventual replacement by maturing stands. On the other hand, in landscapes with lower burn rates, managers and conservationists may find the greatest number of conservation options, as old growth should be plentiful if not dominant (figure 8-3). However, if a change in the dominant forest type (the matrix) leads to negative consequences for some species, managers may be faced with an equally difficult challenge in regions with low burn rates. In regions where forestry practices have already greatly reduced the proportion of boreal old growth, active management in the form of ecological restoration will be required to accelerate the rees-

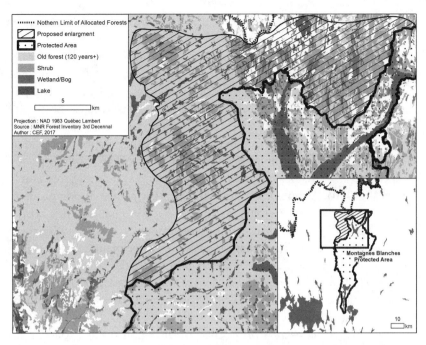

FIGURE 8-4. A map of the White Mountain protected area in Quebec, designed primarily for woodland caribou, but which also protects much old growth. An enlargement of this area (hatched line) would increase habitat and old-growth protection, with minor losses to harvesting potential due to the high proportion of wetlands and unproductive forests. Harvesting forests in this sector would lead to high road-building costs and, thus, low economic viability.

tablishment of old-growth characteristics (e.g., uneven-sized structure) (Bergeron et al. 1999).

Understanding that old growth is not a single entity, but rather a dynamic state that requires promotion and conservation of all of its compositional, structural, and developmental stages, requires thinking in terms of an interconnected network of reserves within vast management areas. Early old growth, in which single-tree replacement of the initial cohort is beginning, would later be old growth added to some reserves. In other cases, intermediate-moderate disturbances (e.g., Hansen and Lorimier 2007; Reyes and Kneeshaw 2008; Shorohova et al. 2011) could revert more advanced old-growth conditions to earlier old-growth stages. This type of network could potentially include some dynamic reserves (that could become available for harvest at some point) but should be based on a large core of permanent reserves that allows all forest development stages to occur.

In terms of the habitat needs of wildlife, managers must consider the species with the largest home range that are dependent (at least in some part of their life cycle) on old-growth forests. One of the most compelling and frequently cited examples is that of the woodland caribou. Old-growth forest reserves greater than 1,000 square kilometers in size would be required to maintain the viability of caribou populations (see Soulé 1987 for discussion of minimum viable populations). For other species, "messy forestry" that leaves structural attributes may be sufficient (Franklin et al. 2000). Although connectivity (i.e., the ability of organisms to move from one site to another) may be an issue for organisms that are dependent on old growth, planning for large distances between old-growth reserves will help to ensure that old growth will be maintained in different boreal regions. For example, in a region where the largest known fire is 1,000,000 hectares, reserves located a minimum of 1,000 kilometers apart should ensure that not all old-growth reserves will be affected by the same disturbance event.

Old-growth conservation in the boreal region needs explicit policies to ensure that, when designated old-growth forests are lost to disturbance (especially fire or logging), an equivalent area of old-growth replacement forest will be protected. Approaches such as dynamic or floating reserves, although creative, are currently risky because of the planning required to ensure that replacement old growth is available in sufficient quantity and quality. Given that old growth encompasses a variety of stand conditions that change through time, ensuring that all old-growth stages are available would be a complex task. Extended rotation forestry—designating a portion of a forest's stands for long rotations that extend to old-growth stages—may, however, move the concept of dynamic reserves to an operationally viable conservation tool (Burton et al. 1999). This change requires an explicit reduction in the annual allowable cut (Fall et al. 2004) and the clear designation of old-growth management areas to ensure that old-growth forests would not disappear due to pressure to harvest "over-mature" old forests. Reduced annual allowable cuts are also an option that could help maintain minimal amounts of old-growth boreal stands if climate change leads to increased burn rates.

Although an increase in the size and number of reserves is required, we can build on the use of habitat that is of high ecological but low economic viability. This could include old-growth forests surrounded by difficult-to-access terrain (i.e., wetlands, rocky outcrops, etc.) or even riparian zones in areas with many lakes or rivers (e.g., figure 8-4). Other reserves should be placed on the edges of unproductive forest land that can serve as buffers or alternative habitat.

## Managing the Matrix between Reserves

Management practices to create elements of old-growth structure should be targeted to the matrix between protected areas. Thus, a zoning approach of different types of management could help the creation of stands with old-growth characteristics in the matrix between old-growth reserves. Areas adjacent to and between reserves would be managed with the objective of maintaining the structural or compositional characteristics of old forests, providing buffers, transition zones, and connectivity for old-growth reserves. This approach is akin to that proposed for the designation of biosphere reserves by Martino (2001) and by Harris (1984) for different zones of forestry activity in growing concentric circles around reserves.

Where old growth is rare or at risk, restorative silvicultural strategies aimed at facilitating the development of old-growth forests or some elements of old-growth structure will be needed. This has been well developed at the stand level for long-lived forests in the Pacific Northwest (e.g., Franklin and Johnson 2012). Chapter 13 explores this topic in detail for a wide range of eastern forests. In the boreal region, when primarily dealing with degraded European forests, Kuuluvainen (2009) as well as Shorohova et al. (2011) suggested that forest management be used to create a range of conditions and structures that are associated with old-growth boreal forests. However, forest management critics note that proposals to use partial harvests may not be economically viable due to the small size of trees in the boreal forest and the low value of individual trees. Furthermore, some partial harvesting operations are based on diameter-limit cutting such as harvesting with advance regeneration protection. These harvests currently remove all stems greater than 14 or 16 centimeters in diameter, thus retaining only the smallest merchantable stems. Consequently, plans for old-growth restoration would have to be modified to encompass the retention of larger trees. Thorpe and Thomas (2007) also note that retention levels are usually low and, subsequently, experience high mortality. Such silvicultural practices may accelerate the time it takes to develop old-growth characteristics over those associated with clear-cutting (Deans et al. 2003), but if no measures are provided to ensure their maintenance once old-growth characteristics develop, then it is unlikely that such diameter-limit partial cuts will maintain habitat for old-growth-associated species.

## Conclusion

The natural proportion of old-growth forest varies across the boreal region depending primarily on regional fire regime. Because of their dynamics, old-growth forest stands are also diverse over space and time. All these characteristics are important for maintaining biodiversity and present challenges to the maintenance of old-growth values. The proposed solutions are difficult to implement as they result in a decreased annual allowable cut in managed forest landscapes, with a concordant cost to forest companies and to the economic activity in forest-dependent communities. However, the benefit of maintaining old growth for its potential role in protecting some species has been well recognized (Gauthier et al. 2009). There is, therefore, a crucial need to develop economic incentives to maintain or restore these dynamic elements of forested landscapes. Efforts should be dedicated to document the multiple economic and ecological benefits (in terms of caribou habitat maintenance, carbon sequestration, biodiversity conservation, etc.) of implementing such a diverse array of solutions. In boreal forests, there is no universal approach to maintain all old-growth forests and dependent species. Rather, conservation of old-growth lies in a concerted application of a large number of approaches in a large number of locations. In other words, due to the high regional variability in natural (e.g., fires and insect outbreaks) and human (e.g., clear-cut logging) disturbances, old-growth conservation must be addressed using multiple techniques at multiple scales in order to achieve overall goals.

## References

Andrew, M. E., M. A. Wulder, and J. A. Cardille. 2014. "Protected areas in boreal Canada: a baseline and considerations for the continued development of a representative and effective reserve network." *Environmental Review* 22: 135–160. dx.doi.org/10.1139/er-2013-0056.

Beaudoin, A., P. Y. Bernier, L. Guindon, P. Villemaire, X. J. Guo, G. Stinson, T. Bergeron, S. Magnussen, and R. J. Hall. 2014. "Mapping attributes of Canada's forests at moderate resolution through kNN and MODIS imagery." *Canadian Journal of Forest Research* 44: 521–532.

Bergeron, Y., D. Cyr, C. R. Drever, M. Flannigan, S. Gauthier, D. Kneeshaw, E. Lauzon, et al. 2006. "Past, current, and future fire frequencies in Quebec's commercial forests: implications for the cumulative effects of harvesting and fire on age-class structure and natural disturbance-based management." *Canadian Journal of Forest Research 36*: 2737–2744.

Bergeron, Y., B. Harvey, A. Leduc, and S. Gauthier. 1999. "Forest management guidelines based on natural disturbance dynamics: stand-and forest-level considerations." *The Forestry Chronicle 75*: 49–54.

Bergeron, Y., D. B. I. P. Vijayakumar, H. Ouzennou, F. Raulier, F., A. Leduc, and S. Gauthier. 2017. "Projections of future forest age class structure under the influence of fire and harvesting: implications for forest management in the boreal forest of eastern Canada." *Forestry 90*: 485–495

Bouchard, M., D. Kneeshaw, and Y. Bergeron. 2006. "Forest dynamics after successive spruce budworm outbreaks in mixedwood forests." *Ecology 87*: 2318–2329.

Boucher, D., L. De Grandpré, D. Kneeshaw, B. St-Onge, J. C. Ruel, K. Waldron, and J. M. Lussier. 2015. "Effects of 80 years of forest management on landscape structure and pattern in the eastern Canadian boreal forest." *Landscape Ecology 30*: 1913–1929.

Boulanger, Y., S. Gauthier, and P. J. Burton. 2014. "A refinement of models projecting future Canadian fire regimes using homogeneous fire regime zones." *Canadian Journal of Forest Research 44*: 365–376.

Bureau du forestier en chef (BFEC). 2015. "État de la forêt publique du Québec et de son aménagement durable Bilan 2008-2013." Roberval, Québec: Gouvernement du Québec.

Burton, P. J., D. D. Kneeshaw, and K. D. Coates.1999. "Managing forest harvesting to maintain old growth in boreal and sub-boreal forests." *The Forestry Chronicle 75*: 623–631.

Burns, R. M., and B. H. Honkala, Compilers. 1990. *Silvics of North America* (Volume 2). Washington, DC: United States Department of Agriculture.

Cumming, S. G., P. J. Burton, and B. Klinkenberg. 1996. "Boreal mixedwood forests may have no "representative" areas: some implications for reserve design." *Ecography 19*: 162–180.

Deans, A. M., J. R. Malcolm, S. M. Smith, and T. J. Carleton. 2003. "A comparison of forest structure among old-growth, variable retention harvested, and clearcut peatland black spruce (*Picea mariana*) forests in boreal northeastern Ontario." *The Forestry Chronicle 79*: 579–589.

De Grandpré, L., S. Gauthier, A. T. Pham, D. Cyr, S. Périgon, D. Boucher, J. Morissette, G. Reyes, T. Aakala, and T. Kuuluvainen. 2009. "Towards an ecosystem approach to managing the boreal forest in the North Shore region: disturbance regime and natural forest dynamics." In *Ecosystem Management in the Boreal Forest*, edited by S. Gauthier, M-A. Vaillancourt, A. Leduc, L. De Grandpré, D. Kneeshaw, H. Morin, P. Drapeau, and Y. Bergeron, 228–256. Québec, Québec: Presses de l'Université du Québec.

Dragotescu, I., and D. D. Kneeshaw. 2012. "A comparison of residual forest following fires and harvesting in boreal forests in Quebec, Canada." *Silva Fennica 46*: 365–376.

Fall, A., M. J. Fortin, D. D. Kneeshaw, S. H. Yamasaki, C. Messier, L. Bouthillier, and C. Smyth. 2004. "Consequences of various landscape-scale ecosystem management strategies and fire cycles on age-class structure and harvest in boreal forests." *Canadian Journal of Forest Research 34*: 310–322.

Franklin, J. F., and K. N. Johnson. 2012. "A restoration framework for federal forests in the Pacific Northwest." *Journal of Forestry 11*: 428–439.

Franklin, J. F., D. Lindenmayer, J. A. MacMahon, A. McKee, J. Magnuson, D. A, Perry, R. Waide, and D. Foster. 2000. "Threads of continuity." *Conservation in Practice 1*: 8–17.

Gauthier, S., P. Bernier, T. Kuuluvainen, A. Z. Shvidenko, and D. G. Schepaschenko. 2015. "Boreal forest health and global change." *Science 349*: 819–822.

Gauthier, S., P. Lefort, Y. Bergeron, and P. Drapeau. 2002. "Time since fire map, age-class distribution and forest dynamics in the Lake Abitibi Model Forest. Information report LAU-X-125E." Laurentian Forestry Centre, Canadian Forest Service. Ste-Foy, Québec: Natural Resources Canada.

Gauthier, S., M. A. Vaillancourt, D. Kneeshaw, P. Drapeau, L. De Grandpré, Y. Claveau, and D. Paré. 2009. "Aménagement forestier écosystémique: origines et fondements." In *Ecosystem Management in the Boreal Forest*, edited by S. Gauthier, M-A. Vaillancourt, A. Leduc, L. De Grandpré, D. Kneeshaw, H. Morin, P. Drapeau, and Y. Bergeron, 13–40. Québec, Québec: Presses de l'Université du Québec.

Girard, F., S. Payette, and R. Gagnon. 2008. "Rapid expansion of lichen woodlands within the closed-crown boreal forest zone over the last 50 years caused by stand disturbances in eastern Canada." *Journal of Biogeography* 35: 529–537.

Hanson, J. J., and C. G. Lorimer. 2007. "Forest structure and light regimes following moderate wind storms: Implications for multi-cohort management." *Ecological Applications* 17: 1325–1340.

Harris, L. D. 1984. *The Fragmented Forest: Island Biogeography Theory and the Preservation of Biotic Diversity*. Chicago, IL: University of Chicago Press.

Johnson, E. A., and S. L. Gutsell. 1994. "Fire frequency models, methods and interpretations." *Advances in Ecological Research* 25: 239–287.

Kasischke, E. S., D. L. Verbyla, T. S. Rupp, A. D. McGuire, K. A. Murphy, R. Jandt, J. L. Barnes, et al. 2010. "Alaska's changing fire regime—implications for the vulnerability of its boreal forests." *Canadian Journal of Forest Research* 40: 1313–1324.

Kneeshaw, D., Y. Bergeron, and T. Kuuluvainen. 2011. "Forest ecosystem structure and disturbance dynamics across the circumboreal forest." In *The Sage Handbook of Biogeography*, edited by A. Millinton, M. Blimler, and U. Schickhoff, 263–280. London, UK: Sage Publications.

Kneeshaw, D., and P. J. Burton. 1998. "Assessment of functional old-growth status: a case study in the sub-boreal spruce zone of British Columbia, Canada." *Natural Areas Journal* 18: 293–308.

Kneeshaw, D., and S. Gauthier. 2003. "Old growth in the boreal forest: a dynamic perspective at the stand and landscape level." *Environmental Reviews* 11(S1): S99–S114.

Kouki, J., S. Löfman, P. Martikainen, S. Rouvinen, and A. Uotila. 2001. "Forest fragmentation in Fennoscandia: linking habitat requirements of wood-associated threatened species to landscape and habitat changes." *Scandinavian Journal of Forest Research* 16(S3): 27–37.

Krawchuk. M. A., S. L. Haire, J. Coop. M-A. Parisien, E. Whitman, G. Chong, and C. Miller. 2016. "Topographic and fire weather controls of fire refugia in forested ecosystems of northwestern North America." *Ecosphere* 7: e01632.

Kuuluvainen, T. 2009. "Forest management and biodiversity conservation based on natural ecosystem dynamics in northern Europe: the complexity challenge." *Ambio* 38: 308–315.

Leroux, S. J., F. K. Schmiegelow, R. B. Lessard, and S. G. Cumming. 2007. "Minimum dynamic reserves: a framework for determining reserve size in ecosystems structured by large disturbances." *Biological Conservation* 138: 464–473.

Lindenmayer, D. B., M. L. Hunter, P. J. Burton, and P. Gibbons. 2009. "Effects of logging on fire regimes in moist forests." *Conservation Letters* 2: 271–277. https://doi.org/10.1111/j.1755-263X.2009.00080.x.

Martino, D. 2001. "Buffer zones around protected areas: a brief literature review." *Electronic Green Journal* 1: 15. http://repositories.cdlib.org/uclalib/egj/vol1/iss15/art2.

Nikolov, N., and H. Helmisaari. 1992. "Silvics of the circumpolar boreal forest tree species." In *A Systems Analysis of the Global Boreal Forest*, edited by H. H. Shugart, R. Leemans, and G. B. Bonan, 13–84. Cambridge, UK: Cambridge University Press.

Oliver, C. D., and B. C. Larson. 1990. *Forest Stand Dynamics*. New York: McGraw-Hill.

Patry, C., D. Kneeshaw, I. Aubin, and C. Messier. 2017. "Intensive forestry filters understory plant traits over time and space in boreal forests." *Forestry 90*: 436–444.

Perron, N., L. Bélanger, and M-A. Vaillancourt. 2008. "Forêt résiduelle sous régimes de feu et de coupes." In *Aménagement Écosystémique en Forêt Boréale,* edited by S. Gauthier, M-A. Vaillancourt, A. Leduc, L. De Grandpré, D. Kneeshaw, H. Morin, P. Drapeau, and Y. Bergeron, 137–164. Québec, Québec: Presses de l'université du Québec.

Pickett, S. T., and J. N. Thompson. 1978. "Patch dynamics and the design of nature reserves." *Biological Conservation 13*: 7–37.

Portier, J., S. Gauthier, A. Leduc, D. Arseneault, and Y. Bergeron. 2016. "Fire regime along latitudinal gradients of continuous to discontinuous coniferous boreal forests in eastern Canada." *Forests 7*: 211.

Rayfield, B., P. M. James, A. Fall, and M. J. Fortin. 2008. "Comparing static versus dynamic protected areas in the Quebec boreal forest." *Biological Conservation 141*: 438–449.

Reyes, G. P., and D. Kneeshaw. 2008. "Moderate-severity disturbance dynamics in *Abies balsamea–Betula* spp. forests: the relative importance of disturbance type and local stand and site characteristics on woody vegetation response." *Ecoscience 15*: 241–249.

Rogers, B. M., A. J. Soja, M. L. Goulden, and J. T. Randerson. 2015. "Influence of tree species on continental differences in boreal fires and climate feedbacks." *Nature Geoscience 8*: 228.

Schneider, R. 2001. *Establishing A Protected Area Network in Canada's Boreal Forest: An Assessment of Research Needs*. Edmonton, Alberta: Alberta Centre for Boreal Studies,

Shorohova, E., D. Kneeshaw, T. Kuuluvainen, and S. Gauthier. 2011. "Variability and dynamics of old-growth forests in the circumbolear zone: implications for conservation, restoration and management." *Silva Fennica 45*: 785–806.

Shugart, H. H., Jr., and D. C. West. 1981. "Long-term dynamics of forest ecosystems." *American Scientist 69*: 647–652.

Soulé, M. E., ed. 1987. *Viable Populations for Conservation*. Cambridge, UK: Cambridge University Press.

Stralberg, D., E. M. Bayne, S. G. Cumming, P. Sólymos, S. J. Song, and F. K. Schmiegelow. 2015. "Conservation of future boreal forest bird communities considering lags in vegetation response to climate change: a modified refugia approach." *Diversity and Distributions 21*: 1112–1128.

Thorpe, H. C., and S. C. Thomas. 2007. "Partial harvesting in the Canadian boreal: success will depend on stand dynamic responses." *The Forestry Chronicle 83*: 318–325.

Tikkanen, O. P., P. Martikainen, E. Hyvärinen, K. Junninen, and J. Kouk. 2006. "Red-listed boreal forest species of Finland: associations with forest structure, tree species, and decaying wood." *Annales Zoologici Fennici 43*: 373–383.

Venier, L. A., I. D. Thompson, R. Fleming, J. Malcolm, I. Aubin, J. A. Trofymow, D. Langor, et al. 2014. "Effects of natural resource development on the terrestrial biodiversity of Canadian boreal forests." *Environmental Review 22*: 457–490. dx.doi.org/10.1139/er-2013-0075.

# Chapter 9

# Forest-Stream Interactions in Eastern Old-Growth Forests

*Dana R. Warren, William S. Keeton, Heather A. Bechtold, and Clifford E. Kraft*

The largescale recovery of eastern forests from historic clearing is a remarkable example of forest ecosystem resilience (Foster and Aber 2004). More than 150 years after the peak of agricultural clearing in eastern North America, many forests across the region have reached maturity and some are progressing toward an old-growth condition (Brooks et al. 2012; figure 9-1). With this forest recovery and an increasing abundance of old-growth stands, we see the recovery of ecosystem functions and ecosystem services not only in the terrestrial environment but also in the streams and rivers that flow through this increasingly complex forested landscape (Warren et al. 2016; Urbano and Keeton 2017).

Since the emergence of stream ecology as a subfield of ecology in the late 1970s and early 1980s, scientists have studied forest-stream interactions. Early studies focused on how the simple presence or absence of forest cover within stream corridors (riparian zones) affected streams (e.g. Burton and Likens 1973; Bisson and Sedell 1984; Bilby and Bisson 1992). But as the field has developed further, we are learning that forest-stream interactions are governed by far more than just whether riparian forests are present. The age, developmental condition, and architecture (or "structure") of streamside forests strongly influence stream ecosystem processes and the resilience of stream corridors to disturbance (Gregory et al. 1991). From recent studies it is clear that streams bordered by old-growth forests, in particular, are profoundly different from those surrounded by younger forests in the eastern United States (Keeton et al. 2007; Valett et al. 2002; Warren et al. 2009; Bechtold et al. 2017). Therefore, as forests along stream corridors (riparian zones) continue to develop in the coming decades, we can expect to see increasing complexity

159

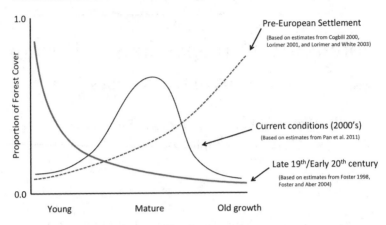

FIGURE 9-1. Stand age class distributions in the past over three periods—preEuro-American settlement, nineteenth- and early twentieth-century (period of greatest open, agricultural, or early successional land in the region), and current conditions. Curves are based on Foster et al. (1998); Cogbill (2000); Lorimer (2001); Lorimer and White (2003); Foster and Aber (2004); and Pan et al. (2011).

not only in riparian forest structural characteristics but also in the streams that flow through those forests.

Assessing how and why streams that flow through old-growth forests might differ from those that flow through younger forests begins with an understanding of the key links between forests and streams. Most research on aquatic-terrestrial linkages has focused on headwater ecosystems (streams generally less than 15 meters wide) where riparian forests have been found to influence many stream features and functions, including physical characteristics, the structure and complexity of food webs, and the cycling and retention of nutrients. For example, the input of wood from riparian forests can influence the width of a stream as well as the size and frequency of pool habitats (Gregory et al. 2003). Forests also influence stream food webs directly through the input of leaf litter and other smaller organic material that provide food for aquatic microbes and larger stream biota (Hall et al. 2000). Indirectly, the forests influence stream food webs by controlling stream light availability which, in turn, regulates in-stream primary productivity and temperature (Hill et al.1995; Wootton 2012; Bilby and Bisson 1992). Through controls on organic matter inputs, large wood structure, and light, riparian forests also strongly influence nutrient cycling in stream environments (Bernot et al. 2010). In addition, the complex structural characteristics that develop together in old-growth forests and the streams flowing through them enhance the resilience of these systems to flood disturbances (Keeton et al. 2017).

In this chapter, we explore key forest-stream interactions and explain how these may be influenced by riparian forest development, age, and structure with an emphasis on old-growth riparian forest conditions. In the final section, we present a conceptual framework to describe how stream function may change over time with the progression of riparian forests from structurally simple younger stands towards more complex old-growth stands.

## Importance of Wood in Streams

In dominant forest types of the eastern United States, streams that flow through old-growth stands generally contain large amounts of dead wood (figure 9-2). Not only is dead wood usually more abundant in late-successional forest streams, the individual logs are, on average, larger than those in streams flowing through younger forests (Warren et al. 2009; Valett et al. 2002; Keeton et al. 2007). The size and abundance of wood in a stream is important because wood is a key structural element that can influence streams in multiple ways. For example, wood inputs can increase favorable habitat for stream fish by creating pools and by enhancing overall habitat complexity (Flebbe 1999; Montgomery et al. 2003). In addition, sediment and organic material retained by wood and debris jams comprise a large carbon sink (Beckman and Wohl 2014), and these structures strongly influence the processing and retention of nutrients like phosphorous and nitrogen (Steinhart et al. 2000; Warren et al. 2007; Valett et al. 2002; Beckman and Wohl 2014). In low gradient streams with soft or highly mobile substrates, wood also provides a critical substrate for many invertebrates (Smock et al. 1989; Lamberti and Berg 1995).

Certainly, there is an overall trend toward more and larger wood in old-growth streams. However, there are also notable exceptions in situations where disturbances kill trees and lead to pulse recruitment of logs into streams, particularly periodic partial or intermediate intensity disturbances (e.g., Meigs and Keeton 2018), such as microburst wind events, ice storms, and some insect outbreaks (see chapter 6). Indeed, periodic or episodic disturbances can lead to enormously high wood vol-

FIGURE 9-2. Large woody debris in two streams running through old-growth riparian forests in the Adirondacks, New York State. Some of the large logs are able to anchor on either side of the stream channel, around which debris dams form. This structure, in turn, creates pool habitats, armors banks, traps sediment, elevates nutrient uptake and spiraling (or processing) rates, and increases "roughness" or the surface area capable of dissipating kinetic energy. Photo credit: W. S. Keeton.

umes in eastern forest streams of all ages (Kraft et al. 2002), and wood recruitment after high intensity, stand-replacing disturbances follows a different dynamic. In these stand-replacing disturbance situations, wood volume and debris-dam frequency are typically high immediately after the disturbance, they decline as logs break down or decay, then they remain relatively low as a young forest regenerates (Valett et al. 2002). Density-dependent mortality (i.e., intertree competition) and density independent events (e.g., windthrow events or ice storms) during the early and middle phases of stand development recruit some logs into the stream channel, but loading rates are generally low and the trees that die tend to be from smaller diameter, less vigorous stems. Wood loading increases later in stand development when both further density-dependent mortality combined with density-independent events and individual tree deaths lead to the mortality of much larger trees (Franklin and Van Pelt 2004). These larger trees not only add more total wood volume, but the larger logs tend to persist longer and generally decay more slowly than smaller wood because, all else being equal, the surface area to volume ratio is lower on bigger trees. Collectively, in relation to forest use, wood volume in the stream takes on a U-shaped distribution when natural disturbances kill trees without removing them from the system (Valett et al. 2002). However, where riparian trees are cleared for timber or agriculture, linear or consistent increases in wood loading over time are likely to occur, with periodic pulses or elevated recruitment following disturbance events (Warren et al. 2009). We focus here on the overall tendency toward an increase in size and abundance of large wood in streams, which is useful in developing rules of thumb and broad hypotheses about ecosystem function. Nevertheless, disturbance history must always be considered when general patterns are applied to understanding a particular forest-stream location.

## Benefits of Large Wood to Stream Fish

Prior to the 1970s, many fisheries managers thought large logs and debris dams were bad for fish, going so far even as to actively clear wood out of streams, thinking this would help fish passage. Fortunately, this practice has been discontinued, and we have learned that wood can be a critical structural feature in headwater streams (Nislow 2005). Large wood enhances stream habitat in a number of ways. Wood is itself an important source of cover for fish, and large stable wood can create or enlarge pool habitat (Riley and Fausch 1995: Flebbe 1999). Most work in eastern forests streams—and across North America—examining the role of wood in creating fish habitat has focused on trout and salmon. Although these salmonid fishes can use other stream habitat features (e.g., rocks and deep water in pools) as cover habitat to avoid predation (Sweka and Hartman 2006), researchers have found that fish use wood as a preferred cover type (Flebbe 1999). In systems with abundant wood and complex habitat architecture, visual isolation can reduce aggressive interactions among fish, allowing for higher densities in a given pool (Sundbaum and Naslund 1998). A number of studies in eastern North America have found increased fish abundance following experimental wood additions (Burgess and Bider 1980; Culp et al. 1996). Overall, increasing habitat complexity and pool size are expected to improve habitat for fish as more wood accumulates in streams bordered by old-growth riparian forests.

Not only do old-growth streams generally have more wood, the wood that enters often comes from larger trees and, therefore, is more stable (Braudrick and Grant 2000; Warren and Kraft 2008). Large, stable wood is also more likely to anchor wood jams, which are collections of multiple pieces of wood that often span the stream channel, sometimes nucleating around boulders (Keeton et al. 2007). Wood jams are a highly complex habitat structure that can increase fish abundance, particularly when they increase pool volume and habitat complexity along channels, resulting in braiding and backwater habitats (Warren and Kraft 2003; Burgess and Bider 1980).

Anglers have long recognized that larger pools generally have more and larger fish, an observation confirmed by many studies (Riley and Fausch 1995; Warren et al. 2010). Because wood is an important pool-forming feature (Montgomery et al. 1995), restoration efforts across eastern forest landscapes and throughout North America often focus on adding large logs to streams. Individual pieces of stable wood that span the wetted area of the stream functionally reduce the active channel dimensions of the stream, as does wood that projects into a channel. This decrease in channel size increases water velocity, thereby increasing local stream energy. Higher energy, in turn, allows for greater scour of stream substrates (Thompson and Hoffman 2001). Scour pools created by wood can be critical habitat for fish as they are often quite deep. Alternatively, many wood jams dissipate energy and create dammed pools upstream of the wood structures. In these cases the wood jams also tend to create plunge pools downstream of the structure (Montgomery et al. 1995). Experimental wood additions in Colorado streams created deep plunge pools where logs were added and then anchored to prevent movement during high flows. This increase in pool area resulted in more fish and, ultimately, greater overall trout production (Riley and Fausch 1995; White et al. 2011).

In eastern forest streams, Keeton et al. (2007) found that more large logs (defined in that study as wood greater than 30 centimeters in diameter) were present in streams flowing through old-growth riparian forests than streams flowing through mature forests. This difference in large-log frequency was notable because it was positively associated with pool habitat in the study reaches; by contrast, total wood abundance and total wood volume were *not* important in that analysis, indicating that the key factor was the presence of larger wood pieces. This highlights how and why streams in old-growth forests may differ from those in younger forests. However, increasing large wood in a stream does not necessarily benefit all fish species or all life stages, as noted in an experimental wood addition study by Langford et al. (2012). Other factors, such as stream temperature, macroinvertebrate abundance (i.e., food), and stream chemical conditions, also affect the number and size of fish in a stream, and wood will not always increase pool volume in a watershed with many boulders (Sweka and Hartman 2006). However, given high recruitment rates for large logs in old-growth streams, and given the role of this wood in creating high-quality trout habitat conditions, our working hypothesis in on-going research is that eastern old-growth riparian forests will foster stream habitats that are favorable to native brook trout (*Salvelinus fontinalis*) and other dominant stream fish.

### *Influence of Large Wood on Aquatic Macroinvertebrates*

Macroinvertebates, such as mayflies (Ephemeroptera), caddisflies (Trichoptera), and stoneflies (Plecoptera), spend most of their lives in their larval aquatic stage, living in a stream for months to years before emerging as adults to breed and then die. These insects play an important role in stream ecosystems, processing nutrients and linking the base of the food web with higher trophic levels. Despite a lack of research on macroinvertebrates in eastern old-growth forests, we expect that invertebrate communities will differ between systems with young, mature, and old-growth riparian forests. As with fish, increasing wood loads over time would be the most likely driver. In low gradient blackwater systems with soft streambeds, wood can provide critical substrate for many macroinvertebrates (Smock et al. 1989; Lamberti and Berg 1995). Similarly, in sand-bed streams and other systems with highly mobile substrates, stable wood offers refuge from scour (Borchardt 1993). Where large logs create jams, accumulated organic material supports macroinvertebrates in the shredder-feeding guild that consume leaves and other coarse organic material (Lemly and Hilderbrand 2000). As with fish habitat, the amount and stability of wood are key and are enhanced in old-growth riparian forests because of higher recruitment rates for large logs (Keeton et al. 2007). To date, there have been no studies of streams in the eastern United States that have explicitly investigated variation in macroinvertebrate communities across stand-development series extending into old growth. This contrasts with western Oregon, where researchers found higher stream macroinvertebrate biomass in old-growth forest reaches as compared to regenerating riparian forest sections of the same stream, with the difference attributed to greater light beneath frequent canopy gaps in old growth forest streams (Kaylor and Warren 2017).

## Terrestrial Carbon Inputs to Streams

The amount, size, and energetic quality of organic matter are of interest to scientists across many ecological disciplines. The fixed carbon in organic matter is, in effect, the fundamental energetic medium of life on earth. Therefore, understanding carbon dynamics is important in understanding how food webs and ecosystems function. Further, sequestration and respiration of organic matter can influence the fate of carbon and whether it remains sequestered in a reduced state or is respired and returned to the atmosphere. Given its relatively slow decay, large wood may accumulate in streams for many years, and when left to accumulate, large wood can represent a sizable carbon pool in streams (Beckman and Wohl 2014). Greater wood volumes have also been found in late successional forest streams relative to streams with mature riparian forest in the eastern and upper Midwestern regions of the United States (Valett et al. 2002). In addition, a space-for-time study in the northeastern United States found a positive relationship between stand age and large-wood accumulation in streams, with little evidence of reduced accumulation rates even when the age of dominant riparian trees exceeded 300 years (Warren et al. 2009).

Slower decay rates for large logs, originating from the predisturbance stand but persisting through secondary succession (termed "biological legacies"), influence carbon accumulation rates in streams and disrupt the straightforward progression by which increasing stand age increases the amount of large wood. As noted above, older forest streams have more wood *on average* than younger forest streams, but slowly decaying logs that carry the legacy of past disturbances alter this relationship and increase carbon storage in streams. For example, one study in the central Appalachians found little evidence for a relationship between stream wood and riparian forest stand age (Hedman et al. 1996). However, this region was heavily impacted by the chestnut blight in the early 1900s. Therefore, many stream reaches retained large dead wood, originating from the slowly decaying old-growth American chestnut trees (*Castanea dentata*) that died almost a century before. When chestnut trees were removed from the assessments, an increase in wood volume over time was well in line with expectations. This highlights the importance of considering wood inputs from future disturbances and mortality agents. These include invasive insect pests such as the hemlock woolly adelgid (*Adelges tsugae*) and the emerald ash borer (*Agrilus planipennis*), as well as changes in physical export processes, such as increased flooding, but reduced ice flows in the northeastern United States as climate change proceeds (see Keeton et al. 2017). These drivers of future wood recruitment and loading dynamics will have important implications for predicting stream carbon dynamics over the coming decades.

Wood—particularly stable wood—dramatically influences annual and seasonal stream carbon dynamics through its effect on organic matter retention. In one of the first studies on this topic, conducted at the Hubbard Brook Experimental Forest in New Hampshire, Bilby (1981) found that wood removal caused up to a five-fold increase in fine particulate carbon export relative to reference conditions. A number of studies since then have highlighted the role of wood, boulders, and other channel roughness elements in the retention of leaves and other particulate organic matter throughout the autumn and winter seasons (Muotka and Laasonen 2002). In the headwater streams of old-growth forests, where large logs are more common and large wood jams occur regularly, more coarse and fine organic material will be retained that will increase carbon storage and whole ecosystem respiration (Bechtold et al. 2017).

Organic material from the riparian forest is particularly important in forested streams because this "allochthonous" (a term meaning "derived from outside the stream") carbon is often the dominant basal resource supporting stream food webs (Wallace et al. 1997; Fisher and Likens 1972; Hall et al. 2000). Input of allochthonous material from the riparian forest affects not only the composition of the aquatic biotic community but also fundamental nutrient cycling along stream networks (Tank et al. 2010; Mulholland and Webster 2010). In a classic experiment conducted in the southern Appalachian Mountains, litter exclusion demonstrated that allochthonous carbon inputs can control stream insect communities and food web structure (Wallace et al. 1997; Hall et al. 2000). The type of litter that enters the stream is also important. Litter from deciduous trees is generally more labile (easy to break down) than that from conifers, but it also predominately enters the stream en masse in the autumn, whereas most conifers tend to distribute their litter input more evenly throughout the year. In riparian forests dominated by deciduous species, stream nutrient concentrations decline and ecosystem respiration spikes upward following leaf fall in autumn (Roberts and Mulholland 2007). Although leaf litter may persist for weeks to months in a stream, an increase in respiration and decline in stream nutrient concentrations are enhanced by the new litter and the highly accessible dissolved organic carbon that rapidly leaches from allochthonous leaves. Organic material from soils also washes into the stream, providing a consistent but low-level carbon source that can be substantial during seasonal or periodic flood events, especially in low gradients in floodplain forests.

## Stream Light and In-Stream Primary Production

In forested headwater ecosystems, the structure of riparian forests directly influences the amount of light reaching a stream (Kaylor et al. 2017; Keeton et al. 2007; Bechtold et al. 2017). But canopy structure in riparian forests is not static or uniform. Instead, it changes dynamically as forests age, develop, and interact with natural disturbances, becoming spatially heterogeneous as canopy gaps develop (Van Pelt and Franklin 2000). The resulting complex light environment within riparian corridors and over stream channels is driven by canopy gap dynamics (Curzon and Keeton 2010). Light is a key limiting factor for primary production on stream substrates (Rosemond 1993; Bilby and Bisson 1992). Yet, only recently have scientists linked the temporal and spatial dynamics of in-stream primary production to the age and canopy architecture of adjoining riparian forests, particularly old-growth forest structure in eastern stream systems. (Curzon and Keeton 2010; Kaylor et al. 2017; Bechtold et al. 2017; Stovall et al. 2009).

While external plant material falling into channels is usually the dominant mass carbon source in forested streams, controls on primary production are important because stream algae provide a higher "quality" carbon source (Cross et al. 2005). Algae also generally have more nitrogen and phosphorous per unit carbon than leaves, branches, and other fine litter (Cross et al. 2005), and studies using isotope and gut analyses of fish and invertebrates have shown that the contributions of algae to stream food webs are disproportionately high relative to the availability of algae (McCutchan and Lewis 2002). Consequently, controls on stream light may affect not only algal production but also the productivity of secondary consumers in a system (Bilby and Bisson 1992; Rosemond 1993; Kaylor and Warren 2017).

From studies of headwater streams in both eastern deciduous forests and the coniferous forests of the Pacific Northwest, scientists have learned that old-growth riparian forests have, on average, more light reaching channels than do young and mature riparian forests (Keeton et al. 2007; Kaylor et al. 2017; Warren et al. 2013). However, this is not to say that streams within old-growth riparian forests receive greater light fluxes uniformly. As in terrestrial ecosystems, light availability is patchy in old-growth forest streams due to their complex canopies, characterized by continuous variation in both vertical (e.g., canopy layering) and horizontal (e.g., tree density, gaps) dimensions. Intense light patches in old forests are interspersed with areas of low light, leading to frequent transitions from light-controlled benthic algae production in the shaded reaches to nutrient regulated algae production in sections with greater light flux (Warren et al. 2017). In contrast, young and mature riparian forests are most often dominated by closed, more uniformly structured canopies, having fewer intense light patches and lower total light flux to the stream benthos. Bechtold et al. (2017) found that light flux to streams exhibited a U-shaped distribution over successional seres, with higher light soon after riparian clearing, lowest light in the middle stages of stand development and higher light later in stand evolution due to the development of canopy gaps. A similar pattern was observed by Kaylor et al. (2017) in a review of stream studies across the Pacific Northwest. Canopy "openness," a metric commonly used as a proxy for potential light exposure, was a dominant predictor of differences in stream vertebrate biomass in a comparison of paired reaches in old-growth and second-growth riparian forests (Kaylor and Warren 2017). In that study, the canopy openness proxy measure of stream light accounted for over 70 percent of the variability in the difference in vertebrate biomass between adjacent reaches with different riparian forest conditions. No comparable paired studies of the influence of riparian forest age class on stream vertebrate biomass have been conducted in eastern forests. However, given the distinct differences in canopy structure and stream light noted by Keeton et al. (2007) and Bechtold et al. (2017), and given the clear increases in algal standing stocks in light patches of eastern old-growth forests observed by Stovall et al. (2009), similar differences are likely in eastern stream systems.

Riparian forests and the shade they create are also important in regulating stream temperature, because solar radiation has the strongest thermal influence on streams (Johnson 2004; Garner et al. 2017). Removal of riparian forests, for instance by logging or natural disturbance, eliminates shade and can lead to substantial increases in stream temperature (Johnson 2004). Although light levels are generally higher in streams bordered by late successional and old-growth forests, a comparable increase in temperature has not been quantified in eastern forest ecosystems. This is likely due to the more moderate increases in light associated with patchy openings in the canopy of late successional forests relative to the larger changes induced by experiments in which forest cover was cleared (Klos and Link 2018; Janisch et al. 2012). Further, the presence of abundant logs and debris dams in old-growth streams may force water into subsurface flowpaths (the *hyporheic zone*) where it may remain cooler—at least in summer—and thereby insulate the stream from warming effects of solar radiation (Arrigoni et al. 2008).

## Nutrient Dynamics in Old-Growth Streams

Medical doctors have shown that, as humans age, their dietary needs change in response to how the body absorbs, processes, and excretes nutrients; that is, the elderly eat less, are inefficient at absorption (due to decreased stomach acid), and, thus, excrete more vitamins and nutritive compounds than young humans. Similarly, a classic paradigm in ecosystem ecology suggests that as forests age, nutrients will be lost or leaked from the terrestrial system at higher rates than for younger forests (Likens et al. 1970; Vitousek and Reiners 1975). Viewing these processes with a holistic watershed-scale perspective requires that we consider the absorption, the processing, and the export of nutrients from both upland and in-stream components of the system. Thus, nutrient loss from terrestrial soils is accompanied by nutrient gain in downstream aquatic systems (Bechtold et al. 2017). We focus here on how streams that flow through aging forests retain, transform, and export nutrients, thereby regulating the interplay of nutrients between aquatic and terrestrial ecosystems (Bernhardt et al. 2003; Mulholland 2004). Most importantly, forest age and structure strongly influence these processes.

Streams have often been thought of as passive pipes flowing out of a watershed (Hotchkiss et al. 2015), where water and nutrients pass through rapidly and ultimately reach an end point. Nutrients are quickly assimilated by microbes and algae when they are available in aquatic systems. Then, after those microbes or algae die, the nutrients are again released back into the water and made available for uptake, thereby creating a cycle of nutrients between organic and inorganic forms. A similar cycle, called a "nutrient spiral," also occurs in terrestrial soils, but the longitudinal nature or downstream flow of water stretches this cycle into a springlike spiral. The spiral can be stretched out or contracted, depending on how quickly nutrients are absorbed from the water column and how long they are retained in stream biota (Mulholland et al. 1985). The average distance that a nutrient particle travels in a stream before being absorbed again can be measured and compared between systems. This value is referred to as the uptake length ($S_w$ or the portion of the spiral in the water column). Thus, streams are not pipes. Instead, they cycle nutrients and are affected by many factors influenced by the structure and age of surrounding forests that, in turn, influence in-stream structural characteristics.

At the stream reach scale, there are many reasons to expect nutrient uptake to be greater in streams with old-growth riparian forests than in streams with young or mature riparian forest. As noted above, headwater streams in old-growth forests on average receive more sunlight than streams flowing through younger forests. This difference in light is important because increasing light in forested streams often leads to large increases in primary production and associated increases in stream nutrient demand. The role of light as a driver of stream nutrient dynamics has been most thoroughly explored in experimental studies in which all or nearly all of the riparian vegetation has been removed (Sabater et al. 2000). While differences in light availability between old-growth and second-growth forests are more subtle, they have been shown to alter nutrient uptake (Sobota et al. 2012). Old-growth forests have more frequent and larger light patches than younger forests, creating a patchwork of light and dark areas along the streambed. This light mosaic creates localized areas (hot spots) of nutrient demand due to greater algal standing stocks (Stovall et al. 2009) and produces in-stream fluctuations between light and nutrient limitation (Warren et al. 2017). Simultaneously, the downstream flowpaths of nutrient spirals elongate and shrink in response to the availability of leaky nutrients from forest soils and patches of sunlight from the old-growth canopy. As a result, we expect autotrophic nutrient demand to be greater in old-growth reaches based solely on differences in light dynamics between old-growth and mature forests.

In addition to greater light levels that can enhance autotrophic uptake, a higher frequency of large wood and wood jams in old-growth forest systems can also enhance nutrient uptake. This anticipated increase in uptake is attributed to heterotrophic fungi and microbes that process litter retained behind dams. These microbes have a high nutrient demand as they process dead wood, leaves, and other litter. Indeed, in a bioassay study comparing nitrogen losses among different stream substrates, Steinhardt (2000) found greater nitrogen loss potential in substrates behind wood jams relative to substrates in the open channel. Nutrient uptake was also found to decrease with the loss of a carbon subsidy in a litter removal study from a headwater stream in the southern Appalachian Mountains (Webster et al. 2000). As a result, we also expect increased nutrient demand in old-growth forest streams due to greater heterotrophic demand.

While the hypothesis that old-growth forest streams have greater nutrient demand than streams flowing through young forest is supported conceptually by the observational and manipulative studies noted above, research quantifying nutrient uptake in streams across a range of stand ages in the eastern United States has yielded surprisingly variable results. There was support for this hypothesis in a study comparing phosphate uptake between streams with late successional versus mature riparian forest stands in the central Appalachian Mountains, where uptake was distinctly higher in the old-growth reaches (Valett et al. 2002). In that study, the relationship between phosphate uptake and stand age was attributed to differences in carbon retention associated with wood and wood jams. The importance of wood jam frequency for stream phosphate uptake in eastern forests was further supported by a subsequent study in central New Hampshire (Warren et al. 2007), yet that study found no strong relationships between stream wood and nitrogen uptake. The absence of a relationship in that study was attributed to alternative factors affecting nitrogen uptake, such as autotrophic demand, use of nitrate in denitrification, or the influence of hydrologic retention times (Bernot and Dodds 2005). More recently, a study exploring nutrient uptake, primary production, and ecosystem respiration in streams across upstate New York and central New Hampshire, revealed inconsistent results when evaluating sites with riparian forests ranging from 10 to at least 360 years in age. Bechtold et al. (2017) did find greater primary production and greater respiration rates in old-growth forest streams. However, nutrient uptake did not follow this pattern. Some of the streams in old-growth reaches did indeed have greater uptake rates, but no consistent relationship was found between stand age and nutrient uptake.

In summary, available evidence provides some support for an increase in stream nutrient concentrations as riparian and upland forests age, but results are more equivocal in assessing stream nutrient cycling over time. Overall, empirical assessments are inconsistent despite credible reasons for expecting that stream nutrient uptake rates will increase as riparian forests progress toward old growth. As a result, we still have a long way to go in determining how, why, and to what degree nutrient cycling changes through the process of stand development in stream riparian zones.

## Conclusion

The high degree of riparian functionality associated with old-growth forests is an ecosystem service that has received little attention, yet riparian influences will assume greater importance as eastern forest landscapes continue to recover and mature from historic land use. While forest development can follow multiple pathways (Lorimer and Halpin 2014; Urbano and Keeton 2017), current research suggests that development of complex late successional characteristics will lead to concurrent shifts in stream ecosystem function independent of the rate or pathway by which it occurs (Warren et al. 2016). Drivers of shifts in stream function as young and ma-

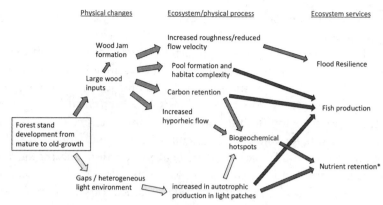

Figure 9-3. Pathways by which the transition from mature to old-growth riparian forests can affect physical and ecological processes in streams, with associated implications for stream ecosystem services. Empirical studies provide support for greater inorganic phosphorous uptake in older forest streams; results are equivocal for inorganic nitrogen uptake.

ture riparian forests progress toward old growth are directly and indirectly associated with increases in two key factors—large wood and stream light (figure 9-3), which influence stream ecosystem processes and stream biota. These, in turn, influence the ecosystem services provided by streams (figure 9-3). Streams with old-growth riparian forests have also been found to be more resilient to disturbance than those with younger riparian forests, resulting, in part, from the greater stand structural complexity of the forests and increased "roughness" and wood loading in the riparian zone (Keeton et al. 2017). Understanding that canopy structure and stream wood are key factors driving the differences in functionality between systems with old-growth versus younger riparian forests informs our projections of future change in eastern forest streams and the ecosystem service consequences of that change. This understanding also raises questions about how best to effectively manage or restore riparian forests to promote old-growth characteristics associated with desirable streams functions, such as flood resilience, nutrient processing, and high-quality fish habitat.

Further, understanding that stream ecosystem function is affected by wood loading, riparian forest canopy structure, and the species composition of riparian forest community allows us to consider how stream function may change in the future not only as a result of anticipated shifts in forest succession and age class distributions, but also as we see changes in forest structure due to increased species invasion, altered climate regimes, and changing land use. For example, the invasion of both the hemlock wooly adelgid and the emerald ash borer (see chapter 12) are both of particular importance to projections of stream ecosystem function in the future as these important riparian trees are lost, altering both wood loading and canopy structure in and along stream corridors. To date, land-use changes across eastern forests have been the driving factor affecting streamside forests and associated stream function. The removal of riparian forests for timber and agriculture in the nineteenth and early twentieth century fundamentally altered stream function, and the subsequent regeneration of riparian forests has led to a slow recovery toward historic function. But the remarkable redevelopment of secondary forest cover alone does not constitute a full recovery of riparian functionality. Important differences remain between streams that run through old-growth forests versus those that run through young and mature forests. And with few riparian forest stands currently in an old-growth condition, most forest headwaters have likely not fully returned to their historic level of functionality. With careful stewardship and adaptive management, we can hope that this will change in the future.

# References

Arrigoni, A. S., G. C. Poole, L. A. K. Mertes, S. J. O'Daniel, W. W. Woessner, and S. A. Thomas. 2008. "Buffered, lagged, or cooled? Disentangling hyporheic influences on temperature cycles in stream channels." *Water Resources Research* 44: Artn W09418.

Bechtold, H. A., E. J. Rosi, D. R. Warren, and W. S. Keeton. 2017. "Forest age influences in-stream ecosystem processes in Northeastern US." *Ecosystems 20*: 1058–1071.

Beckman, N. D., and E. Wohl. 2014. "Carbon storage in mountainous headwater streams: The role of old-growth forest and logjams." *Water Resources Research 50*: 2376–2393.

Bernhardt, E. S, G. E Likens, C. T. Driscoll, and D. C. Buso. 2003. "In-stream uptake dampens effects of major forest disturbance on watershed nitrogen export." *Proceedings of the National Academy of Sciences of the United States of America 100*: 10304–10308.

Bernot, M. J., and W. K. Dodds. 2005. "Nitrogen retention, removal, and saturation in lotic ecosystems." *Ecosystems 8*: 442–453.

Bernot, M. J., D. J. Sobota, R. O. Hall, P. J. Mulholland, W. K. Dodds, J. R. Webster, J. L. Tank, et al. 2010. "Inter-regional comparison of land-use effects on stream metabolism." *Freshwater Biology 55*: 1874–1890.

Bilby, R. E. 1981. "Role of organic debris dams in regulating the export of dissolved and particulate matter from a forested watershed." *Ecology 62*: 1234–1243.

Bilby, R. E., and P. A. Bisson. 1992. "Allochthonous versus autochthonous organic-matter contributions to the trophic support of fish populations in clear-cut and old-growth forested streams." *Canadian Journal of Fisheries and Aquatic Sciences 49*: 540–551.

Bisson, P. A., and J. R. Sedell. 1984. "Salmonid poulations in streams in clearcut vs. old-growth forests of western Washington." In *Fish and Wildlife relationships in old-growth forests*, 121–129. Juneau, AK: American Institute of Fishery Research Biologists.

Borchardt, D. 1993. "Effects of Flow and Refugia on Drift Loss of Benthic Macroinverte-brates - Implications for Habitat Restoration in Lowland Streams." *Freshwater Biology* *29:* 221–227.

Braudrick, C. A., and G. E. Grant. 2000. "When do logs move in rivers?" *Water Resources Research 36:* 571–583.

Brooks, R. T., K. H. Nislow, W. H. Lowe, M. K. Wilson, and D. I. King. 2012. "Forest succession and terrestrial-aquatic biodiversity in small forested watersheds: a review of principles, relationships and implications for management." *Forestry 85*: 315–327.

Burgess, S. A., and J. R. Bider. 1980. "Effects of stream habitat improvements on inver-tebrates, trout populations, and mink activity." *Journal of Wildlife Management 44*: 871–880.

Burton, T. M., and G. E. Likens. 1973. "Effect of Strip-Cutting on Stream Temperatures in Hubbard Brook Experimental Forest, New-Hampshire." *Bioscience 23*: 433–435.

Cross, W. F., J. P. Benstead, P. C. Frost, and S. A. Thomas. 2005. "Ecological stoichiometry in freshwater benthic systems: recent progress and perspectives." *Freshwater Biology 50*: 1895–1912.

Culp, J. M., G. J. Scrimgeour, and G. D. Townsend. 1996. "Simulated fine woody debris accu-mulations in a stream increase rainbow trout fry abundance." *Transactions of the American Fisheries Society 125*: 472–479.

Curzon, M. T., and W. S. Keeton. 2010. "Spatial characteristics of canopy disturbances in riparian old-growth hemlock - northern hardwood forests, Adirondack Mountains, New York, USA." *Canadian Journal of Forest Research 40*: 13–25.

Fisher, S. G., and G. E. Likens. 1972. "Stream Ecosystem - Organic Energy Budget." *Bioscience 22*: 33–35.

Flebbe, P. A. 1999. "Trout use of woody debris and habitat in Wine Spring Creek, North Carolina." *Forest Ecology and Management 114*: 367–376.

Foster, D. R., and J. D. Aber, eds. 2004. *Forests In Time: The Environmental Consequences of 1,000 Years of Change.* New Haven, CT: Yale University Press.

Franklin, J. F., and R. Van Pelt. 2004. "Spatial aspects of structural complexity in old-growth forests." *Journal of Forestry 102*: 22–28.

Garner, G., L. A. Malcolm, J. P. Sadler, and D. M. Hannah. 2017. "The role of riparian veg-etation density, channel orientation and water velocity in determining river temperature dynamics." *Journal of Hydrology 553*: 471–485.

Gregory, S., K. Boyer, and A. Gurnell, eds. 2003. *The Ecology and Management of Wood in World Rivers.* Bethesda, MD: American Fisheries Society.

Gregory, S. V., F. J. Swanson, W. A. McKee, and K.W. Cummins. 1991. "An ecosystem perspective on riparian zones focus on links between land and water." *Bioscience 41*: 540–551.

Hall, R. O., J. B. Wallace, and S. L. Eggert. 2000. "Organic matter flow in stream food webs with reduced detrital resource base." *Ecology 81*: 3445–3463.

Hedman, C. W., D. H. VanLear, and W. T. Swank. 1996. "In-stream large woody debris load-ing and riparian forest seral stage associations in the southern Appalachian Mountains." *Canadian Journal of Forest Research 26*: 1218–1227.

Hill, W. R., M. G. Ryon, and E. M. Schilling. 1995. "Light Limitation in a Stream Ecosystem - Responses by Primary Producers and Consumers." *Ecology 76*: 1297–1309.

Hotchkiss, E. R., R. O. Hall, R. A. Sponseller, D. Butman, J. Klaminder, H. Laudon, M. Ros-vall, and J. Karlsson. 2015. "Sources of and processes controlling $CO_2$ emissions change with the size of streams and rivers." *Nature Geoscience 8*: 696–699.

Janisch, J. E., S. M. Wondzell, and W. J. Ehinger. 2012. "Headwater stream temperature: Interpreting response after logging, with and without riparian buffers, Washington, USA." *Forest Ecology and Management 270*: 302–313.

Johnson, S. L. 2004. "Factors influencing stream temperatures in small streams: substrate effects and a shading experiment." *Canadian Journal of Fisheries and Aquatic Sciences 61*: 913–923.

Kaylor, M. J., and D. R. Warren. 2017. "Linking riparian shade and the legacies of forest management to fish and vertebrate biomass in forested streams." *Ecosphere 8* (6): e01845.

Kaylor, M. J., D. R. Warren, and P. M. Kiffney. 2017. "Long-term effects of riparian forest harvest on light in Pacific Northwest (USA) streams." *Freshwater Science 36*: 1–13.

Keeton, W. S., E. M. Copeland, S. M. P. Sullivan, and M. C. Watzin. 2017. "Riparian forest structure and stream geomorphic condition: implications for flood resilience." *Canadian Journal of Forest Research 47*: 476–487.

Keeton, W. S., C. E. Kraft, and D. R. Warren. 2007. "Mature and old-growth riparian forests: Structure, dynamics, and effects on Adirondack stream habitats." *Ecological Applications 17*: 852–868.

Klos, P. Z., and T. E. Link. 2018. "Quantifying shortwave and longwave radiation inputs to headwater streams under differing canopy structures." *Forest Ecology and Management 407*: 116–124.

Kraft, C. E., R. L. Schneider, and D. R. Warren. 2002. "Ice storm impacts on woody debris and debris dam formation in northeastern US streams." *Canadian Journal of Fisheries and Aquatic Sciences 59*: 1677–1684.

Lamberti, G. A., and M. B. Berg. 1995. "Invertebrates and Other Benthic Features as Indicators of Environmental-Change in Juday-Creek, Indiana." *Natural Areas Journal 15*: 249–258.

Langford, T. E. L., J. Langford, and S. J. Hawkins. 2012. "Conflicting effects of woody debris on stream fish populations: implications for management." *Freshwater Biology 57*: 1096–1111.

Lemly, A. D., and R. H. Hilderbrand. 2000. "Influence of large woody debris on stream insect communities and benthic detritus." *Hydrobiologia 421*: 179-185.

Likens, G. E., F. H. Bormann, N. M. Johnson, D. W. Fisher, and R. S. Pierce. 1970. "Effects of Forest Cutting and Herbicide Treatment on Nutrient Budgets in Hubbard Brook Watershed-Ecosystem." *Ecological Monographs 40*: 23–47.

Lorimer, C. G., and C. R. Halpin. 2014. "Classification and dynamics of developmental stages in late-successional temperate forests." *Forest Ecology and Management 334*: 344–357.

McCutchan, J. H., and W. M. Lewis. 2002. "Relative importance of carbon sources for macroinvertebrates in a Rocky Mountain stream." *Limnology and Oceanography 47*: 742–752.

Meigs, G. W., and W. S. Keeton. 2018. "Intermediate-severity wind disturbance in mature temperate forests: effects on legacy structure, carbon storage, and stand dynamics." *Ecological Applications*. In Press. doi:10.1002/eap.1691.

Montgomery, D. R., J. M. Buffington, R. D. Smith, K. M. Schmidt, and G. Pess. 1995. "Pool spacing in forest channels." *Water Resources Research 31*: 1097–1105.

Montgomery, D. R., T. M. Massong, and S. C. S. Hawley. 2003. "Influence of debris flows and log jams on the location of pools and alluvial channel reaches, Oregon Coast Range." *Geological Society of America Bulletin 115*: 78–88.

Mulholland, P. J. 2004. "The importance of in-stream uptake for regulating stream concentrations and outputs of N and P from a forested watershed: evidence from long-term chemistry records for Walker Branch Watershed." *Biogeochemistry 70*: 403–426.

Mulholland, P. J., J. D. Newbold, J. W. Elwood, L. A. Ferren, and J. R. Webster. 1985. "Phosphorus Spiraling in a Woodland Stream - Seasonal-Variations." *Ecology 66*: 1012–1023.

Mulholland, P. J., and J. R. Webster. 2010. "Nutrient dynamics in streams and the role of J-NABS." *Journal of the North American Benthological Society 29*: 100–117.

Muotka, T., and P. Laasonen. 2002. "Ecosystem recovery in restored headwater streams: the role of enhanced leaf retention." *Journal of Applied Ecology 39*:1 45–156.

Nislow, K. H. 2005. "Forest change and stream fish habitat: lessons from 'Olde' and New England." *Journal of Fish Biology 67*: 186–204.

Riley, S. C., and K. D. Fausch. 1995. "Trout population response to habitat enhancement in six northern Colorado streams." *Canadian Journal of Fisheries and Aquatic Science 52*: 34–53.

Roberts, B. J., and P. J. Mulholland. 2007. "In-stream biotic control on nutrient biogeochemistry in a forested stream, West Fork of Walker Branch." *Journal of Geophysical Research-Biogeosciences 112*: G04002.

Rosemond, A. D. 1993. "Interactions among irradiance, nutrients, and herbivores constrain a stream Algal community." *Oecologia 94*: 585–594.

Sabater, F., A. Butturini, E. Marti, I. Munoz, A. Romani, J. Wray, and S. Sabater. 2000. "Effects of riparian vegetation removal on nutrient retention in a Mediterranean stream." *Journal of the North American Benthological Society 19*: 609–620.

Smock, L. A., G. M. Metzler, and J. E. Gladden. 1989. "Role of debris dams in the structure and function of low-gradient headwater streams." *Ecology 70*: 764–775.

Sobota, D. J., S. L. Johnson, S. V. Gregory, and L. R. Ashkenas. 2012. "A stable isotope tracer study of the influences of adjacent land use and riparian condition on fates of nitrate in streams." *Ecosystems 15*: 1–17.

Steinhart, G. S., G. E. Likens, and P. M. Groffman. 2000. "Denitrification in stream sediments in five northeastern (USA) streams." *Verhandlungen des Internationalen Verein Limnologie 27*: 1331–1336.

Stovall, J. P., W. S. Keeton, and C. E. Kraft. 2009. "Late-successional riparian forest structure results in heterogeneous periphyton distributions in low-order streams." *Canadian Journal of Forest Research 39*: 2343–2354.

Sundbaum, K., and I. Naslund. 1998. "Effects of woody debris on the growth and behavior of brown trout in experimental stream channels." *Canadian Journal of Zoology 76*: 56–61.

Sweka, J. A., and K. J. Hartman. 2006. "Effects of large woody debris addition on stream habitat and brook trout populations in Appalachian streams." *Hydrobiologia 559*: 363–378.

Tank, J. L., E. J. Rosi-Marshall, N. A. Griffiths, S. A. Entrekin, and M. L. Stephen. 2010. "A review of allochthonous organic matter dynamics and metabolism in streams." *Journal of the North American Benthological Society 29*: 118–146.

Thompson, D. M., and K. S. Hoffman. 2001. "Equilibrium pool dimensions and sediment-sorting patterns in coarse-grained, New England channels." *Geomorphology 38*: 301–316.

Urbano, A. R., and W. S. Keeton. 2017. "Carbon dynamics and structural development in recovering secondary forests of the northeastern US." *Forest Ecology and Management 392*: 21–35.

Valett, H. M., C. L. Crenshaw, and P. F. Wagner. 2002. "Stream nutrient uptake, forest succession, and biogeochemical theory." *Ecology 83*: 2888–2901.

Van Pelt, R., and J. F. Franklin. 2000. "Influence of canopy structure on the understory environment in tall, old-growth, conifer forests." *Canadian Journal of Forest Research-Revue Canadienne De Recherche Forestiere 30*: 1231–1245.

Vitousek, P. M., and W. A. Reiners. 1975. "Ecosystem succession and nutrient retention - hypothesis." *Bioscience 25*: 376–381.

Wallace, J. B., S. L. Eggert, J. L. Meyer, and J. R. Webster. 1997. "Multiple trophic levels of a forest streams linked to terrestrial litter inputs." *Science 277*: 102–104.

Warren, D. R., E. S. Bernhardt, R. O. Jr. Hall, and G. E. Likens. 2007. "Forest age, wood, and nutrient dynamics in headwater streams of the Hubbard Brook Experimental Forest, NH." *Earth Surface Processes and Landforms 32*: 1154–1163.

Warren, D. R., S. M. Collins, E. M. Purvis, M. J. Kaylor, and H. A. Bechtold. 2017. "Spatial variability in light yields colimitation of primary production by both light and nutrients in a forested stream Ecosystem." *Ecosystems 20*: 198–210.

Warren, D. R., W. S. Keeton, H. A. Bechtold, and E. J. Rosi-Marshall. 2013. "Comparing streambed light availability and canopy cover in streams with old-growth versus early-mature riparian forests in western Oregon." *Aquatic Sciences 75*: 547–558.

Warren, D. R., W. S. Keeton, P. M. Kiffney, M. J. Kaylor, H. A. Bechtold, and J. Magee. 2016. "Changing forests-changing streams: riparian forest stand development and ecosystem function in temperate headwaters." *Ecosphere 7* (8). doi:10.1002/ecs2.1435.

Warren, D. R., and C. E. Kraft. 2003. "Brook trout (*Salvelinus fontinalis*) response to wood removal from high-gradient streams of the Adirondack Mountains (NY, USA)." *Canadian Journal of Fisheries and Aquatic Sciences 60*: 379–389.

Warren, D. R., and C. E. Kraft. 2008. "Dynamics of large wood in an eastern U.S. mountain stream." *Forest Ecology and Management 256*: 808–814.

Warren, D. R., C. E. Kraft, W. S. Keeton, J. S. Nunery, and G. E. Likens. 2009. "Dynamics of wood recruitment in streams of the northeastern U.S." *Forest Ecology and Management 258*: 804-813.

Warren, D. R., M. M. Mineau, E. J. Ward, and C. E. Kraft. 2010. "Relating fish biomass to habitat and chemistry in headwater streams of the northeastern United States." *Environmental Biology of Fishes 88*: 51–62.

Webster, J. R., J. L Tank, J. B. Wallace, J. L. Meyer, S. L. Eggert, T. P. Ehrman, B. R. Ward, B. L. Bennet, P. F. Wagner, and M. E. McTammy. 2000. "Effects of litter exclusion and wood removal on phosphorous and nitrogen retention in a forest stream." *Verhandlungen des Internationalen Verein Limnologie 27*: 1337–1340.

White, S. L., C. Gowan, K. D. Fausch, J. G. Harris, and W. C. Saunders. 2011. "Response of trout populations in five Colorado streams two decades after habitat manipulation." *Canadian Journal of Fisheries and Aquatic Sciences 68*: 2057–2063.

Wootton, J. T. 2012. "River food web response to large-scale riparian zone manipulations." *PLOS One 7* (12): ARTN e51839.

# Chapter 10

# Belowground Ecology and Dynamics in Eastern Old-Growth Forests

*Timothy J. Fahey*

Eastern old-growth forests possess features that distinguish them from human-disturbed stands, especially in terms of the structure, diversity, and function of the ecosystem. To the forest visitor, the distinctive aboveground structure of the old-growth forest is most visible and striking: towering trees, multiple canopy layers, large snags, and coarse woody debris. Hidden from view is an equally complex belowground world that has largely escaped study by ecologists simply because access and observation are so difficult. Are the complex structures and distinctive function aboveground in eastern old-growth forests mirrored in the soil? In this chapter, I summarize current evidence about features of the belowground dynamics of eastern old-growth forests that are characteristic of these spectacular ecosystems.

As summarized by Burrascano et al. (2013) and in chapters 1, 7, and 11 of this volume, the key structural features that distinguish old-growth forests from younger, second-growth stands are: (1) abundance of large living trees; (2) high volume of coarse woody debris in varying stages of decay; (3) vertical heterogeneity, including large canopy gaps created by the mortality of large trees; (4) pronounced microtopography created by tree uprooting; and (5) species composition dominated by understory tolerant species. What features of belowground forest dynamics would reflect these aboveground characteristics in eastern old-growth forests?

In terms of land-use history in eastern North America, one pervasive anthropogenic effect on belowground dynamics requires special attention: Except in the most rugged and remote landscapes, a period of agricultural activity followed European settlement. The legacy of agricultural activity on forest soils and their properties is profound and persistent, typically much more so than forest harvest. In Europe, for example, Depouey et al. (2002) observed that effects of agricultural land use on forest soil properties persisted for millennia. In northeastern forests, the nitrogen dynamics of post-agricultural soils were strikingly different from primary forest more than a century after abandonment and reversion to forest (Compton and Boone 2000). In contrast, after logging, most biogeochemical features of soil recover more quickly, although Latty et al. (2004) were able to detect more subtle but significant differences in soil carbon and nitrogen stocks between mature, postlogging stands compared to virgin forest in the Adirondack Mountains, New York. In this chapter, I will not focus on the belowground legacy of agriculture, which has received considerable attention.

## Overview of the Effects of Features of Old-Growth Forest on Belowground Dynamics

I begin by briefly considering each of the five distinctive aboveground features of old-growth forest, listed earlier, and how they might affect belowground dynamics, especially pedodiversity, the highly variable morphology and other properties of old-growth forest soils.

### Large and Old Living Trees

It has long been known that individual trees can profoundly influence soil properties beneath their crowns, including pH, carbon, nitrogen, and base cations (Zinke 1962). The pattern and intensity of this influence will depend on the crown dimensions of individual trees as well as how long the tree lives. One important source of single-tree influence, particularly in humid, eastern broadleaf forests with decurrent crowns (spreading or rounded), is the funneling of precipitation to the base of the tree by "stem flow." This mechanism is less important in excurrent (cone-shaped crowns) conifer forests where rainwater drips from drooping branches. The stem flow effect is further extended belowground by "double-funneling" whereby the roots induce preferential flow of intense rains through soils, bypassing the micropores of the soil matrix and delivering water to deep soil layers (Johnson and Lehman 2006). Another pervasive influence of big trees on soils resulting from tree death is the formation of large gaps and, in the case of windthrow, the formation of pit-and-mound topography.

## Coarse Woody Debris (CWD)

Accumulation of dead logs on the forest floor is a defining feature of old-growth forests that influences numerous ecosystem patterns, processes, and functions (Harmon et al. 1986). Of course, most CWD occurs at the aboveground-belowground interface and as such can profoundly influence belowground dynamics. In fact, in some forests, especially in cold boreal zones, a considerable proportion of the CWD actually accumulates within the soil matrix, often as a result of preservation from rapid decay by insulating bryophytes (e.g., mosses). However, a recent summary indicates that buried wood is only a minor component in eastern deciduous forests in part due to limited development of bryophyte cover (Moroni et al. 2015) and also because of the more rapid and complete decay of angiosperm than gymnosperm wood (Cornwell et al 2009).

## Canopy Gaps

Death of large, old trees in old-growth forests leads to the formation of large gaps with consequences for belowground dynamics. For example, Dahir and Lorimer (1996) observed that mean gap area was fourfold greater in old-growth than mature northern hardwoods and hemlock stands in upper Michigan. Belowground responses to large gaps are likely to be greater than for small gaps because of the potential to form a "root gap" where few roots of surrounding edge trees immediately colonize the gap (figure 10-1); however, the evidence for formation of root gaps is mixed, as detailed later. Nevertheless, there is some clear evidence for soil nutrient responses to canopy gaps, and the implications for the nutrient balance of eastern old-growth forest landscapes is intriguing. Perhaps gaps contribute to net losses of limiting nutrients (McGee et al. 2007; see nitrogen dynamics section).

## Microtopography

Among the most distinctive effects of agricultural land use on forest soils is the elimination of pit-and-mound microtopography. Conversely, the formation of highly developed microtopography is favored in eastern old-growth forests by the uprooting of large trees that are particularly susceptible to windthrow in part because of their tall stature and broad crowns

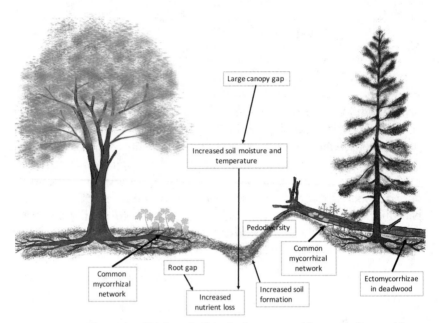

FIGURE 10-1. Elements of high pedodiversity in eastern old-growth forests: blow-down of big trees leads to large canopy gaps and tip-up mounds and pits. The largest gaps may contain "root gaps" in soils that result in nutrient losses as well as renewed soil profile development, especially in pits. Mycorrhizal associations and common mycorrhizal networks may play an important role in tree regeneration in old-growth forests.

(figures 10-1, 10-2). Of course, the pits and mounds formed by big trees are more accentuated and persistent than for smaller trees (Sobhani et al. 2014), contributing to the high pedodiversity observed in eastern old-growth forests (Scharenbroch and Bockheim 2007).

### Forest Composition

As noted earlier, soil properties like pH, nitrogen, and base cations can be influenced by individual trees, and species effects at the stand scale also are common. Thus, the composition of eastern old-growth forests, dominated by highly-tolerant understory species, can regulate belowground processes. The spatial dynamics of changes in species composition in old-growth forest may exhibit intriguing self-organization. For example, Frelich et al. (1993) suggested that the formation of a mosaic of discrete patches of

FIGURE 10-2. Two photos showing tip-up mound size and related gap structural complexity in an old-growth hemlock-northern hardwood stand in the Adirondack State Park, New York State. Photo credits: W. S. Keeton.

hardwood (maple) and conifer (hemlock) stands in an old-growth upper Michigan forest resulted from strong negative reciprocal association between these two dominant species, which could further result in distinctive soil chemistry feedbacks (pH, calcium; Fujinuma et al. 2005). Moreover, this sort of pattern also could be reinforced by the mix of tree species characterized by associations with ectomycorrhizal fungi (Pinaceae, Fagaceae, Betulaceae, etc.) versus arbuscular mycorrhizal fungi (Sapindaceae, Rosaceae, Magnoliaceae, etc.). Phillips et al. (2013) suggested that these classes of mycorrhizae can have an overriding effect on carbon-nutrient coupling in temperate forests so that nutrient dynamics are highly dependent on forest composition. Thus, belowground dynamics in eastern old-growth forests are distinct from those in successional stands even though few of these features will be obvious to the casual visitor.

## Seven Key Features of Belowground Dynamics in Eastern Old-Growth Forests

In this section I provide a detailed overview of seven features of the patterns and mechanisms of belowground dynamics of eastern old-growth forests: canopy gap effects, pit-and-mound microtopography, roots and mycorrhizae, soil carbon, nitrogen cycling, terrestrial salamanders, and invasive earthworms.

### Gaps in Eastern Old-Growth Forests

As forest stands mature following largescale disturbances like hurricanes, fires, or clear-cutting, in the transition to old-growth status, the overstory canopy may break up as some overstory trees senesce. The death of overstory trees results in canopy gaps, and these tend to be larger in old-growth forests because some of the trees are bigger, causing more collateral damage when they fall. Nevertheless, even in old-growth stands, the mean size of gaps is usually relatively small ($\bar{x}$ = 44 m$^2$; Dahir and Lorimer 1996) except when intermediate-severity disturbance events like microburst windstorms cause more extensive damage (Papaik and Canham 2006). Typical small gaps caused by single-tree deaths can result in increased insolation (i.e, sunlight) and consequently higher surface soil temperatures, but these gaps are usually too small to cause a "root gap" because the horizontal extent of neighboring trees will completely encompass the gap; that is, the root systems of canopy trees extensively overlap (Buttner and Leuschner 1994). Nevertheless, root gaps have sometimes been observed in deciduous forests either as reduced fine root biomass or growth (Bauhus and Bartsch 1996). In some cases, however, fine root growth has been observed to increase in old-growth forest gaps (Battles and Fahey 1996), possibly reflecting increased soil resource availability and increased root growth of advance regeneration. The timing and duration of the gap influence on root growth also varies, but some observations indicate rapid (less than one year) and persistent (several years) effects (Bauhus and Bartsch 1996). In sum, a suite of factors will influence the dynamics of root gaps within canopy gaps, including 1) gap size, 2) size and species of neighboring trees, and 3) abundance of advance regeneration.

The existence of a root gap together with changes in the soil environment (e.g., higher temperature, moisture; figure 10-1) could result in increased soil nutrient availability and possibly consequent nutrient losses in dissolved or gaseous forms. Indeed, Scharenbroch and Bockheim (2007) provided evidence for increased leaching losses in canopy gaps in old-growth northern hardwood forests in upper Michigan, and McGee et al. (2007) noted similar slight increases in Adirondack northern hardwood forest gaps. Especially large nitrogen leaching losses were reported by Ritter and Vesterdal (2006) for gaps in mature Danish beech forests on nutrient-rich soils, suggesting dependence on site fertility (see nitrogen cycling below). Complex interactions between gap formation and coincident pedoturbation was documented in eastern old-growth forests by McGee et al. (2007).

## Pit-and-Mound Microtopography

In eastern North America there has been a gradual reduction in the extent of pit-and-mound topography over centuries as a result of agricultural land use, repeated logging, and a decline in the average size of trees, as old-growth forests clearly retain more and larger pits and mounds (Samonil et al. 2010). The principal effect of pit-and-mound formation is to increase the local pedodiversity and spatial heterogeneity of soils rather than larger-scale averages. For example, Liechty et al. (1997) found similar total carbon and nitrogen stocks in soils of old-growth forest stands with and without abundant pit-and-mound topography in upper Michigan. However, spatial heterogeneity is increased by tree uprooting as forest floor organic matter is mixed into the mineral soil of mounds and plant litter accumulates in pits. In humid climates and on poorly drained soils, wet conditions in pits can suppress root growth, litter decomposition, and soil invertebrate activity.

The extent and depth of pit-and-mound microtopography obviously is dependent upon the size of the uprooted trees. The persistence of the resulting microtopography also depends upon tree size, so that old-growth forests would be expected to retain more pedodiversity. Pit-and-mound microtopography can persist for surprisingly long periods, but recent observations indicate that the persistence of pits and mounds depends primarily on soil texture and porosity (Samonil et al. 2010). At the extreme, Samonil et al. (2016) observed that, on sandy outwash soils in upper Michigan with very low innate erodibility, pits and mounds can still be detected over 6,000 years after formation, whereas 500 to 2,000 years is more typical on finer soils. Taking into account the frequency of tree uprooting and the area disturbed, Samonil et al. (2010) concluded that the turnover time of soils in virgin temperate forests is on the order of 1,000 years. In general, pits fill in more quickly than mounds erode (Samonil et al. 2010). Plotkin et al. (2017) observed that 11 percent of the mounds formed from blowdowns in the 1938 hurricane at the Pisgah old-growth forest in New Hampshire were still over one meter high half a century later!

The deep soil mixing from uprooting of large trees eliminates surface horizons and mixes forest floor organic matter into mineral soil. One effect of this pedoturbation may be the interruption of paludification, the accumulation of mineral nutrients in slowly decaying organic matter that can result in forest "retrogression" due to nutrient limitation. Although this process has been observed in wet temperate conifer forests (Bormann et al. 1995), it has not been demonstrated in eastern old-growth forests. Consequently, the importance of pedoturbation for forest health deserves further study. Notably, pedoturbation resets surface soil horizon formation, a process that appears to proceed more rapidly in pits than mounds, at least in Spodosols where spodic horizon formation is particularly rapid (Samonil et al. 2016).

### Roots and Mycorrhizae

There is little basis for concluding that systematic differences exist between the root systems of eastern old-growth forests and younger stands. Some reasons for this conclusion are 1) few measurements of old-growth forest roots have been reported owing to the difficulty of measurement; 2) forest tree root systems exhibit high spatial variability making the detection of patterns challenging; and 3) factors other than forest age may override disturbance history, especially climate, soil fertility, depth, stoniness, etc. However, some evidence does exist for trends in forest root systems across younger stand ages, but even here the patterns appear to vary across sites and regions. Considering first the small feeder roots that provide the trees with soil resources, certainly this fine root biomass increases during early stages of stand development; however, the timing of a peak in fine root biomass sometimes coincides with the early peak in stand leaf area (Claus and George 2005), whereas in other cases, it may continue to increase for decades longer (Yanai et al. 2006). Some evidence suggests that the ratio of fine-root biomass to leaf area remains constant across later stages of stand development (Bauhus and Bartsch 1996), but in other cases, an increase in this ratio may coincide with decreased soil nutrient availability as would be expected from a functional standpoint (i.e., more fine roots needed to obtain soil resources). The depth distribution of fine roots also does not appear to differ systematically between old-growth and younger forests (Bauhus and Bartsch 1996), presumably because the depth distribution exhibits idiosyncratic patterns related to forest floor development, soil mixing, profile development, texture, hydrology, parent material, and other factors. Although we might expect that the high pedodiversity in old-growth forests should lead to high spatial variation in fine-root biomass, no studies have demonstrated such a pattern in eastern old-growth forests.

Obviously, the biomass of coarse, woody roots increases with stand age, probably roughly in parallel with aboveground biomass, but again few measurements are available. Based on the extraordinary allometric measurements at Hubbard Brook Experimental Forest by Whittaker et al. (1974), who excavated the root systems of large trees up to 63 centimeters DBH "with the encouragement of dynamite" (p. 235), the ratio of total belowground biomass to aboveground biomass of northern hardwood trees appears to increase slightly for larger trees as would be expected from the standpoint of wind firmness of tall trees.

Mycorrhizal associations are an ubiquitous feature of all forest trees and are critically important for forest productivity. These associations are symbiotic relationships between mycorrhizal fungi and tree roots, helping trees acquire water and nutrients, and sometimes protection from root pathogens, in exchange for photosynthate supplied to the fungi. There is some evidence for differences in mycorrhizal associations between young and more mature forest stands (Twieg et al. 2007), but this has not been reported for eastern old-growth forests. In theory, either changing forest composition (e.g., arbuscular versus ectomycorrhizal associated trees) or shifts in nutrient availability (e.g., nitrogen versus phosphorus, organic nitrogen; Lilleskov and Bruns 2003) and its spatial variation would be expected to cause successional changes in mycorrhizal communities. Moreover, Johnson et al. (2005) concluded that, in boreal forests, plant species richness is related to mycorrhizal species richness. Nevertheless, Dickie et al. (2013) recently concluded that mycorrhizal communities do not exhibit consistent trends with ecosystem development for many of the same reasons listed for fine roots.

A prominent and interesting feature of belowground dynamics is the so-called "wood-wide web" (Helgason et al. 1998) in which trees are linked together through the mycorrhizal fungal network where carbon and nutrients can be transferred between trees—even those of different species (Figure 10-1). Perhaps this web of interaction becomes more complex in old-growth forests coinciding with structural complexity of the ecosystem; however, this subject has received less attention in eastern as compared to western old-growth forests.

One important feature of old-growth forests that certainly influences the distribution of fine roots and mycorrhizae is the abundance of coarse woody debris. Although fine roots often proliferate in decaying wood in eastern forests, the abundance of CWD is not sufficient to support more than a few percent of the forest fine-root system (Arthur et al. 1993). However, from a diversity standpoint, CWD does provide a niche for particular species of ectomycorrhizal fungi that colonize decaying wood (Tedersoo et al. 2003; figure 10-1).

The woody root system of large trees in old-growth forests also could play a significant role in forest hydrology. As mentioned earlier, woody roots act as a conduit for rapid deep percolation of rainwater as indicated by using dye tracers (Schwarzel et al. 2012). Moreover, dead roots form channels in soil that act as pipes for deep routing of water; these root channels can persist in fine, highly structured soils and, thus, would be expected to accumulate with increasing stand age, though no such evidence has been reported.

## Soil Carbon Cycle

Forests contain the largest terrestrial carbon stock on Earth, and the majority of that carbon is stored in the soil. Forest harvest often results in a decrease of soil carbon storage, mostly from the surface horizons (Nave et al. 2010); thus, protection of old-growth forests supplies a global ecosystem service in the form of soil carbon retention. Do old-growth forest soils also sequester additional carbon in parallel with the aboveground carbon accumulation described in chapter 14? In theory, soils of old-growth forests should be at or near carbon saturation because nearly all the sites where carbon is stabilized should already be occupied. Nevertheless, some empirical observations suggest that carbon may be accumulating in surface soil (Zhou et al. 2006) or deep soil (Tang et al. 2009) of some old-growth forests. What could explain such unexpected observations?

Soil carbon stocks represent the balance between inputs of carbon from plant detritus and outputs by heterotrophic respiration (decomposition), leaching, and erosion. Increased inputs of detrital carbon to old-growth forest soils might accompany continued aboveground biomass accumulation. However, most of the carbon that is stabilized in soils (i.e., extremely slowly decomposed) is actually derived from microbial residues so that increased inputs to the stabilized carbon pool requires a mechanism that causes greater microbial biomass turnover. Perhaps global change drivers, such as climate change, increasing atmospheric carbon dioxide or nitrogen deposition, have stimulated a higher supply of microbial residues in some old-growth forests. However, the principal mechanisms by which soil carbon is stabilized against microbial degradation are strong interactions with mineral surfaces (clay, silt, amorphous metal oxides) and physical protection within soil microaggregates. These sites of stabilization are thought to be near saturation in most forest soils so that even increased inputs of microbial residues would not be expected to result in soil carbon accumulation (Wiesmeier et al. 2014). One recent suggestion is that carbon may be accumulating (at least temporarily) in some soils in the form of unprotected particulate organic matter (Castellano et al. 2015), and perhaps that could contribute to observations of soil carbon accumulation in old-growth forest soils. In any case, more and better measurements of soil carbon stocks and their temporal changes are needed to determine how widely this phenomenon may be occurring in mature forests worldwide.

## Nitrogen Cycling

The nitrogen dynamics of eastern old-growth forests have received considerable study because nitrogen is often the most limiting nutrient in temperate forests, and atmospheric pollution has greatly elevated the inputs of this key element, leading to concerns about ecosystem nitrogen saturation. In theory, retention of nitrogen in forests is expected to peak early in succession, before declining in old age as live biomass reaches a maximum, and some evidence supports this theory (Vitousek and Reiners 1975). Studies of old deciduous forests suggest that large forest gaps could be the source of nitrogen losses (Ritter and Vesterdal 2006; McGee et al. 2007). However, strong retention of nitrogen has been observed in many eastern old-growth forests, even in the face of high anthropogenic inputs via atmospheric deposition (Goodale et al. 2003).

One possible explanation of this anomalous observation is that high accumulation of carbon-rich detritus in old-growth forests could act as a sink for nitrogen. In effect, microbial heterotrophs can immobilize lots of mineral nitrogen in their biomass, and microbial turnover could result in the formation of stable soil organic matter with high nitrogen content. Fisk et al. (2002) tested this hypothesis by comparing several adjacent old-growth and second-growth northern hardwood forests in upper Michigan. Although they found evidence of higher soil detrital pools and microbial nitrogen immobilization in the old-growth stands, this pattern was not correlated with soil nitrogen retention, which was similarly high across ages. Based on analysis using a simulation model for the Hubbard Brook ecosystem in New Hampshire, Aber et al. (2002) concluded that complex patterns of forest nitrogen retention could only be explained in light of the full spectrum of disturbance events (e.g., blowdowns, defoliation by insects, tree harvest) and changes in environmental conditions (e.g., climate, various pollutants), influencing forest ecosystem dynamics.

## Invasive Earthworms

A pervasive influence on belowground dynamics of northern forests has been the invasion of large lumbricid earthworms. These are native to Europe but have widely colonized northern landscapes, which lacked these ecosystem engineers prior to European settlement (Bohlen et al. 2004). The effects of earthworms on forest soils result primarily from their feeding and burrowing activities: They can rapidly eliminate the surface organic horizons (forest floor) that accumulate in most cold temperate forests and mix the organic matter into the mineral soil. These changes in the soil profile influence many aspects of belowground dynamics, including roots and mycorrhizae, microbial communities, litter decay and soil organic matter stability, mineral nutrient availability and leaching, and soil structure and aggregation (Bohlen et al. 2004). Some consequences of earthworm invasions for forest health and diversity also have been indicated by comparisons of adjacent invaded and earthworm-free forests.

Although no comprehensive studies of the occurrences of earthworm invasion of old-growth forests have been reported, some local studies suggest that many eastern old-growth forests have so far escaped colonization by European lumbricid earthworms. For example, Gundale et al. (2005) observed that old-growth northern hardwood forests in upper Michigan contained smaller and less diverse invasive earthworm communities than nearby second-growth stands. They suggested that earthworms had less opportunity for introduction or dispersal into more remote wilderness areas by such vectors as recreational vehicles, logging machinery, and anglers.

Observations from old-growth beech forests within the native range of lumbricid earthworms in Europe suggest that both forest development stage and soil characteristics influence earthworm communities and that these fluctuate spatially and temporally over long timescales (Ponge et al. 1999). Most importantly, many lumbricid earthworm species are sensitive to low soil pH and calcium availability, which may limit their ability to invade many eastern old-growth forests located on acidic soils. In the future, colonization by additional exotic earthworms, especially the Asian taxa *Amynthas*, could result in still greater effects on belowground dynamics of northern forests (Greiner et al. 2012).

## Plethodontid Salamanders

Lungless terrestrial salamanders from the family Plethodontidae are typically the most abundant soil-dwelling vertebrates in eastern North American broadleaf forests, and the southern Appalachian Mountains comprise the global center of their diversity. These woodland salamanders have been proposed as ideal indicators of forest ecosystem health (Welsh and Droege 2001). By profoundly changing the habitat of these surface-soil-dwelling predators, intensive forest harvest can drastically reduce the density of plethodontid salamanders (Petranka et al. 1993). The recovery of salamander populations following logging also can be slow, limited by their low reproductive rate, short-range dispersal, and the delayed recovery of their forest floor habitat; however, deMaynadier and Hunter (1995) concluded that effects of logging can be mitigated if microhabitat structure is maintained. Old-growth forests provide particularly suitable microhabitats for plethodontid salamanders in the form of well-developed forest-floor horizons, complex soil structures, pedodiveristy, and especially large decaying logs. Several studies have noted higher salamander abundance in old-growth than in mature, second-growth forests (deMaynadier and Hunter 1995; Hicks and Pearson 2003) or where coarse woody debris is silviculturally enhanced to emulate old-growth conditions (McKenny et al. 2006). These salamanders spend most of their time in the soil and forest floor, feeding on microarthropods (mites and springtails) and hiding in cool, moist microhabitats beneath rocks and dead logs. Notably, invasion of northern forests by lumbricid earthworms could decrease habitat quality by eliminating the forest floor and consequently reducing the abundance of soil microarthropods.

Given their low reproductive rate, narrow thermal tolerance, and limited dispersal ability, amphibians may be particularly sensitive to rapid climate change. For this reason, old-growth forests that provide particularly high-quality microhabitat conditions could serve as valuable climate refugia for plethodontid salamanders. A long-term re-survey of salamander populations in the southern Appalachians suggested that increase in abundance at both low and high elevations was associated primarily with forest maturation rather than climate warming; at low-elevation sites salamanders were near the limit of their thermal tolerance but forest recovery from early twentieth century logging apparently compensated for temperature effects (Moskwik 2014).

## Conclusion

The hidden belowground dynamics of eastern old-growth forests are a reflection of the distinctive structural complexity of these ecosystems aboveground. Large trees, gaps, and microtopography lead to greater pedodiversity than for younger forests. However, further research is needed to demonstrate whether and how such pedodiversity influences functional characteristics and biodiversity of eastern old-growth forests.

## References

Aber, J. D., S. V. Ollinger, C. T. Driscoll, G. E. Likens, R. T. Holmes, R. J. Freuder, and C. L. Goodale. 2002. "Inorganic nitrogen losses from a forested ecosystem in response to physical, chemical, biotic, and climatic perturbations." *Ecosystems* 5: 0648–0658.

Arthur, M. A., L. M. Tritton, and T. J. Fahey. 1993. "Dead bole mass and nutrients remaining 23 years after clear-felling of a northern hardwood forest." *Canadian Journal of Forest Research* 23: 1298–1305.

Battles, J. J., and T. J. Fahey. 1996. "Spruce decline as a disturbance event in the subalpine forests of the northeastern United States." *Canadian Journal of Forest Research* 26: 408–421.

Bauhus, J., and N. Bartsch. 1996. "Fine-root growth in beech (*Fagus sylvatica*) forest gaps." *Canadian Journal of Forest Research 26*: 2153–2159.

Bohlen, P. J., S. Scheu, C. M. Hale, M. A. McLean, S. Migge, P. M. Groffman, and D. Parkinson. 2004. "Non-native invasive earthworms as agents of change in northern temperate forests." *Frontiers in Ecology and the Environment 2*: 427–435.

Bormann, B. T., H. Spaltenstein, M. H. McClellan, F. C. Ugolini, K. Cromack Jr., and S. M. Nay. 1995. "Rapid soil development after windthrow disturbance in pristine forests." *Journal of Ecology 83*: 747–757.

Burrascano, S., W. S. Keeton, F. M. Sabatini, and C. Blasi. 2013. "Commonality and variability in the structural attributes of moist temperate old-growth forests: a global review." *Forest Ecology and Management 291*: 458–479.

Büttner, V., and C. Leuschner. 1994. "Spatial and temporal patterns of fine root abundance in a mixed oak-beech forest." *Forest Ecology and Management 70*: 11–21.

Castellano, M. J., K. E. Mueller, D. C. Olk, J. E. Sawyer, and J. Six. 2015. "Integrating plant litter quality, soil organic matter stabilization, and the carbon saturation concept." *Global Change Biology 21*: 3200–3209.

Claus, A., and E. George. 2005. "Effect of stand age on fine-root biomass and biomass distribution in three European forest chronosequences." *Canadian Journal of Forest Research 35*: 1617–1625.

Compton, J. E., and R. D. Boone. 2000. "Long-term impacts of agriculture on soil carbon and nitrogen in New England forests." *Ecology 81*: 2314–2330.

Cornwell, W. K., J. H. Cornelissen, S. D. Allison, J. Bauhus, P. Eggleton, C. M. Preston, F. Scarff, J. T. Weedon, C. Wirth, and A. E. Zanne. 2009. "Plant traits and wood fates across the globe: rotted, burned, or consumed?" *Global Change Biology 15*: 2431–2449.

Dahir, S. E., and C. G. Lorimer. 1996. "Variation in canopy gap formation among developmental stages of northern hardwood stands." *Canadian Journal of Forest Research 26*: 1875–1892.

deMaynadier, P. G., and M. L. Hunter, Jr. 1995. "The relationship between forest management and amphibian ecology: a review of the North American literature." *Environmental Reviews 3*: 230–261.

Dickie, I. A., L. B. Martínez-García, N. Koele, G. A. Grelet, J. M. Tylianakis, D. A. Peltzer, and S. J. Richardson. 2013. "Mycorrhizas and mycorrhizal fungal communities throughout ecosystem development." *Plant and Soil 367*: 11–39.

Dupouey, J. L., E. Dambrine, J. D. Laffite, and C. Moares. 2002. "Irreversible impact of past land use on forest soils and biodiversity." *Ecology 83*: 2978–2984.

Fisk, M. C., D. R. Zak, and T. R. Crow. 2002. "Nitrogen storage and cycling In old- and second-growth northern hardwood forests." *Ecology 83*: 73–87.

Frelich, L. E., R. R. Calcote, M. B. Davis, and J. Pastor. 1993. "Patch formation and maintenance in an old-growth hemlock-hardwood forest." *Ecology 74*: 513–527.

Fujinuma, R., J. Bockheim, and N. Balster. 2005. "Base-cation cycling by individual tree species in old-growth forests of Upper Michigan, USA." *Biogeochemistry 74*: 357–376.

Goodale, C. L., J. D. Aber, and P. M. Vitousek. 2003. "An unexpected nitrate decline in New Hampshire streams." *Ecosystems 6*: 0075–0086.

Greiner, H. G., D. R. Kashian, and S. D. Tiegs. 2012. "Impacts of invasive Asian (*Amynthas hilgendorfi*) and European (*Lumbricus rubellus*) earthworms in a North American temperate deciduous forest." *Biological Invasions 14*: 2017–2027.

Gundale, M. J., W. M. Jolly, and T. H. Deluca. 2005. "Susceptibility of a northern hardwood forest to exotic earthworm invasion." *Conservation Biology 19*: 1075–1083.

Harmon, M. E., J. F. Franklin, F. J. Swanson, P. Sollins, S. V. Gregory, J. D. Lattin, N. H. Anderson, et al. 1986. "Ecology of coarse woody debris in temperate ecosystems." In *Advances in Ecological Research 15*: 133–302. San Diego, CA: Academic Press.

Helgason, T., T. J. Daniell, R. Husband, A. H. Fitter, and J. P.W. Young. 1998. "Ploughing up the wood-wide web?" *Nature 394*: 431.

Hicks, N. G., and S. M. Pearson. 2003. "Salamander diversity and abundance in forests with alternative land use histories in the Southern Blue Ridge Mountains." *Forest Ecology and Management 177*: 117–130.

Johnson, D., M. IJdo, D. R. Genney, I. C. Anderson, and I. J. Alexander. 2005. "How do plants regulate the function, community structure, and diversity of mycorrhizal fungi?" *Journal of Experimental Botany 56*: 1751–1760.

Johnson, M. S., and J. Lehmann. 2006. "Double-funneling of trees: stemflow and root-induced preferential flow." *Ecoscience 13*: 324–333.

Latty, E. F., C. Canham, and P. L. Marks. 2004. "The effects of land-use history on soil properties and nutrient dynamics in northern hardwood forests of the Adirondack Mountains." *Ecosystems 7*: 193–207. doi:10.1007/s10021-003-0157-5.

Liechty, H. O., M. F. Jurgensen, G. D. Mroz, and M. R. Gale. 1997. "Pit and mound topography and its influence on storage of carbon, nitrogen, and organic matter within an old-growth forest." *Canadian Journal of Forest Research 27*: 1992–1997.

Lilleskov, E. A., and T. D. Bruns. 2003. "Root colonization dynamics of two ectomycorrhizal fungi of contrasting life history strategies are mediated by addition of organic nutrient patches." *New Phytologist 159*: 141–151.

McGee, G. G., M. J. Mitchell, D. J. Leopold, and D. J. Raynal. 2007. "Comparison of soil nutrient fluxes from tree-fall gap zones of an old-growth northern hardwood forest." *The Journal of the Torrey Botanical Society 134*: 269–280.

McKenny, H. C., W. S. Keeton, and T. M. Donovan. 2006. "Effects of structural complexity enhancement on eastern red-backed salamander (*Plethodon cinereus*) populations in northern hardwood forests." *Forest Ecology and Management 230*: 186–196.

Moroni, M. T., D. M. Morris, C. Shaw, J. N. Stokland, M. E. Harmon, N. J. Fenton, K. Merganičová, J. Merganič, K. Okabe, and U. Hagemann. 2015. "Buried wood: a common yet poorly documented form of deadwood." *Ecosystems 18*: 605–628.

Moskwik, M. 2014. "Recent elevational range expansions in plethodontid salamanders (Amphibia: Plethodontidae) in the southern Appalachian Mountains." *Journal of Biogeography 41*: 1957–1966.

Nave, L. E., E. D. Vance, C. W. Swanston, and P. S. Curtis. 2010. "Harvest impacts on soil carbon storage in temperate forests." *Forest Ecology and Management 259*: 857–866.

Papaik, M. J., and C. D. Canham. 2006. "Species resistance and community response to wind disturbance regimes in northern temperate forests." *Journal of Ecology 94*: 1011–1026.

Petranka, J. W., M. E. Eldridge, and K. E. Haley. 1993. "Effects of timber harvesting on southern Appalachian salamanders." *Conservation Biology 7*: 363–370.

Phillips, R. P., E. Brzostek, and M. G. Midgley. 2013. "The mycorrhizal-associated nutrient economy: a new framework for predicting carbon-nutrient couplings in temperate forests." *New Phytologist 199*: 41–51.

Plotkin, A. B., P. Schoonmaker, B. Leon, and D. Foster. 2017. "Microtopography and ecology of pit-mound structures in second-growth versus old-growth forests." *Forest Ecology and Management 404*: 14–23.

Ponge, J. F., N. Patzel, L. Delhaye, E. Devigne, C. Levieux, P. Beros, and R. Wittebroodt. 1999. "Interactions between earthworms, litter and trees in an old-growth beech forest." *Biology and Fertility of Soils 29*: 360–370.

Ritter, E., and L. Vesterdal. 2006. "Gap formation in Danish beech (*Fagus sylvatica*) forests of low management intensity: soil moisture and nitrate in soil solution." *European Journal of Forest Research 125*: 139–150.

Šamonil, P., K. Král, and L. Hort. 2010. "The role of tree uprooting in soil formation: a critical literature review." *Geoderma 157*: 65–79.

Šamonil, P., M. Valtera, R. J. Schaetzl, D. Adam, I. Vašíčková, P. Daněk, D. Janík, and V. Tej-necký. 2016. "Impacts of old, comparatively stable, treethrow microtopography on soils and forest dynamics in the northern hardwoods of Michigan, USA." *Catena 140*: 55–65.

Scharenbroch, B. C., and J. G. Bockheim. 2007. "Pedodiversity in an old-growth northern hardwood forest in the Huron Mountains, Upper Peninsula, Michigan." *Canadian Journal of Forest Research 37*: 1106–1117.

Schwärzel, K., S. Ebermann, and N. Schalling. 2012. "Evidence of double-funneling effect of beech trees by visualization of flow pathways using dye tracer." *Journal of Hydrology 470*: 184–192.

Sobhani, V. M., M. Barrett, and C. J. Peterson. 2014. "Robust prediction of treefall pit and mound sizes from tree size across 10 forest blowdowns in eastern North America." *Ecosystems 17*: 837–850.

Tang, J., P. V. Bolstad, and J. G. Martin. 2009. "Soil carbon fluxes and stocks in a Great Lakes forest chronosequence." *Global Change Biology 15*: 145–155.

Tedersoo, L., U. Kõljalg, N. Hallenberg, and K. H. Larsson. 2003. "Fine scale distribution of ectomycorrhizal fungi and roots across substrate layers including coarse woody debris in a mixed forest." *New Phytologist 159*: 153–165.

Twieg, B. D., D. M. Durall, and S. W. Simard. 2007. "Ectomycorrhizal fungal succession in mixed temperate forests." *New Phytologist 176*: 437–447.

Vitousek, P. M., and W. A. Reiners. 1975. "Ecosystem succession and nutrient retention: a hypothesis." *BioScience 25*: 376–381.

Welsh, H. H., and S. Droege. 2001. "A case for using plethodontid salamanders for monitoring biodiversity and ecosystem integrity of North American forests." *Conservation Biology 15*: 558–569.

Whittaker, R. H., F. H. Bormann, G. E. Likens, and T. G. Siccama. 1974. "The Hubbard Brook ecosystem study: forest biomass and production." *Ecological Monographs 44*: 233–254.

Wiesmeier, M., R. Hübner, P. Spörlein, U. Geuß, E. Hangen, A. Reischl, B. Schilling, M. Lützow, and I. Kögel-Knabner. 2014. "Carbon sequestration potential of soils in southeast Germany derived from stable soil organic carbon saturation." *Global Change Biology 20*: 653–665.

Yanai, R. D., B. B. Park, and S. P. Hamburg. 2006. "The vertical and horizontal distribution of roots in northern hardwood stands of varying age." *Canadian Journal of Forest Research 36*: 450–459.

Zhou, G., S. Liu, Z. Li, D. Zhang, X. Tang, C. Zhou, J. Yan, and J. Mo. 2006. "Old-growth forests can accumulate carbon in soils." *Science 314*: 1417–1417.

Zinke, P. J. 1962. "The pattern of influence of individual forest trees on soil properties." *Ecology 43*: 130–133.

# Chapter 11

# Biological Diversity in Eastern Old Growth

*Gregory G. McGee*

The objective of this chapter is to update the current understanding of the unique role of eastern old-growth forests in supporting biological diversity, a topic that was first addressed for vertebrates and understory herbs by various contributors to Davis (1996). Therefore, this treatment will focus on diversity of invertebrates, bryophytes, lichens, fungi, vascular plants, and their respective interactions. The biological significance of eastern old-growth forests is linked, in large part, to heterogeneous habitats provided by dead trees (snags), rotting logs, and old live trees. Therefore, it is necessary to first describe eastern old-growth forest structure and contrast it with that of variously disturbed and managed forests in order to establish context for the ecological significance of the structure-diversity relationships.

## Structural Characteristics of Eastern Old-Growth Forests

Old growth emerges in physiographic regions where disturbances recur at intervals that are infrequent enough, intensities low enough, and extents local enough that trees are able to approach their maximum life spans. Under these conditions, trees attain old age, senesce, and eventually die due to defects or weaknesses associated with age, rather than catastrophic, stand-replacing disturbances such as wind, fire, or insect outbreaks. Therefore, by definition, old-growth forests are characterized by having relatively high densities of large-diameter trees of advanced age, and large volumes of coarse woody debris (CWD), although there are alternative views on this issue (chapters 3 and 4). Beginning in the 1980s, considerable research effort was devoted to understanding the ecological value of old-growth forest structure in the Pacific Northwest. Forest ecologists working in eastern North America soon followed suit (chapter 1).

While numerous studies have characterized various aspects of eastern old-growth forest structure, synthesizing this literature presents challenges since all relevant structural characteristics (e.g., large tree densities, CWD abundance) are not reported in all studies. Further, sampling inclusion thresholds (i.e., diameter or size limits) and units of measurement (CWD volume versus biomass) differ among studies, thereby limiting comparisons. Consequently, a perusal of Tyrrell et al.'s (1998) compilation of old-growth structural data across eastern forest types reveals numerous missing data points of relevant forest structural components. Still, in some eastern forest types, sufficient information has been published to approximate the range of forest structural conditions in old growth and provide comparisons to younger and variously managed forests.

Figures 11-1 and 11-2 present available structural data for some of the more thoroughly studied eastern old-growth forest types. Total live basal area and large tree (greater than or equal to 50 centimeters DBH) densities across select eastern forest types are summarized in figure 11-1. Although there is substantial variation in the structure of any given forest type, eastern old-growth forests generally carry about 30 to 45 square meters per hectare of live basal area and include more than 30 trees per hectare that are larger than 50 centimeters DBH (that is, on average, a tree larger than 50 centimeters DBH occurs approximately every 21 meters). The northern pine forests are unique in their capacity to carry up to 100 large trees per hectare, and this reflects the tendency of long-lived eastern white pine to establish, following catastrophic disturbances, and to form 300-year-old emergent canopies under which subcanopies of other long-lived shade-tolerant species can establish (e.g., Abrams and Orwig 1996).

Figure 11-2 summarizes available data for downed and standing CWD greater than 10 centimeters in diameter. In most forest types, median downed CWD volumes range from about 40 to 110 cubic meters per hectare in old-growth forests. The notable exception for this metric is the spruce-fir forest type, which averages 450 cubic meters per hectare and exhibits a substantial range (66 to 951 cubic meters per hectare). These high values may reflect pulsed inputs of CWD that are long-lived due to slow decay rates in high-elevation, cold, nutrient-poor environments. Average standing dead basal areas range from approximately 2.0 square meters per hectare in mesic oak and beech-maple-basswood forests to 6.0 to 8.5 square meters per hectare in mixed-conifer hardwood and spruce-fir types, and represent 5 to 7 percent of the live basal areas in the oak and beech-maple-

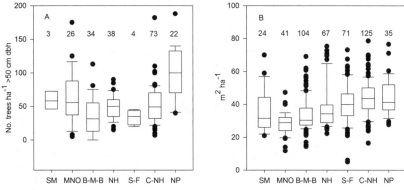

FIGURE 11-1. Summaries of (A) live tree (greater than 10 centimeters DBH) basal areas and (B) large tree (greater 50 centimeters DBH) densities in old-growth stands of select eastern forest types. Forest types are: Southern Mixed Mesophytic, Mesic Northern Oak, Beech-Maple-Basswood, Northern Hardwood, Spruce-Fir, Conifer-Northern Hardwood, and Northern Pine. Boxplot values are 10th, 25th, 50th, 75th, 90th percentiles, and outliers. Numbers of observations used in analyses are given above box plots. Data are compiled primarily from Tyrrell, et al. 1998, along with Martin 1975; Mroz et al. 1985; Greenidge 1987; Palmer 1987; Abrams and Orwig 1996; Roovers and Shifley 1997; Dodds and Smallidge 1999; Forrester and Runkle 2000; Ziegler 2000; Chokkalingam and White 2001; Crow et al. 2002; Jenkins 2007; Kincaid 2007; D'Amato et al. 2008; Vanderwel et al. 2008; Fraver and White 2009; Keeton et al. 2011; D'Amato et al. 2017.

basswood types, and 10 to 21 percent in the other forest types. In northern hardwoods, recent mortality of American beech due to beech bark disease (chapter 12) has probably elevated CWD loads in old-growth forests by about 20 percent (McGee 2000).

In order to evaluate the significant ecological role of old-growth forest structure in supporting biological diversity, we need to contextualize these parameters by comparing them with those of disturbed and managed forests that typify the region's contemporary landscapes. While some studies have directly compared stand structure between old-growth and younger or variously managed forests, these studies have in no way exhausted consideration of the variety of management practices that do or can exist, and that may enhance structural habitat in managed forests (chapter 13). In traditionally managed forests, large trees are purposefully removed using empirically derived cutting guides that serve to maximize wood biomass or sawtimber production. Further, smaller stems are thinned in well-tended stands to reduce crowding and to focus growth on dominant or

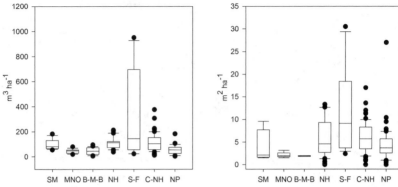

FIGURE 11-2. Summaries of (A, *left*) downed coarse woody debris (greater than 10 centimeters in diameter) volume and (B, *right*) standing coarse woody debris (greater than 10 centimeters DBH) basal area in select eastern forest types. Boxplot values are 10th, 25th, 50th, 75th, 90th percentiles, and outliers. Data are compiled primarily from Tyrrell, et al. 1998, along with Gregory and Lienkaemper (unpublished) as reported by Harmon et al. 1986; Hardt 1993; Muller and Liu 1991; Hardt and Swank 1997; Batista and Platt 1997; Dodds and Smallidge 1999; Chokkalingam and White 2001; Hura and Crow 2004; Jenkins 2007; D'Amato et al. 2008; Vanderwel et al. 2008; Fraver and White 2009; Keeton et al. 2011.

desired stems, thus removing biomass that would otherwise contribute to the CWD loads. For instance, widely applied cutting guides in northern hardwood forests recommend maximum diameter limits of 45 to 60 centimeters DBH for sawtimber production and 40 to 45 centimeters DBH for fiber production (see McGee et al. 1999 and references therein). Thus, when landowners seek to maximize fiber and sawtimber production, forest structure becomes homogenized. Applying prevailing cutting guides in Adirondack northern hardwood forests yields an average of 18 square meters per hectare live basal area (range: 16 to 20) with 5 trees per hectare greater than 50 centimeters DBH, snag basal areas of 1.2 square meters per hectare (range: 0.6 to 2.5), and downed CWD volumes of 69 cubic meters per hectare (range: 55 to 101). This contrasts with 34 square meters per hectare live basal area, 55 trees per hectare greater than 50 centimeters DBH, snag basal areas of 8.6 square meters per hectare, and downed CWD volumes of 139 cubic meters per hectare in comparable old-growth stands (McGee et al. 1999). Similarly, in beech-maple-basswood forests of Minnesota, old-growth stands carry greater live basal area than unevenaged managed stands (31 versus 25 square meters per hectare), double the proportion of more than 50 centimeters DBH trees, greater snag volumes

(27 versus 8 cubic meters per hectare), and greater log volumes (55 versus 40 cubic meters per hectare) (Hale et al. 1999). Available information for southern forests (McMinn and Hardt 1996) indicates greater volumes of downed CWD in late successional forests (82 cubic meters per hectare) versus younger or managed forests (40 cubic meters per hectare). Collectively, current research reveals overall greater live basal areas, greater densities of large-diameter trees, and greater volumes and/or biomasses of CWD in old-growth versus younger or managed forests (Hale et al. 1999; McGee et al. 1999; Ziegler 2000; Hura and Crow 2004; D'Amato et al. 2008; Keeton et al. 2011; but see Vanderwel et al. 2008).

## Ecological Roles of Old-Growth Keystone Structures

Structural features provide the environmental habitat heterogeneity that enhances and maintains biodiversity (chapters 9 and 13). When these structures are unique, at a regional or microscale, such that they provide habitat or resources necessary to support the presence of certain taxa, when those taxa would not otherwise be present, they are referred to as "keystone structures" (sensu Tews et al. 2004). For instance, regional biodiversity is enhanced by the presence of unique conditions associated with limestone or serpentine outcrops, alluvial fans or kettle ponds. The presence of vernal pools within upland forests, grasslands, or even agricultural systems support diverse plants and animals that otherwise would not be present locally. Similarly, the variety of organisms associated with late successional and old-growth forests appear to rely on the presence, abundance, and proximity of the keystone structures that define old-growth forests (large trees and coarse woody debris).

The magnitude of structural differences between managed and old-growth forests can be quite consequential for the diversity of organisms that are adapted to use those keystone structures. For sessile organisms, these structures represent habitat "islands" that directly limit their distributions. It is to these patches that they immigrate (usually by chance), in and upon which they must acquire all resources necessary to grow and complete their entire life cycles, and then from which their offspring or diaspores must emigrate before conditions become unfavorable. Therefore, environmental heterogeneity within and among these habitat islands, and the proximity of like habitat islands, regulate the abundance and community composition of these organisms. The following consideration of the ecological roles of old-growth structures is presented through the lens of the structures serving as necessary resource patches for some organisms. The objective of this review is to draw upon literature from eastern forests to present our current understanding of these relationships. An attempt was made to exhaustively review the eastern North American literature, but in several instances when that literature was scant, the analysis was supplemented with information from temperate spruce, pine, and mixed-deciduous forest types of northern and central Europe.

## Large Old Trees

To state the obvious, trees do not live forever. Those rare individuals that outcompete thousands of other stems during their emergence to canopy dominance eventually senesce and die after reaching relatively predictable sizes and ages. Gower et al. (1996) reviewed the physiological mechanisms leading to the relatively predictable "pathological rotation" of trees—that is, the approximate age and/or diameter of trees when the rate of wood decay exceeds the rate of new wood production, leading to their senescence and eventual death. An understanding of pathological rotations informs the empirically derived diameter limits (described above) that are applied to optimize long-term wood production in managed forests. However, as trees grow large and old, their morphological changes result in unique physical structures that provide habitat for other organisms. For instance, old canopy-dominant trees develop large, twisting, spreading upper branches (Pederson 2010) that provide substrate for arboreal organisms. The bark of many tree species becomes fissured or platelike with age, and these textures provide protected microsites, in and under which many invertebrates shelter and diaspores (spores and vegetative propagules) establish. Chemical and physical conditions of bark also change as trees grow and age. In sugar maples, bark of older, large trees hold twice as much moisture per unit surface area as younger, small trees, and mineral nutrient concentrations of bark leachate increase about five-fold with increasing trunk diameter (McGee et al. in press). These moisture and nutrient gradients may be consequential for partitioning bark habitat among arboreal organisms.

## Epiphytes

Epiphytes are organisms, such as bryophytes (nonvascular plants that include mosses and liverworts), lichens (symbiotic associations of fungi and algae or cyanobacteria), slime molds, and free-living algae that inhabit the surfaces of live plants. Epiphytes are not parasitic on their host plant. Instead, they receive all necessary water and mineral nutrients from the environment on and around the tree surface. Consequently, epiphytes are quite sensitive to variation in bark chemistry and the availability of moisture, light, and atmospheric humidity.

Several studies in eastern forests (Selva 1994; Cooper-Ellis 1998; Mc-Gee and Kimmerer 2002; Root et al. 2007b; Garmendia-Zapata 2015) and other northern temperate forest regions (e.g., Nascimbene et al. 2013) report increased epiphyte abundance and diversity, and compositional turnover with increasing tree diameter. The percent of cover by epiphytic macrolichens increases from roughly 12 percent on small (15 to 35 centimeters DBH) to roughly 25 percent on large (more than 50 centimeters DBH) sugar maples (Root et al. 2007b), as does the cover of mosses and liverworts at the base (0 to 1.5 meters above the ground) of several tree species in northern hardwoods (McGee and Kimmerer 2002). However, these studies found no differences in total epiphytie-bryophyte cover among old-growth, maturing 100-year-old, partially cut Adirondack northern hardwood forests (range: 90 to 130 square meters per hectare), or in total epiphytie-lichen cover on sugar maples among old-growth, reserve shelterwood, and selection stands (range: 240 to 340 square meters per hectare). Species richness of basal bryophytes on sugar maple increased two-fold as trees increased in diameter over a range of 10 to 80 centimeters DBH (Garmendia-Zapata 2015), and while that study found no differences in diameter-richness relationships among stand types or different regions of New York State, Cooper-Ellis (1998) did detect a two-fold greater richness of bryophytes on trees in old growth compared to trees of equal sizes in secondary forests in Massachusetts.

To date only one study has identified any epiphyte species restricted to eastern old growth. McMullin et al. (2008) detected 16 lichens present only in "advanced" old-growth stands in Nova Scotia. However, several epiphytes exhibit affinities for large-diameter (e.g., more than 50 centimeters DBH) trees. Several mesophytic (moisture-requiring) and calciphilic (calcium-requiring) bryophytes occur with greater cover and frequency on large-diameter sugar maples (plate 6). Given the greater abundance of large-tree substrate in old-growth forests (207 square meters per hectare in old-growth versus 22 to 32 square meters per hectare in partially cut and maturing stands), these mesophytes and calcicoles achieve greater overall cover in old growth (McGee and Kimmerer 2002). Similarly, studies in northern hardwood and spruce-fir forests (Selva 1994; Richardson and Cameron 2004; Root et al. 2007b; McMullin et al. 2008) have recognized some cyanolichens (e.g., *Lobaria* spp.), some fruticose green-algal hair lichens (e.g., *Bryoria, Alectoria, Usnea*), and the crustose order Caliciales to be indicators of large trees and old-growth conditions (plate 6, photo d).

The relationship between epiphyte communities and tree age and/or diameter appears to be more complex than just straightforward affinities for large tree substrates. Root et al. (2007a) found not only greater abundance of several arboreal lichen species on large versus small sugar maples, but also detected an interaction between tree size and forest disturbance history. *Lobaria pulmonaria* was more frequent and abundant on large trees in old-growth and selection system stands cut from old growth than on similarly sized trees in reserve shelterwoods that had been more heavily cut from old growth 25 to 30 years prior. Additionally, small sugar maples retained in the shelterwoods also supported fewer of the large-tree lichens than were present on trees of similar sizes in old and selection system stands.

Several possible mechanisms may account for the unique epiphyte communities observed on large trees and in old-growth forests. First, as described earlier, habitat conditions on large, old trees (physical and chemical conditions of bark) change through time in a manner that meets habitat requirements of mesophytic and calciphilic lichen species. Further, the microenvironmental conditions in the canopies of undisturbed forests may remain more protected. For instance, Root et al. (2007b) speculated that the relative rarity of *L. pulmonaria* in shelterwoods, along with characteristically different communities on small-diameter trees in shelterwoods, may be the result of exposure of branches and boles to elevated light, heat and desiccating wind currents in the heavily disturbed shelterwood stands. In a review of northern-European literature, Nascimbene et al. (2013) also noted negative impacts of intensive canopy disturbances to microenvironmental conditions required by epiphytic lichens.

But this explanation is confounded by the fact that large-tree substrate is also long-lived and persistent. Therefore, the unique epiphyte composition of old-growth forests may alternatively be a function of colonization probabilities for dispersal-limited organisms on these ephemeral habitat islands (Lobell et al. 2006). That is, the greater surface areas of large trees that persist for long periods in relative close proximity to other large trees are more likely to intercept rare diaspores of poorly dispersed epiphytes. For instance, a study in European forests found that *Neckera pennata* (one of the eastern large-tree indicator species) requires 10 years to grow from a spore to a 2-square centimeter colony, and initial sexual reproduction does not occur until 19 to 29 years (Wiklund and Rydin 2004). The dispersal distances of epiphytic bryophyte spores, which generally occur close to the ground, isolated from winds above forest canopies, are typically on the order of a few meters. While vegetative propagules (gemmae, brood branches, stem fragments) may experience infrequent long-distance dispersal by birds and mammals, these too generally do not disperse far from the parent plant.

Lichen dispersal is more complicated than that of bryophytes since the mycobiont (the fungus) and photobiont (the photosynthetic alga or cyanobacterium) must both disperse. Sexual reproduction requires fungal spores to alight upon substrate where the photobiont is already present in a free-living state. Therefore, lichen establishment may be limited if the fungus requires a specific photobiont partner. Conversely, generalists are more successful in long-distance dispersal by spores. Vegetative reproduction of lichens occurs through dispersal of soredia (photobiont cells surrounded by fungal hyphae), but when wind-dispersed, the larger soredia have shorter dispersal distances than spores. Soredia are also dispersed passively by birds and mammals (Richardson and Cameron 2004) and mites (Root et al. 2007a). Lobell et al. (2006) concluded that many asexually dispersed epiphytic lichens tend to be more dispersal-limited and to have more patchy distributions in forests than sexually dispersed generalists, thereby requiring long-lived substrates (old trees) for local persistence of metapopulations. Distinguishing the effects of habitat uniqueness from those of habitat longevity for explaining abundant and diverse old-growth epiphyte communities remains problematic because of the difficulty of experimentally testing these factors or accounting for them in observational studies.

### Arboreal Invertebrates

Arboreal invertebrate diversity and communities change with the age and size of trees due to the development of textured bark and epiphytic mats of lichens and bryophytes (Ulyshen 2011). Where arboreal arthropods have been studied in eastern North America, abundance of several taxa are directly and positively related to foliose lichen abundance, and these lichens provide food, shelter, and oviposition sites for a variety of micro- and macroinvertebrates, including aquatic fauna such as nematodes, rotifers, and tardigrades, as well as for terrestrial gastropods and arthropods, including spiders, mites, insects, and millipedes (Stubbs 1989). In a quantitative assessment of arboreal oribatid mite communities, Root et al. (2007a) extracted 25 species, including three previously undescribed species, from a mere 0.26 square meter of lichen sampled from twelve sugar maple trees in Adirondack northern hardwoods in New York State. They noted differences in oribatid mite communities within the same species of lichen substrate between old-growth and reserve shelterwood stands, and suggested these differences may be due to the greater occurrence of poorly dispersed taxa in the old growth.

The ecological consequences of epiphyte-invertebrate linkages are poorly understood for eastern forests. However, springtails (Collembola) are known to depend on epiphytic bryophytes for shelter and food. In turn, their populations support predatory arboreal spiders, the abundance of which has been shown to decline in direct relation to reduced springtail abundance and bryophyte cover on trees retained within experimental gaps (Miller et al. 2008). Studies in European temperate forests have shown that the abundance of arboreal invertebrates, including those large enough to be consumed by passerine birds, is positively associated with lichen biomass. In turn, old-growth forest canopies support greater invertebrate densities than managed forests due to the greater abundance of lichens (Pettersson et al. 1995). Old-growth Douglas-fir-western hemlock forest canopies support greater codominance of predatory and detritivorous arthropods than those of young, even-aged forests, which are dominated by leaf- and needle-feeding (phytophagous) species (Schowalter 1995). This evidence suggests that old-growth canopies, which support greater abundance of epiphytic lichens and bryophytes, harbor a greater abundance and diversity of arthropods, which, in turn, lead to more complex trophic interactions that maintain higher-order predators and provide biological regulation of potentially destructive insect herbivores.

## Downed Coarse Woody Debris

### Vascular Plants

Decaying logs provide important regeneration sites for several woody and herbaceous vascular plants in eastern forests. For instance, yellow birch, eastern hemlock, red spruce, and balsam fir germinants all occur with greater densities on decayed logs versus undifferentiated forest floor microsites (McGee 2001; Caspersen and Saprunoff 2005; Marx and Walters 2008). Wood fern, Christmas fern (McGee 2001; Flinn 2007), and some club mosses (Scheller and Mladenoff 2002) establish in greater densities on decaying wood and humus microsites compared to hardwood leaf litter. As described previously, these more favorable substrates are more common in old-growth forests compared to younger and managed ones.

The ecological mechanisms controlling these associations are still not clearly understood. The upper surfaces and crevasses of decaying logs appear to serve as "safe sites" where small seeds and spores escape smothering by hardwood leaf litter, while larger seeds, such as those of maples and American beech, are more readily shed from convex log surfaces. While well-decomposed logs provide moist substrate that favors early establishment of seedlings and ferns, past studies (Harmon et al. 1986; Arthur et al. 1992) indicate that coarse woody debris (CWD) is generally nutrient-poor compared to mineral soil, and therefore likely cannot support tree growth without facilitation by mycorrhizal fungi (Harmon et al. 1986). In eastern forests, Marx and Walters (2006) determined that only 4 percent of hemlock and yellow birch seedlings less than 2 years old and 0 to 12 percent of older seedlings on CWD were involved in mycorrhizal associations, but the few that did form mycorrhizal associations grew faster. They further reported that eastern hemlock and yellow birch seedlings tended to establish on hemlock and birch logs but not on sugar maple logs, and suggested that this substrate selectivity may be mediated by the mycorrhizal fungi present in different species of decaying wood.

Even though seeds or spores may germinate more readily on decaying logs, the young root systems must access nutrient-rich mineral soil to sustain long-term growth, and light levels greater than those experienced under closed canopies appear to be necessary for germinants to grow into larger regeneration classes. For example, while McGee (2001) reported greater basal areas of yellow birch, balsam fir, and eastern hemlock regeneration on logs versus the forest floor in old-growth, partially cut, shelterwood sites, these species did not necessarily develop to dominant sapling stages at these sites. Even in old growth, where decaying logs were most abundant and seeds readily available, yellow birch sapling regeneration greater than 2 millimeters in diameter at the root collar was absent. This lack of regeneration was probably due to deep shading in these stands. However, under the moderate light conditions of partial-cuts, larger saplings (up to 125 millimeters in diameter) occurred on logs. Others (Caspersen and Saprunoff 2005; Shields et al. 2007; Bolton and D'Amato 2011) have similarly concluded that yellow birch, hemlock, and balsam fir regeneration does not attain dominance, even in small (0.04 hectares) gaps, without CWD or scarified mineral soil substrate.

Canopy gaps exert other important influences on ground-layer vegetation. Any evaluation of the significance of old-growth forest conditions needs to consider how overall forest canopy structure influences herbaceous diversity and abundance. McCarthy (2003) provides a thorough review of herbaceous vegetation in eastern old growth and discusses how factors such as site conditions, aspect, past land use, and canopy development confound conclusions regarding the significance of old-growth forest conditions for herbaceous communities. In recent studies that controlled for these extraneous factors in northern hardwoods-hemlock (D'Amato et al. 2009) and southern cove forests (Wyatt and Silman 2010), 100- to 150-year-old, even-aged, aggrading stands, that established after clear-cutting, exhibited lower cover and richness of ground-layer vegetation and lower seedling and sapling densities than comparable old-growth forests. Together, these studies concluded that the differences were largely due to low dispersal rates of the herbaceous plants and lack of canopy gaps and well-decayed CWD substrate in the closed-canopy aggrading stands. When canopy gaps are created periodically in uneven-aged, managed forests, few if any differences with old-growth conditions have been detected in herbaceous layer diversity, richness, or indicator species (e.g., Metzger and Schultz 1984; Goebel et al. 1999; Scheller and Mladenoff 2002). However, successive partial cuts that increase and homogenize light levels on the forest floor over extended periods can create conditions favored by ruderal and competitive vascular plants. With higher and sustained light levels, herbaceous plants having the capacity to propagate rapidly by seed or by vegetative means can establish large, more continuous understory patches, which, through time, may lead to eventual declines in local vascular plant richness and diversity (Scheller and Mladenoff 2002).

## Bryophytes and Lichens

Coarse woody debris provides structure that maintains local diversity of epixylic (inhabiting wood surfaces) bryophytes and lichens (Cooper-Ellis 1998; Cole et al. 2008). As with epiphytes, there is a lack of understanding of the degree to which abundance and diversity of these organisms in old-growth forests is limited by habitat availability per se, by local immigration or extinction patterns of potentially dispersal-limited species on short-lived habitat patches, or by more favorable overall microclimatic conditions in old growth. Cooper-Ellis (1998) found no difference in average percent bryophyte cover or bryophyte richness on decaying logs between old-growth and secondary northern hardwood forests. Similarly, McGee (1998) detected no species that were limited to old growth and no differences in community composition or percent cover of bryophytes on logs that occurred in old-growth, partially cut, even-aged, aggrading forests. However, there was 2- to 4-times greater total cover of epixylic bryophytes in old versus partially cut or aggrading secondary forests due, in part, to 2-times greater log surface area in the old stands. However, studies conducted elsewhere (reviewed by Frego 2007) suggest that bryophyte communities in mixed conifer-hardwood forests are sensitive to changing understory microclimates following a range of canopy disturbances. In particular, Shields et al. (2007) reported that ground-layer bryophyte cover declined in experimental gaps ranging from 200 to 1,460 square meters and concluded that frequently recurring and expansive disturbances more than 200 square meters would be detrimental to ground-layer bryophytes.

No substantial research has summarized epixylic lichen communities in eastern forest types, but a review of lichens in temperate forests of the Pacific Northwest and northern Europe (Spribille et al. 2008) determined that 43 percent of 1,271 lichen species occur on woody substrates, with 10 percent being exclusively lignicolous (wood-dependent), with most obligate lignicoles being sexually reproducing crustose lichens. Spribille et al. (2008) suggest that the strict lignicolous association is likely due to dispersal abilities to ephemeral habitats. Further, the distributions of several lichen species are restricted to particular tree species and decay stages of wood substrate.

## Fungi

Fungi are highly specialized in their use of specific types of organic matter. Mycorrhizal fungi form symbiotic relationships with the roots of plants. They derive their energy from stored carbohydrates in roots and, in turn, their filamentous hyphae extend the functional root system of the plant to enhance acquisition of mineral nutrients and water. Saprotrophic fungi are decomposers that acquire energy from carbon compounds in dead wood (lignicolous) or leaf litter and soil (humicolous). It is these saprotrophs that have been the subject of most research in temperate old-growth systems given their associations with woody debris. However, fungal inventories are extremely challenging because the spore-bearing structures used for identification are frequently short-lived and because of the problematic nature of drawing inference regarding abundances from observing the spore-bearing structures.

Few studies have directly assessed fungal diversity in eastern old growth. Lindner et al. (2006) found no differences in plot-level lignicolous fungal richness between old-growth and managed northern hardwoods, but observation frequency (abundance) of spore-bearing structures was positively correlated with substrate volume, which was greater in the old growth versus managed stands. Further Lindner et al. (2006) found that several species were restricted to certain diameter classes of woody debris (both small and large classes), and some were possible indicators of old growth. Other studies have focused directly on the role of CWD volume and quality in maintaining fungal diversity. In northern hardwood forests, Brazee et al. (2014) detected greater species richness and observation frequency of lignicolous fungi in study plots that received experimental CWD additions meant to mimic CWD loads and canopy conditions of an old-growth forest. They further determined that large diameter logs (more than 40 centimeters) were important for enhancing local diversity of decay fungi and supported 24 percent of all species encountered. Rubino and McCarthy (2003) detected positive correlations between lignicolous fungus richness and the size (volume) of oak logs, as well as the volume of logs in the surrounding study plot.

Substantially more research has been conducted on fungal diversity in the temperate forests of Europe where numerous studies confirm the positive relationships between lignicolous fungal diversity and variation in tree species, size class, and decay stage of wood substrate (e.g., Norden et al. 2004; Odor et al. 2006; Stokland and Larsson 2011). These studies emphasize the conservation value of old-growth forests. In particular, Stokland and Larson (2011) found that old-growth forests not only carried greater richness of wood decay fungi than managed forests, due to greater volumes of CWD substrate, but they also found that logs of equal quality (species, size, decay) carried greater richness in old-growth versus managed forests, especially for logs more than 30 centimeters DBH. In other words, a lower proportion of available habitat patches were occupied in the managed forests, perhaps due to excessive isolation of relatively rare logs. Rydin et al. (1997) further concluded that a large proportion of threatened macrofungal species in northern European temperate forests are associated with CWD and suggested that the threatened status of macrofungi is a consequence of reduced CWD loads in managed forests.

## Invertebrates

Approximately 20 percent of woodland arthropods are saproxylic and, thus, rely on decaying wood or other wood-inhabiting organisms (e.g., decay fungi, bacteria, and algae) at some point in their life cycles (Ulyshen 2011). Therefore, the abundance and diversity of such saproxylic organisms would be expected to be closely related to the availability of coarse woody debris. Chandler (1991) determined that densities of some beetles associated with fungi dependent on decaying wood were greater in old-growth, aggrading forests versus young, aggrading forests. In both the conifer- and the hardwood-dominated forests of Nova Scotia, Kehler et al. (2004) concluded that decaying log volume was the best habitat predictor of beetle (Coleoptera) species richness. However, Zeran et al. (2007) detected no difference in richness or diversity of fungivorous Coleoptera in southern Ontario when they compared isolated, old-growth stands with similarly aged but selectively cut northern hardwood-hemlock forest fragments.

A more comprehensive body of research on old-growth forest invertebrate communities exists from Europe. Martikainen et al. (2000) trapped 553 species of beetles in old-growth and managed spruce forests, of which roughly 40 percent were saproxylic, and of the saproxylic species, nearly 80 percent were most abundant in old-growth stands. Further, Berg et al. (1994) reported that, of the more than 730 threatened forest invertebrate species in Sweden on The IUCN Red List of Threatened Species, about 25 and 35 percent rely on logs and snags, respectively, as important habitat elements. Terrestrial snail (Gastropoda) diversity and abundance is correlated with coarse woody debris (CWD) volume (Kappes et al. 2006). That study further reported that while snails were not using CWD fragments per se, both saprophagous and predaceous snail abundance was greater on forest floor microsites near (less than 0.1 meters) CWD fragments than away (greater than 2 meters) from the fragments. They determined that leaf litter tended to accumulate to greater depths near CWD, and most likely provided more favorable microenvironmental conditions and foraging opportunities. In an allied study, Topp et al. (2006) determined that millipede (Diplopod) and woodlice (Isopod) abundance and richness were also several times greater in microsites near CWD fragments versus away from the fragments

## Conclusion

Current research indicates that temperate old-growth forests support diverse biological communities that rely on persistent keystone habitat structures or that are sensitive to microenvironmental changes caused by canopy disturbances. This literature review highlights the need for future research that elucidates old growth structure-diversity relationships in eastern North America in order to bring the state of knowledge on a par with that of

Europe and the Pacific Northwest. In addition to completing basic biotic surveys, future research should include systems-level investigations that demonstrate consequences of old-growth keystone structures on organisms occupying higher trophic levels. For instance, some evidence suggests that epiphytic and/or epixylic lichen and bryophyte mats can support more complex trophic systems that, in turn, may provide for internal natural biological controls or support greater abundance and diversity of higher-order predators. Finally, integrated stand-level experiments that account for the effects of old-growth keystone structures versus old-growth conditions, per se, are needed to determine the efficacy of management systems aimed at enhancing late successional structural habitat (chapter 13).

## References

Abrams, M. D., and D. A. Orwig. 1996. "A 300-year history of disturbance and canopy recruitment for co-occurring white pine and hemlock on the Allegheny Plateau, USA." *Journal of Ecology 84*: 353–363.

Arthur, M. S., L. M. Tritton, and T. J. Fahey. 1992. "Dead bole mass and nutrients remaining 23 years after clear-felling a northern hardwood forest." *Canadian Journal of Forest Research 23*: 1298–1305.

Batista, W. B., and W. J. Platt. 1997. "An old-growth definition for southern mixed hardwood forests." General Technical Report SRS-9. Asheville, NC: USDA Forest Service.

Berg, A., B. Ehnstrom, L. Gustafsson, T. Hallingback, M. Jonsell, and J. Weslien. 1994. "Threatened plant, animal, and fungus species in Swedish forests: distribution and habitat associations." *Conservation Biology 8*: 718–731.

Bolton, N. W., and A. W. D'Amato. 2011. "Regeneration responses to gap size and coarse woody debris within natural disturbance-based silvicultural systems in northeastern Minnesota, USA." *Forest Ecology and Management 262*: 1215–1222.

Brazee, N. J., D. L. Lindner, A. W. D'Amato, S. Fraver, J. A. Forrester, and D. J. Mladenoff. 2014. "Disturbance and diversity of wood-inhabiting fungi: effects of canopy gaps and downed woody debris." *Biodiversity and Conservation 23*: 2155–2172.

Caspersen, J. P., and M. Saprunoff. 2005. "Seedling recruitment in a northern temperate forest: the relative importance of supply and establishment limitation." *Canadian Journal of Forest Research 35*: 978–989.

Chandler, D. S. 1991. "Comparison of some slime-mold and fungus-feeding beetles (Coleoptera: Eucinetoidea, Cucujoidea) in an old-growth and 40-year-old forest in New Hampshire." *Coleopterists Bulletin 45*: 239–256.

Chokkalingam, U., and A. White. 2001. "Structural and spatial patterns of trees in old-growth northern hardwood and mixed forests of northern Maine." *Plant Ecology 156*:1311–160.

Cole, H. A., S. G. Newmaster, F. W. Bell, D. Pitt, and A. Stinson. 2008. "Influence of microhabitat on bryophyte diversity in Ontario mixedwood boreal forest." *Canadian Journal of Forest Research 38*: 1867–1876.

Cooper-Ellis, S. 1998. "Bryophytes in old-growth forests of western Massachusetts." *Journal of the Torrey Botanical Society 125*: 117–132.

Crow, T. R., D. S. Buckley, E. A. Nauertz, and J. C. Zasada. 2002. "Effects of management on the composition and structure of northern hardwood forests in Upper Michigan." *Forest Science 48*: 129–145.

D'Amato, A. W., D. A. Orwig, and D. R. Foster. 2008. "The influence of successional processes and disturbance on the structure of *Tsuga canadensis* forests." *Ecological Applications* 18: 1182–1199.

D'Amato, A. W., D. A. Orwig, and D. R. Foster. 2009. "Understory vegetation in old-growth and second-growth *Tsuga canadensis* forests in western Massachusetts." *Forest Ecology and Management 257*: 1043–1052.

D'Amato, A. W., D. A. Orwig, D. R. Foster, A. B. Plotkin, P. K. Schoonmaker, and M. R. Wagner. 2017. "Long-term structural and biomass dynamics of virgin *Tsuga canadensis-Pinus strobus* forests after hurricane disturbance." *Ecology 98*: 721–733.

Davis, M. B., ed. 1996. *Eastern Old-Growth Forests: Prospects for Discovery and Recovery*. Washington, DC: Island Press.

Dodds, K. J., and P. J. Smallidge. 1999. "Composition, vegetation and structural characteristics of a presettlement forest in western Maryland." *Castanea 64*: 337–345.

Flinn, K. M. 2007. "Microsite-limited recruitment controls fern colonization of post-agricultural forests." *Ecology 88*: 3103–3114.

Forrester, J. A., and J. R. Runkle. 2000. "Mortality and replacement patterns of an old-growth *Acer-Fagus* woods in the Holden Arboretum, northeastern Ohio." *The American Midland Naturalist 144*: 227–242.

Fraver, S., and A. S. White. 2007. "Disturbance dynamics of old-growth *Picea rubens* forests of northern Maine." *Journal of Vegetation Science 16*: 597–610.

Frego, K. A. 2007. "Bryophytes as potential indicators of forest integrity." *Forest Ecology and Management 242*: 65–75.

Garmendia-Zapata, M. A. 2015. "Influence of environmental variation on epiphytic bryophyte assemblages on sugar maple (*Acer saccharum* Marsh.) in northern and central New York." Master's thesis. Syracuse, NY: State University of New York–College of Environmental Science and Forestry.

Goebel, P. C., D. M. Hix, and A. M. Olivero. 1999. "Seasonal ground-flora patterns and site factor relationships of second-growth and old-growth south-facing ecosystems, southern Ohio." *Natural Areas Journal 19*: 12–29.

Gower, S. T., R. E. McMurtrie, and D. Murty. 1996. "Aboveground net primary production decline with stand age: potential causes." *Trends in Ecology and Evolution 11*: 378–382.

Greenidge, K. N. H. 1987. "Compositional-structural relations in old-growth forests, Cape Breton Island." *Rhodora 89*: 279–297.

Hale, C. M., J. Pastor, and K. A. Rusterholz. 1999. "Comparison of structural and compositional characteristics in old-growth and mature, managed hardwood forests of Minnesota, U.S.A." *Canadian Journal of Forest Research 29*: 1471–1489.

Hardt, R. A. 1993. "Characterization of old-growth forests in the southern Appalachian region of the United States and implications for their management." PhD diss. Athens, GA: University of Georgia.

Hardt, R. A., and W. T. Swank. 1997. "A comparison of structural and compositional characteristics of southern Appalachian young second-growth, maturing second-growth and old-growth stands." *Natural Areas Journal 17*: 42–52.

Harmon, M. E., J. F. Franklin, F. J. Swanson, P. Sollins, S. V. Gregory, J. D. Lattin, N. H. Anderson, et al. 1986. "Ecology of coarse woody debris in temperate ecosystems." *Advances in Ecological Research 15*: 133–302.

Hura, C. E., and T. R. Crow. 2004. "Woody debris as a component of ecological diversity in thinned and unthinned northern hardwood forests." *Natural Areas Journal 24*: 57–64.

Jenkins, M. A. 2007. "Vegetation communities of Great Smoky Mountains National Park." *Southeastern Naturalist 6*: 35–56.

Kappes, H., W. Topp, P. Zach, and J. Kulfan. 2006. "Coarse woody debris, soil properties and snails (Mollusca: Gastropoda) in European primeval forests of different environmental conditions." *European Journal of Soil Biology 42*: 131–146.

Keeton, W. S., A. A. Whitman, G. G. McGee, and C. L. Goodale. 2011. "Late-successional biomass development in northern hardwood-conifer forests of the northeastern United States." *Forest Science 57*: 489–505.

Kehler, D., S. Bondrup-Nielsen, and C. Corkum. 2004. "Beetle diversity associated with forest structure including deadwood in softwood and hardwood stands in Nova Scotia." *Proceedings of the Nova Scotia Institute of Science 42*: 227–239.

Kincaid, J. A. 2007. "Compositional and environmental characteristics of *Tsuga canadensis* (L.) Carr. forests in the southern Appalachian Mountains, USA." *Journal of the Torrey Botanical Society 134*: 479–488.

Lindner, D. L., H. H. Burdsall, Jr., and G. R. Stanosz. 2006. "Species diversity of Polyporoid and Corticioid fungi in northern hardwood forests with differing management histories." *Mycologia 98*: 195–217.

Lobell, S., T. Snall, and H. Rydin. 2006. "Species richness patterns and metapopulation processes – evidence from epiphyte communities in boreo-nemoral forests." *Ecography 29*: 161–182.

Martikainen, P., J. Siitonen, P. Punttila, L. Kaila, and J. Rauh. 2000. "Species richness of Coleoptera in mature managed and old-growth boreal forests in southern Finland." *Biological Conservation 94*: 199–209.

Martin, W. H. 1975. "The Lilley Cornett Woods: a stable mixed mesophytic forest in Kentucky." *Botanical Gazette 136*: 171–183.

Marx, L., and M. B. Walters. 2006. "Effects of nitrogen supply and wood species on *Tsuga canadensis* and *Betula alleghaniensis* seedling growth on decaying wood." *Canadian Journal of Forest Research 36*: 2873–2884.

Marx, L., and M. B. Walters. 2008. "Survival of tree seedlings on different species of decaying wood maintains tree distribution in Michigan hemlock-hardwood forests." *Journal of Ecology 96*: 505–513.

McCarthy, B. C. 2003. "The herbaceous layer of eastern old-growth deciduous forests." In *The Herbaceous Layer in Forests of Eastern North America*, edited by F. S. Gilliam, and M. R. Roberts, 163–176. New York: Oxford University Press.

McGee, G. G. 1998. "Structural characteristics of Adirondack northern hardwood forests: implications for ecosystem management." PhD diss. Syracuse, N.Y: State University of New York–College of Environmental Science and Forestry.

McGee, G. G. 2000. "The contribution of beech bark disease-induced mortality to coarse woody debris loads in northern hardwood stands of Adirondack Park, New York, U.S.A." *Canadian Journal of Forest Research 30*: 1453–1462.

McGee, G. G. 2001. "Stand-level effects on the role of decaying logs as vascular plant habitat in Adirondack northern hardwood forests." *Journal of the Torrey Botanical Society 128*: 370–380.

McGee, G. G., M. Cardon, and D. Kiernan. 2018. "Variation in sugar maple (*Acer saccharum* Marshall) bark and stemflow characteristics: implications for epiphytic bryophyte communities." *Northeastern Naturalist*. In Press.

McGee, G. G., D. J. Leopold, and R. D. Nyland. 1999. "Structural characteristics of old-growth, maturing and partially cut northern hardwood forests." *Ecological Applications 9*: 1316–1329.

McGee, G. G., and R. W. Kimmerer. 2002. "Forest age and management effects on epiphytic bryophyte communities in Adirondack northern hardwood forests, New York, USA." *Canadian Journal of Forest Research 32*: 1562–1576.

McMinn, J. W., and R. A. Hardt. 1996. "Accumulations of coarse woody debris in southern forests." In *Biodiversity and Coarse Woody Debris in Southern Forests*, edited by J. W.

McMinn, and D. A. Crossley, Jr., 1–9. General Technical Report SE-94. Southeast Forest Research Station, USDA Forest Service.

McMullin, R. T., P. N. Duinker, R. P. Cameron, D. H. S. Richardson, and I. M. Brodo. 2008. "Lichens of coniferous old-growth forests of southwestern Nova Scotia, Canada: diversity and present status." *The Bryologist 111*: 620–637.

Metzger, F., and J. Schultz. 1984. "Understory response to 50 years of management of a northern hardwood forest in Upper Michigan." *American Midland Naturalist 112*: 209–223.

Miller, K. M., R. G. Wagner, and S. A. Woods. 2008. "Arboreal arthropod associations with epiphytes following gap harvesting in the Acadian forest of Maine." *The Bryologist 111*: 424–434.

Mroz, G. D., M. R. Gale, M. F. Jurgensen, D. J. Frederick, and A. Clark III. 1985. "Composition, structure, and above-ground biomass of two old-growth northern hardwood stands in Upper Michigan." *Canadian Journal of Forest Research 15*: 78–82.

Muller, R. N., and Y. Liu. 1991. "Coarse woody debris in an old-growth deciduous forest on the Cumberland Plateau, southeastern Kentucky." *Canadian Journal of Forest Research 21*: 1567–1572.

Nascimbene, J., G. Thor, and P. L. Nimis. 2013. "Effects of forest management on epiphytic lichens in temperate deciduous forests of Europe – a review." *Forest Ecology and Management 298*: 27–38.

Norden, B., M. Ryberg, F. Gotmark, and B. Olausson. 2004. "Relative importance of coarse and fine woody debris for the diversity of wood-inhabiting fungi in temperate broadleaf forests." *Biological Conservation 117*: 1–10.

Odor, P., J. Heilmann-Clausen, M. Christensen, E. Aude, K. W. van Dort, A. Piltaver, I. Siller, et al. 2006. "Diversity of dead wood inhabiting fungi and bryophytes in semi-natural beech forests in Europe." *Biological Conservation 131*: 58–71.

Palmer, M. W. 1987. "Diameter distributions and the establishment of tree seedlings in the Henry M. Wright Preserve, Macon County, North Carolina." *Castanea 52*: 87–94.

Pederson, N. 2010. "External characteristics of old trees in eastern deciduous forest." *Natural Areas Journal 30*: 396–407.

Pettersson, R. B., J. P. Ball, K-E. Renhorn, P-A. Esseen, and K. Sjoberg. 1995. "Invertebrate communities in boreal forest canopies as influenced by forestry and lichens with implications for passerine birds." *Biological Conservation 74*: 57–63.

Richardson, D. H. S., and R. P. Cameron. 2004. "Cyanolichens: their response to pollution and possible management strategies for their conservation in northeastern North America." *Northeastern Naturalist 11*: 1–22.

Root, H. T., G. G. McGee, and R. A. Norton. 2007a. "Arboreal mite communities on epiphytic lichens of the Adirondack Mountains of New York." *Northeastern Naturalist 14*: 425–438.

Root, H. T., G. G. McGee, and R. D. Nyland. 2007b. "Effects of two silvicultural regimes with large tree retention on epiphytic macrolichen communities in Adirondack northern hardwoods, New York, USA." *Canadian Journal of Forest Research 37*: 1854–1866.

Roovers, L. M., and S. R. Shifley. 1997. "Composition and dynamics of Spitler Woods, an old-growth remnant forest in Illinois (USA)." *Natural Areas Journal 17*: 219–232.

Rubino, D. L., and B. C. McCarthy. 2003. "Composition and ecology of macrofungal and myxomycete communities on oak woody debris in a mixed-oak forest of Ohio." *Canadian Journal of Forest Research 33*: 2151–2163.

Rydin H., M. Diekmann, and T. Hallingback. 1997. "Biological characteristics, habitat associations, and distribution of macrofungi in Sweden." *Conservation Biology 11*: 628–640.

Scheller, R. M., and D. J. Mladenoff. 2002. "Understory species patterns and diversity in old-growth and managed northern hardwood forests." *Ecological Applications 12*: 1321–1343.

Schowalter, T. D. 1995. "Canopy arthropod communities in relation to forest age and alternative harvest practices in western Oregon." *Forest Ecology and Management 78*: 115–125.

Selva, S. B. 1994. "Lichen diversity and stand continuity in northern hardwoods and spruce-fir forest of northern New England and western New Brunswick." *Bryologist 97*: 424–429.

Shields, J. M., C. R. Webster, and L. M. Nagel. 2007. "Factors influence tree species diversity and *Betula alleghaniensis* establishment in silvicultural openings." *Forestry 80*: 293–307.

Spribille, T., G. Thor, F. L. Bunnell, T. Goward, and C. R. Bjork. 2008. "Lichens on dead wood: species-substrate relationships in the epiphytic lichen floras of the Pacific Northwest and Fennoscandia." *Ecography 31*: 741–750.

Stubbs, C. S. 1989. "Patterns of distribution and abundance of corticolous lichens and their invertebrate associates on *Quercus rubra* in Maine." *The Bryologist 92*: 453–460.

Stokland, J. N., and K-H. Larsson. 2011. "Legacies from natural forest dynamics: different effects of forest management on wood-inhabiting fungi in pine and spruce forests." *Forest Ecology and Management 261*: 1707–1721.

Tews, J., U. Brose, V. Grimm, K. Tielborger, M. C. Wichmann, M. Schwager, and F. Jeltsch. 2004. "Animal species diversity driven by habitat heterogeneity/diversity: the importance of keystone structures." *Journal of Biogeography 31*: 79–92.

Topp, W., H. Kappes, J. Kulfan, and P. Zach. 2006. "Distribution pattern of woodlice (Isopoda) and millipedes (Diplopoda) in four primeval forests of the western Carpathians (central Slovakia)." *Soil Biology and Biogeochemistry 38*: 43–50.

Tyrrell, L. E., G. J. Nowacki, T. R. Crow, D. S. Buckley, E. A. Nauertz, J. N. Niese, J. L. Rollinger, and J. C. Zasada. 1998. "Information about old growth for selected forest type groups in the eastern United States." General Technical Report NC-197. North-Central Forest Research Station. USDA Forest Service.

Ulyshen, M. D. 2011. "Arthropod vertical stratification in temperate deciduous forests: implications for conservation-oriented management." *Forest Ecology and Management 261*: 1471–1489.

Vanderwel, M. C., H. C. Thorpe, J. L. Shuter, J. P. Caspersen, and S. C. Thomas. 2008. "Contrasting downed woody debris dynamics in managed and unmanaged northern hardwood stands." *Canadian Journal of Forest Research 38*: 2850–2861.

Wiklund, K., and H. Rydin. 2004. "Colony expansion of *Neckera pennata*: modelled growth rate and effect of microhabitat, competition, and precipitation." *Bryologist 107*: 293–301.

Wyatt, J. L., and M. R. Silman. 2010. "Centuries-old logging legacy on spatial and temporal patterns in understory herb communities." *Forest Ecology and Management 260*: 116–124.

Zeran, R. M., R. S. Anderson, and T. A. Wheeler. 2007. "Effects of small-scale forest management on fungivorous Coleoptera in old-growth forest fragments in southeastern Ontario, Canada." *Canadian Entomologist 139*: 118–130.

Ziegler, S. S. 2000. "A comparison of structural characteristics between old-growth and post-fire second-growth hemlock-hardwood forests in Adirondack Park, New York, U.S.A." *Global Ecology & Biogeography 9*: 373–389.

# Chapter 12

# Eastern Old-Growth Forests under Threat: Changing Dynamics due to Invasive Organisms

*John S. Gunn and David A. Orwig*

One of the most serious threats facing forests in North America is the introduction of plants, pests, and pathogens from other countries (Dukes et al. 2009). This chapter will discuss the mechanisms and impacts of the invasion of these organisms into old forest stands and summarize the stand and landscape factors that make old-growth forests susceptible to these threats. Due to increases in the global movement of products around the world, introductions of plants, pests, and pathogens are increasing and can significantly alter the structure, composition, function, and aesthetics of forests and the habitat they provide (Lovett et al. 2016). Many introduced species pose little or no threat to forested ecosystems, but there is a growing number of species that have already caused serious damage or pose serious threats to our old-growth forests (tables 12-1 and 12-2). While we recognize that native pests and pathogens also exist and can negatively impact forests, we focus here on introduced species that will or could directly impact eastern old-growth forests and the common tree species found therein.

Late successional and old-growth forests typically have been thought to be resistant to nonnative plant invasions because many of the most common invasive plant species are not tolerant of the shady conditions prevalent in older forest stands. However, of the nonnative plants considered to be invasive in New England, 63 percent are considered shade tolerant (Martin et al. 2009). Many more species have moderate shade tolerance (table 12-1) and could present a risk to disturbed old-growth forests where treefall gaps or other small openings are created through tree mortality due to disease, wind, and insect damage. While stands of old growth may have shady, closed-canopy conditions, much of the remaining old-growth forests exist as relict tracts in fragmented landscapes that create conditions favorable to

TABLE 12-1. Important nonnative invasive plant species for old-growth forest impacts in the eastern United States.

| Species | Common Name(s) | Eastern US Extent[1] | Shade Tolerance[2] | Habit[1] | Impact |
|---|---|---|---|---|---|
| *Acer platanoides* | Norway maple | Northern Virginia to Maine | 4.2 | Hardwood Tree | Overstory competition, allelopathic exclusion of native understory and regeneration |
| *Alliaria petiolata* | Garlic mustard | Western North Carolina to Maine | Tolerant (Rodgers et al. 2008) | Forbs/ Herbs | Exclusion of native understory and regeneration |
| *Berberis thunbergii, Berberis vulgaris* | Japanese barberry; common barberry | Georgia to Maine; Northern Virginia to Maine | 1.5; 1.93 (but both species are typically considered by others as tolerant to very tolerant (Swearingen 2008) | Shrub or Subshrub | Exclusion of native understory and regeneration |
| *Celastrus orbiculatus* | Oriental bittersweet | Georgia to Maine | Tolerant (National Park Service) | Vine | Damage to native plants by constricting and girdling stems; exclusion of native understory and regeneration |
| *Elaeagnus umbellate; Eleagnus angustifolia* | Autumn olive; Russian olive | Northern Florida to Maine (more common than E. angustifolia in the eastern US); Florida to Maine (highest frequency is New Jersey to New Hampshire) | 1.35 (*E. angustifolia*) | Shrub or Subshrub; Shrub or Subshrub Hardwood Tree | Exclusion of native understory and regeneration |
| *Euonymus alatus* | Burning bush | Western North Carolina to Maine | 4.33 | Shrub or Subshrub | Exclusion of native understory and regeneration |
| *Lonicera japonica; Lonicera maakii* | Japanese honeysuckle; Amur honeysuckle | Florida to Southern Maine; Northern Virginia to Maine (less common than *L. japonica*) | 3.3–3.75 (other nonnative *Lonicera* spp.) | Shrub or Subshrub | Exclusion of native understory and regeneration |

*continued on next page*

TABLE 12-1. *continued*

| Species | Common Name(s) | Eastern US Extent[1] | Shade Tolerance[2] | Habit[1] | Impact |
|---|---|---|---|---|---|
| *Rhamnus cathartica; Rhamnus frangula* | Common (European) buckthorn; Glossy buckthorn | Delaware to Maine; Delaware to Maine | 1.93 (*R. cathartica*) (Zouhar 2011 Tolerant but grows more quickly in intermediate sunlight); Intolerant (Gucker 2008) | Shrub or Sub-shrub; Shrub or Sub-shrub Hard-wood Tree | Exclusion of native understory and regeneration |
| *Rosa multiflora* | Multi-flora rose | Northern Florida to Maine | Moderately Tolerant (Dlugo et al. 2015) | Shrub or Sub-shrub | Exclusion of native understory and regeneration |

[1] Swearingen, J., and C. Bargeron. 2016. "Invasive Plant Atlas of the United States." University of Georgia Center for Invasive Species and Ecosystem Health. http://www.invasive plantatlas.org/

[2] Numerical values are from: Ninemets, Ülo, and Fernando Valladares. 2006. "Tolerance to Shade, Drought, and Waterlogging of Temperate Northern Hemisphere Trees and Shrubs." *Ecological Monographs* 76: 521–547. www.jstor.org/stable/27646060. The five-level scale used for shade tolerance (1 = very intolerant; 2 = intolerant; 3 = moderately tolerant; 4 = tolerant; 5 = very tolerant). Citations for other shade tolerance classifications are provided.

Gucker, C. L. 2008. "*Frangula alnus*." Fire Effects Information System [Online]. Fire Sciences Laboratory (Producer), Rocky Mountain Research Station, U.S. Department of Agriculture, Forest Service. Accessed October 13, 2017. http://www.fs.fed.us/database/feis.

National Park Service. "Oriental Bittersweet." Accessed October 13, 2017. https://www.nps .gov/plants/alien/pubs/midatlantic/ceor.htm.

Rodgers, V. J., K. A. Stinson, A. C. Finzi. 2008. "Ready or not, garlic mustard Is moving in: *Alliaria petiolata* as a member of eastern North American forests." *BioScience* 58: 426–436. https://doi.org/10.1641/B580510.

Swearingen, J. 2009. "Japanese Barberry (*Berberis thunbergii*)." Weed US Database of Plants Invading Natural Areas in the United States. Accessed October 13, 2017. http://www.inva sive.org/weedus/subject.html?sub = 3010.

Zouhar, K. 2011. "*Rhamnus cathartica, R. davurica*." Fire Effects Information System [Online]. Fire Sciences Laboratory (Producer), Rocky Mountain Research Station, U.S. Department of Agriculture, Forest Service. Accessed October 13, 2017. http://www.fs.fed .us/database/feis.

invasion (chapters 2, 3, 4, 6, and 7). Within intact forest landscapes, likely future increases in temperatures and frequency of stand-altering disturbances put old-growth stands at greater risk of invasion by nonnative plant species that could alter forest structure and threaten the older-age classes that create old-growth conditions.

## Impacts of Invasive Plants on Old-Growth Stands

Nonnative invasive plant species pose a threat to the persistence and recruitment of old-growth forests. The primary mechanisms for these impacts are through the suppression of regeneration of native species, alteration of forest structure, delay of forest succession, and reduced fitness and growth of native species (Vilà et al. 2011). Persistence of an old-growth forest community requires new cohorts of trees to be established in the understory to eventually replace members of the oldest cohorts. This is necessary to meet the process-based "true old growth" definition of Oliver and Larson (1996) where none of the trees in a stand remain from the first cohort initiated by a stand-replacing disturbance. Without replacement from below, the species composition of the old-growth stand would not be maintained, and the forest structural elements accrued through the stand development process would be diminished. Therefore, maintaining the capacity for a stand to regenerate itself, particularly through small gap dynamics in eastern forests, is critical for maintaining old-growth forest structure. Establishment of nonnative invasive plant species in the understory of these stands presents a long-term risk to the persistence of existing old-growth forest communities.

There is a great deal of evidence that shows many nonnative invasive plants do indeed suppress regeneration and can exclude native species altogether. For example, in New Hampshire, Lee et al. (2016) found that glossy buckthorn (*Rhamnus frangula*) inhibits regeneration of white pine, causes a decline in seedling density, and can suppress ground-level plant species abundance (Frappier et al. 2003). Burning bush (*Euonymus alata*), which was introduced to North America from Asia in the 1860s, poses a threat to old-growth forest stands because it tolerates deep shade and spreads through root suckers and bird-dispersed seeds (Martin et al. 2009). Similarly, multiflora rose (*Rosa multiflora*) can invade interior forests and form dense thickets that preclude regeneration of native tree species (Kurtz 2013). Herbaceous species like garlic mustard (*Alliaria petiolate*) can also inhibit germination of native seeds through allelopathic contamination

TABLE 12-2. Introduced forest insects and pathogens in eastern North America that could impact old-growth forests, with potential hosts and impacts. Included are the common name, scientific name, date introduced, host tree(s), and observed or predicted impacts.

| Common name | Scientific name | Date intro'd | Hosts | Impacts |
|---|---|---|---|---|
| **Insect pests** | | | | |
| European gypsy moth | *Lymantria dispar dispar* L. | 1869 | Many hosts, including oaks, aspens, willows, and birches | Periodic outbreaks cause defoliations and can sometimes kill hosts |
| Emerald ash borer (EAB) | *Agrilus planipennis* Fairmaire | 2002 | All North American *Fraxinus* species | Most ash trees succumb. Some species of ash appear to have limited resistance. |
| Hemlock woolly adelgid (HWA) | *Adelges tsugae* Annand | 1950s | Eastern and Carolina hemlock | 90+ percent mortality in most affected stands |
| Balsam woolly adelgid (BWA) | *Adelges piceae* Ratzeburg | 1900 | Most true fir species (*Abies*) in North America | Widespread impacts on firs; severe mortality of Fraser fir on southern Appalachian mountaintops |
| Asian long-horned beetle (ALB) | *Anoplophora glabripennis* Motschulsky | 1990s | Woody vegetation in 15 families, especially maples, elms, and willows | Severe impacts possible in forest landscapes; eradication being attempted |
| Winter moth | *Operophtera brumata* L. | 1940s | Many species including oaks, maples, cherries, | Severe impacts on hosts in southeastern New England |
| European wood wasp | *Sirex noctilio* | 2004 | *Pinus* spp. | Modest impacts so far in United States |
| **Pathogens** | | | | |
| Chestnut blight | *Cryphonectria parasitica* (Murrill) Barr. | 1904 | American chestnut, chinkapin | Virtually eliminated mature chestnuts |
| Beech bark disease (scale insect + fungus) | *Cryptococcus fagisuga* Lindinger + *Nectria coccinea* var. *faginata* (Pers.) Fr. | 1890s | American beech | Severely reduces mature beeches; often replaced by dense thickets of root sprouts |

*continued on next page*

TABLE 12-2. *continued*

| Common name | Scientific name | Date intro'd | Hosts | Impacts |
|---|---|---|---|---|
| | | **Pathogens** | | |
| Dutch elm disease | *Ophiostoma ulmi* (Buisman) Nannf. & *O. novo-ulmi* Brasier; vectored by several insects including *Scolytus multi-striatus* and *S. schevyrewi* | 1933 | American elm; other native elms, (e.g., red or slippery elm are more resistant) | Most large elms killed; some elms remain—although reduced in number and size—in riparian woodlands |
| White pine needle damage | *Mycosphaerella dearnessii, Lophopha-cidium dooksii* and *Bifusella linearis.* | Some early 1900s, others 2006 | Various pine hosts, especially eastern white pine | Crown decline |
| Sudden oak death (SOD) | *Phytophthora ramorum* S. Werres, A.W.A.M. de Cock | 1990s | >100 spp., some eastern oaks vulnerable | Vulnerable hosts often succumb, while other hosts show minor impacts |

of the soil substrate (Prati and Bossdorf 2004). There are many other examples, but the outcomes are the same: the old-growth forest community composition will be altered over time as regeneration of tree species cannot occur in the presence of nonnative invasive understory species.

The lack of regeneration can prevent the persistence of old-growth species, but there are other invasive species that physically alter forests over time. An important example is oriental bittersweet (*Celastrus orbiculatus*), a vine which is another high-risk species to eastern old-growth forests because it can tolerate shade, overtop trees, and increase susceptibility of canopy trees to ice damage and windthrow (Dukes et al. 2009). Oriental bittersweet, which can easily reach heights of 40 to 60 feet, typically establishes on forest edges but also tolerates shade (Leicht and Silander 2006). Over time it can cause significant damage and mortality on edge trees, eventually becoming the de facto tree canopy. The species is currently common throughout eastern North America, except for the most northern portion of the northeastern United States and Canada. However, climate change projections show a high likelihood that favorable conditions for oriental bittersweet will develop in northern New England (Leicht 2005). Again, given the fragmented nature of old-growth forests in the east, species such as oriental bittersweet present a significant threat to remaining stands.

## Impacts of Invasive Pests on Old-Growth Stands

Currently, eastern old-growth forests are infested or threatened with many invasive insect pests and pathogens (table 12-2). Nearly a century ago, we glimpsed the changes a single pathogen can create as the chestnut blight (*Cryphonectria parasitica*) dramatically altered the structure and composition of forests throughout the eastern United States. For example, a reconstruction of the history of old-growth stands in North Carolina showed that more than 20 percent of the trees greater than 15 centimeters in diameter were American chestnut (*Castanea dentata*) prior to the blight (Lorimer 1980).

As invasive pests and pathogens spread across the landscape, even what appear like isolated stands far away from development or population centers are at risk of becoming infested. Insects can reach new forests by flying to them, being transported by wind, or hitching a ride on a bird's feather or piece of logging equipment. In addition, some of the most damaging insects that bore into wood can be moved into a new region from elsewhere by being transported as larvae in firewood. If the wood is left behind at a campground or cabin and not burned, the developing larvae can indeed eat their way out and find new trees nearby to infest.

Once invasive insects reach an old-growth forest, the time it takes until they impact individual trees or an entire stand will vary with the insect type (defoliator, wood borer, sap sucker), host specificity (generalist, attacking several different host trees, or preying on a single host type), and virulence (Lovett et al. 2006). Eastern old-growth forests consist of a combination of mainly deciduous hardwood and evergreen conifer species, so an entire forest is unlikely to be attacked or killed by a single pest. The diversity of these old-growth mixes, however, means that a greater variety of insects and pathogens have potential hosts living in these forests.

Throughout the eastern United States, a high percentage of remaining old-growth forest consists of shade-tolerant eastern hemlock (*Tsuga canadensis*) (Davis 1996; see previous chapters). This long-lived, iconic evergreen, which can reach ages of up to 500 years, is currently under assault from an invasive insect from Japan, the hemlock woolly adelgid (HWA; *Adelges tsugae*). This tiny insect, half a millimeter in size, has two prolific generations per year and feeds on all sizes and age classes of hemlock (Mc-Clure 1987). Within a span of a decade, HWA can bring down the largest and oldest hemlock on the landscape by essentially depleting the tree's reserves. "Gray ghosts" of standing dead and dying hemlocks dot the hillsides of southern Appalachian and mid-Atlantic forests (plate 7), attesting to the effectiveness of HWA at eliminating its host (and the important information contained within their annual ring-widths). As hemlock declines and dies, it is often replaced by a suite of hardwood species like black birch (*Betula lenta*), red maple (*Acer rubrum*), and red oak (*Quercus rubra*), as well as white pine (*Pinus strobus*) (Orwig et al. 2012).

In addition to tree mortality, the elimination of trees by HWA also leads to a cascade of other short- and long-term impacts on structure and function of these ecosystems. The cool, dark microenvironment created by hemlock is transformed into brighter, warmer conditions that often lead to changes in soil respiration, decomposition, and nutrient cycling (Orwig et al. 2013). Since hemlock often grows along streams, the removal of hemlock by HWA alters riparian areas (see chapter 9), leading to warmer stream temperatures, changes in litter inputs, and new assemblages of macroinvertebrates feeding on the litter (Ellison et al. 2005). Hemlock loss is also likely to affect a wide variety of wildlife species that spend a portion of their life cycle in or near hemlock habitat, including birds like Black-throated Green Warblers (*Dendroica virens*), Blackburnian Warblers (*Dendroica fusca*), Acadian Flycatchers (*Empidonax virescens*) (Tingley et al. 2002), several owl species, Northern Goshawks (*Accipiter gentiles*), white-tailed deer (*Odocoileus virginanus*), porcupines (*Erethizon dorsatum*) (Ward et al. 2004), aquatic invertebrates, and brook trout (*Salvelinus fontinalis*) (Snyder et al. 2002).

American beech (*Fagus grandifolia*) is another common old-growth, late successional tree species being negatively impacted by an insect or pathogen pest in the northeastern United States, in this case by the non-native beech bark disease (BBD), which involves an introduced scale insect (*Cryptococcus fasigua*) and a pathogenic fungus (*Neonectria* spp.) (Houston 1994). The feeding of the scale insect on the thin tree bark creates entry points for the fungus, which leads to bark cankers, reduced tree growth, and tree mortality within a decade or so (Houston 1994). Unlike HWA, BBD does not typically eliminate all beech from forests, but usually the largest individuals, shifting the size distribution towards smaller stems, including dense thickets resulting from vegetative sprouting (Forrester et al. 2003). In the Catskill Mountains of New York State, for example, most large beech are now dead and trees greater than 50 centimeters in diameter are rare (Griffin et al. 2003). A recent study revealed that large beech, greater than 90 centimeters DBH, common previously only in old-growth stands, are virtually absent from the 2.3 million square kilometers range of the American beech (Garnas et al. 2011). Beech declines are also sometimes accompanied by shifts in species composition, with sugar maple (*Acer saccharum*) increasing on sites with higher soil calcium, leading to long-term shifts in carbon and nitrogen cycling (Lovett et al. 2006). Beech bark disease, combined with drought in 2002–2003, led to a significant mortality event in old-growth stands previously dominated by American beech in the Big Reed Forest Reserve owned by The Nature Conservancy in northern Maine (Gunn et al. 2014).

An extremely damaging insect that recently arrived in eastern forests is the emerald ash borer (EAB; *Agrilus planipennis*), which has spread rapidly across the eastern United States since being introduced from Asia in the early 2000s and has killed millions of ash in the process (Poland and McCullough 2006). This pest, which feeds on the food-transporting layer of phloem just under the bark, specializes on ash species (*Fraxinus*) and kills them within several years of initial infestation. White ash (*F. americana*) is found in several old-growth forest types and is highly susceptible to EAB. Since stand-level mortality rates can be as high as 99 percent (Knight et al. 2013), old-growth forests containing ash are highly threatened. The recent discovery of extremely old white ashes in the Adirondack Mountains of New York State (290 to 305 years old; Neil Pederson, pers. comm.) highlights that we have much to learn about ash dynamics in old forests, giving an urgency to old-growth studies due to impending EAB infestations.

Eastern old-growth forests are threatened by another invader from China, which has a voracious appetite for a variety of hardwood species: the Asian longhorned beetle (ALB; *Anoplophora glabripennis*). Currently relegated to quarantined areas of New York State, Massachusetts, and Ohio, this large beetle prefers various maple (*Acer*), birch (*Betula*), elm (*Ulmus*), and poplar (*Populus*) species (Eyre and Haack 2017). With the exception of the Massachusetts infestation, ALB has been found in mostly urban to suburban trees. However, Dodds and Orwig (2011) showed that ALB could indeed infest and migrate into forested locations, setting the stage for potential invasion into intact old-growth, maple-dominated forests. Since both ALB and EAB larvae can survive for more than a year within a tree bole or branch, the movement of firewood is a constant threat for new infestations and should be prevented at all costs. The good news is that the Canadian Forest Service has demonstrated that aggressive ALB eradication programs following detection can be successful (Smith et al. 2009).

One of the most important functions of old-growth forests that will be lost with the elimination of dominant trees by any of these plant, insect, or fungal invasive species is the storage of large amounts of carbon (Gunn et al. 2014). The ability to store high levels of carbon will likely be altered for a century or more, as younger replacement species will take that long to accumulate carbon at those levels (chapter 14).

## Ecological Mechanisms of Invasion and Resistance

Some old-growth stands are much more vulnerable to damage by invasive species than others. A combination of abiotic and biotic factors drives invasion and resistance dynamics in old-growth forests (Lodge 1993). First, environmental conditions dictated by climate need to be favorable for invasive plants in terms of temperature and soil moisture (Ibáñez et al. 2009) and for invasive insects in terms of overwinter survivorship (Dukes et al. 2009). Second, the number of invasive insects is highly correlated with the number of host species (i.e., propagule pressure) in the northeastern United States (Liebhold et al. 2013), and species composition varies across sites. Finally, abiotic factors such as road networks and human settlement patterns are also important factors affecting invasive plant establishment (Allen et al. 2013). Thus, the biotic resistance of a community will be driven by characteristics at both the *landscape* and *stand* scales (Johnson et al. 2006; Allen et al. 2013).

### Landscape Structure

Spatial patterns on the landscape are crucial to whether an old-growth forest stand is resistant or vulnerable to plant invasion. Landscape composition (the elements in the landscape) and configuration (the arrangement of these elements), collectively referred to as landscape structure, can influence the movement and spread of invasive plants (Rodewald and Arcese 2016). The concept of "landscape resistance" of forests is a useful way to discuss conditions in which some invasive plants species are less likely to spread because of biotic or anthropogenic barriers (Gonzales and Gergel 2007). The influence of landscape structure on invasive plant movement and incursion into forested areas will likely vary with human settlement patterns. This is a result of the historical context since many invasive plants were historically brought in as ornamentals in a developing landscape, but it is also a pattern that continues to the current day, as patterns of human demography are strongly linked to forest cover loss.

Reduced forest cover is associated with increased human population density, which generally follows a north (low density) to south (high density) gradient in the northeastern United States. Changes in forest cover expose new forest areas to invasive plant species when these changes involve activities such as ground disturbance for road and home building. In a New Hampshire study of early successional habitat, Johnson et al. (2006) found that the amount of human-altered land, particularly land used for agriculture in the surrounding landscape, was a strong factor in colonization and spread of several invasive species. While this is not specific to old-growth forests, the pattern has been observed by others in New England (McDonald et al. 2008; Allen et al. 2013) and is a consistent pattern globally (Rodewald and Arcese 2016). Invasive woody plant species richness is usually highest along the edges that are created when forests become fragmented (Johnson et al. 2006; Allen et al. 2013; Chapman et al. 2015), which likely reflects the general shade intolerance of many invasive plants and the use of edges by vectors such as birds. This dynamic makes it important to understand what constitutes an "edge" in a managed forest context, where the landscape contrasts are not as stark as between a forest and an agricultural field. In a study of US Forest Service Forest Inventory and Analysis plots, Schulz and Gray (2013) found greater occupancy of invasive plant species at forest edge plots versus intact forest plots in the northeastern United States. In a study of hardwood forests in Wisconsin, Watkins et al. (2003) found that nonnative plants were more abundant within 15 meters of unpaved roads than in interior forest. Flory and Clay (2006) found similar results for hardwood sites in Indiana. In addition to providing suitable habitat for the establishment of invasive plants, roads can facilitate movement of invasive plant species across the landscape and create opportunities for incursion into relatively intact forest areas (Parendes and Jones 2000).

Much recent work has focused on invasive species dynamics in forest patches embedded within an agricultural context (e.g., Johnson et al. 2006; Rodewald and Arcese 2016). This is appropriate considering the linkage between human settlement, agricultural land use, and invasive species establishment (Allen et al. 2013). But little is known about patterns of forest invasion in managed forest landscapes where human settlements have a smaller footprint. Invasion dynamics are likely to be different in these less developed landscapes of eastern North America where timber harvest and related activities are the primary disturbance factors. Dynamics in these landscapes are the most relevant to old-growth forests, which generally occur as part of larger forest tracts. As an example of such landscape dynamics, power line rights-of-way (Wagner et al. 2014) and unconventional gas development (Barlow et al. 2017) can act as sources of invasion and a vector for accessing interior forest areas where old-growth forests could be exposed to these new threats. It is also possible that riparian corridors are susceptible to invasion and can act as vectors for invasive species to gain access to interior forests (Liendo et al. 2015).

Like plants, the spread of pests and pathogens across a region is strongly influenced by the arrangements of forest stands on the landscape, consisting of various species that may be hosts to one or more of these invasives. Entire hillsides and valleys dominated by a host species will be much more susceptible to spread by an invasive insect or pathogen than if a few isolated host trees in a ravine were surrounded by miles of nonhost species. Since some pests like HWA are spread by wind and birds (McClure 1990), the location of a stand and host species with respect to prevailing winds and migratory bird flyways is important in determining how rapidly a new forest will be infested. Cold temperatures below -30°C are often limiting to overwinter survivorship of insects like HWA, EAB, ALB and scale insects that transmit BBD (Dukes et al. 2009; DeSantis et al. 2013), and, therefore, habitats and microenvironmental locations on the landscape that experience colder winter temperatures (high elevation, higher latitude, inland locations, or cold valleys) may be refuges for old-growth forests that avoid invasive insect arrival longer than in other areas where cold temperatures are less limiting.

## Stand Structure

The abundance of a particular invasive species within a landscape will obviously influence the invasion risk for every old-growth stand in that landscape. Therefore, the landscape context of forest stands is crucial to long-term susceptibility to invasion. The structure of individual forest stands, however, will have a more proximate influence on invasion resistance of forests. The biological characteristics of invasive plant species, such as the degree of shade tolerance, regeneration requirements, and dispersal mechanisms, interact with existing native forest communities at a stand level. As described earlier, intact forests with closed canopies have generally been thought to be highly resistant to invasion by nonnative plants because of the tendency for invasives to be shade intolerant (Martin et al. 2009). In fact, a large proportion of New England's known invasive plants are very shade tolerant and capable of invading stands under dense forest canopy cover (Martin et al. 2009). New England's late successional and old-growth forests typically have a closed-canopy structure with gaps created by individual or small group tree mortality (Gunn et al. 2014). Burnham and Lee (2010) found the invasive shrub, glossy buckthorn (*Frangula alnus*), in New Hampshire was 96 times more abundant in forest canopy gaps than in undisturbed forest. However, McCarthy et al. (2001) found that old-growth stands with low levels of disturbance in the understory community (e.g., because of individual trees falling) were resistant to invasion, even for small patches embedded in a matrix of land cover types with abundant invasive plant species. Since forest management activities (e.g. harvesting) create canopy gaps, understanding and predicting the susceptibility of eastern old-growth forest communities to invasion may help prioritize management efforts (Stotz et al. 2016).

Davis et al. (2000) theorized that plant communities will be "more susceptible to invasion whenever there is an increase in the amount of unused resources". This suggests that, because old-growth forests are defined essentially by a lack of "stand-replacing" or other significant disturbances, they should inherently be less susceptible to invasion. However, even single tree or slightly larger gaps created by natural mortality events can create enough light and nutrient availability for invasion in old-growth stands. Creating more edge structure in this way increases resource availability (generally in the form of light) within a certain distance of the edge. This then creates an opportunity for invasive plants to become established and potentially expand through processes such as allelopathy, vining, and rapid height growth. In small old-growth forest remnants where the edge-to-area ratio would be greater than within an extensive forest matrix, these stands would clearly be more susceptible to invasion, particularly in more developed landscapes where invasive plant density and diversity are greatest (Allen et al. 2013). Old-growth forest conservation efforts that address threats from invasive plants will need to identify invasibility thresholds in varying disturbance contexts, such as single or multiple tree canopy gaps, whether natural or the result of timber harvests. But these thresholds may vary, depending upon the forest type and the characteristics of the invasive plant posing the threat.

The influence of stand composition and structure varies widely in determining whether an invasive insect or pathogen can become established and/or spread and negatively impact the forest. Certain insects like HWA can feed on all tree sizes and ages, so the actual composition of the forest matters little once the HWA reaches a particular area. In contrast, BBD affects mainly the larger size classes of its beech host (Garnas et al. 2011). In between are insects like ALB, which preferentially infest larger trees (Dodds and Orwig 2011) but then feed on smaller individuals once firmly established in the stand. Thus, the ages and sizes of trees will influence forest impact for some but not all invasive pests. Variability in other stand factors, such as bark chemistry, also plays an important role with BBD, as trees with lower bark phosphorous are more susceptible to the damaging pathogenic fungi (Cale et al. 2015).

## Climate Change

Climate change will likely influence the frequency, duration, and intensity of natural disturbances, such as hurricanes and other wind events, native and exotic pathogen, and insect spread, and the frequency of ice storms (Dale et al. 2001). Such disturbances can increase the likelihood of invasion by nonnative plants if they occur near where such plants are already established. As with native species, there will be winners and losers as the climate warms. Recent work by Merow et al. (2017) has shown that some invasive species, like Japanese barberry (*Berberis thunbergii*), which can take over an understory, will be enhanced; others, like garlic mustard (*Alliaria petiolate*), that seem to be thriving currently, will not necessarily continue this trend—except perhaps in far northern New England and southern Canada.

A warming climate will also continue to facilitate the movement of nonnative insect pests such as the hemlock woolly adelgid, whose main limiting factor is extreme winter cold (Skinner et al. 2003). Warmer winters not only increase the number of surviving HWA, they can also extend the season for egg laying and subsequent dispersal by up to several months (Leppanen and Simberloff 2016). Warming will also likely lead to higher overwinter survival of the scale insect (*Cryptococcus fagisuga*) that helps to transmit BBD. Increases of both BBD and HWA could have large negative impacts on the remaining northern old-growth forests in eastern

North America, as hemlock and beech dominate many of them (Davis 1996; Gunn et al. 2014). Where such pests result in significant disturbance events, invasive plants will likely benefit and move further into intact forest systems. Such looming synergies highlight the need for early detection and removal efforts for invasive plant species.

## Conclusion

Without question, invasive plants, pests, and pathogens represent a significant threat to remaining old-growth forests and create challenges for efforts to restore old forest structure. The best strategy to minimize damage associated with invasive plants, pests, and pathogens is to prevent them from getting there in the first place. This is easier said than done. Once an invasive plant or pest is firmly established in an area, it can be close to impossible to eliminate. Chemically treating individual trees or small cohorts to protect against invasive insects is possible for the examples listed above but is too costly and logistically challenging to protect larger groups of trees at the stand level or larger. Biocontrol efforts, for the major insects affecting forests previously discussed, are ongoing, but they face steep challenges and the potential for negative, unintended consequences.

We know that disturbances create opportunities for invasive plant species to colonize new areas. We also know that proximity to existing propagules heightens the risk of invasion following a disturbance. Using these two factors, old-growth forest and preserve managers must be able to identify the invasive species within the region that are most likely to present a risk. Online mapping tools, such as the Invasive Plant Atlas of New England, with data derived from formal assessments and citizen science, can give an initial perspective on potential risks, but these assessments are not systematic and must only be the starting point. Inventory and monitoring efforts should include at least a qualitative assessment of understory species present within the management unit. If timber harvesting or other management activities that create vegetation and soil disturbance are part of a management plan for old-growth and adjacent forests (see chapter 13), managers must account for the proximity of invasive plant species and manage disturbed sites, such as access roads and log landings, accordingly. This may mean replanting native species in disturbed areas or planning activities during snow cover to avoid soil disturbance. If invasive plants, pests, or pathogens are already present within or adjacent to old-growth stands, then appropriate control efforts may be needed to prevent their further spread. Significant resources have been developed that provide professionals with information to help them control invasive plant species. In the United States, these resources are typically available at the state level through Cooperative Extension, Natural Resources Conservation Service, and state departments of agriculture. In Canada, resources are available through the Invasive Species Centre and at the provincial level (e.g., Early Detection and Rapid Response (EDRR) Network Ontario).

## References

Allen, J. M., T. J. Leininger, J. D. Hurd, D. L. Civco, A. E. Gelfand, and J. A. Silander. 2013. "Socioeconomics drive woody invasive plant richness in New England, USA through forest fragmentation." *Landscape Ecology 28*: 1671–1686. doi:10.1007/s10980-013-9916-7.

Barlow, K. M., D. A. Mortensen, P. J. Drohan, and K. M. Averill. 2017. "Unconventional gas development facilitates plant invasions." *Journal of Environmental Management 202*: 208–216. doi:10.1016/j.jenvman.2017.07.005.

Burnham, K. M., and T. D. Lee. 2010. "Canopy gaps facilitate establishment, growth, and reproduction of invasive *Frangula alnus* in a *Tsuga canadensis* dominated forest." *Biological Invasions 12*: 1509–1520. doi:10.1007/s10530-009-9563-8.

Cale, J. A., S. A. Teale, M. T. Johnston, G. L. Boyer, K. A. Perri, and J. D. Castello. 2015. "New ecological and physiological dimensions of beech bark disease development in aftermath forests." *Forest Ecology and Management 336*: 99–108.

Chapman, J. I., A. L. Myers, A. J. Burky, and R. W. McEwan. 2015. "Edge effects, invasion, and the spatial pattern of herb-layer biodiversity in an old-growth deciduous forest fragment." *Natural Areas Journal 35*: 439–451. doi:10.3375/043.035.0307.

Dale, V. H., L. A. Joyce, S. Mcnulty, R. P. Neilson, M. P. Ayres, M. D. Flannigan, P. J. Hanson, et al. 2001. "Climate change and forest disturbances." *BioScience 51*. doi:10.1641/0006-3568(2001)051[0723:CCAFD]2.0.CO;2.

Davis, M. A., J. P. Grime, and K. E. N. Thompson. 2000. "Fluctuating resources in plant communities : a general theory of invasibility." *Journal of Ecology 88*: 528–534.

Davis, M. B., ed. 1996. *Eastern Old-Growth Forests: Prospects for Rediscovery and Recovery*. Washington, DC: Island Press.

DeSantis, R. D., W. K. Moser, D. D. Gormanson, M. G. Bartlett, and B. Vermunt. 2013. "Effects of climate on emerald ash borer mortality and the potential for ash survival in North America." *Agricultural and Forest Meteorology 178–179*: 120–128.

Dodds, K. J., and D. A. Orwig. 2011. "An invasive urban forest pest invades natural environments—Asian longhorned beetle in northeastern US hardwood forests." *Canadian Journal of Forest Research 41*: 1729–1742.

Dukes, J. S., J. Pontius, D. Orwig, J. R. Garnas, V. L. Rodgers, N. Brazee, B. Cooke, et al. 2009. "Responses of insect pests, pathogens, and invasive plant species to climate change in the forests of northeastern North America: what can we predict? *Canadian Journal of Forest Research 39*: 231–248. doi:10.1139/X08-171.

Ellison, A. M., M. S. Bank, B. D. Clinton, E. A. Colburn, C. R. Ford, D. R. Foster, et al. 2005. "Loss of foundation species: consequences for the structure and dynamics of forested ecosystems." *Frontiers in Ecology and the Environment 3*: 479–486.

Eyre, D., and R. A. Haack, 2017. "Invasive cerambycid pests and biosecurity measures." In *Cerambycidae of the World: Biology and Pest Management*, edited by Q. Wang, 563–607. Boca Raton, FL: CRC Press.

Flory, S. L. 2010. "Non-native grass invasion suppresses forest succession." *Oecologia 164*: 1029–1038. doi:10.1007/s00442-010-1697-y.

Flory, S. L., and K. Clay. 2006. "Invasive shrub distribution varies with distance to roads and stand age in eastern deciduous forests in Indiana, USA." *Plant Ecology 184*: 131–141. doi:10.1007/s11258-005-9057-4.

Forrester, J. A., G. G. McGee, and M. J. Mitchell. 2003. "Effects of beech bark disease on aboveground biomass and species composition in a mature northern hardwood forest, 1985 to 2000." *The Journal of the Torrey Botanical Society 130*: 70–78.

Frappier, B., T. D. Lee, K. F. Olson, and R. T. Eckert. 2003. "Small-scale invasion pattern, spread rate, and lag-phase behavior of *Rhamnus Frangula* L." *Forest Ecology and Management 186*: 1–6. doi:10.1016/S0378-1127(03)00274-3.

Garnas, J. R., M. P. Ayres, A. M. Liebhold, and C. Evans. 2011. "Subcontinental impacts of an invasive tree disease on forest structure and dynamics." *Journal of Ecology 99*: 532–541.

Gonzales, E. K., and S. E. Gergel. 2007. "Testing assumptions of cost surface analysis—a tool for invasive species management." *Landscape Ecology 22*: 1155–1168. doi:10.1007/s10980-007-9106-6.

Griffin, J. M., G. M. Lovett, M. A. Arthur, and K. C. Weathers. 2003. "The distribution and severity of beech bark disease in the Catskill Mountains, N.Y." *Canadian Journal of Forest Research 33*: 1754–1760.

Gunn, J. S., M. J. Ducey, and A. A. Whitman. 2014. "Late-successional and old-growth forest carbon temporal dynamics in the northern forest (northeastern USA)." *Forest Ecology and Management 312*: 40–46. doi:10.1016/j.foreco.2013.10.023.

Houston, D. 1994. "Major new tree disease epidemics: beech bark disease." *Annual Review of Phytopathology 32*: 75–87.

Ibáñez, I., J. A. Silander, J. M. Allen, S. A. Treanor, and A. Wilson. 2009. "Identifying hotspots for plant invasions and forecasting focal points of further spread." *Journal of Applied Ecology 46*: 1219–1228. doi:10.1111/j.1365-2664.2009.01736.x.

Johnson, V. S., J. A. Litvaitis, T. D. Lee, and S. D. Frey. 2006. "The role of spatial and temporal scale in colonization and spread of invasive shrubs in early successional habitats." *Forest Ecology and Management 228*: 124–134. doi:10.1016/j.foreco.2006.02.033.

Knight, K. S., J. P. Brown, and R. P. Long. 2013. "Factors affecting the survival of ash (*Fraxinus* Spp.) trees infested by emerald ash borer (*Agrilus planipennis*)." *Biological Invasions 15*: 371–383.

Kurtz, C.M., and M. H. Hansen. 2013. "An assessment of multiflora rose in northern U.S. Forests." Vol. Res. Note NRS-182. https://www.fs.fed.us/nrs/pubs/rn/rn_nrs182.pdf.

Lee, T., S. Eisenhaure, and I. Gaudreau. 2016. "Pre-logging treatment of invasive glossy buckthorn (*Frangula alnus* Mill.) promotes regeneration of eastern white pine (*Pinus strobus* L.)." *Forests 8*: 16. doi:10.3390/f8010016.

Leicht, S. A. 2005. "The comparative ecology of an invasive bittersweet species (*Celastrus orbiculatus*) and its native congener (*C. scandens*)." PhD diss.. Storrs, CT: University of Connecticut. http://digitalcommons.uconn.edu/dissertations/AAI3193727.

Leicht, S. A., and J. A Silander. 2006. "Differential responses of invasive *Celastrus orbiculatus* (Celastraceae) and native *C. scandens* to changes in light quality." *American Journal of Botany 93*: 972–977. doi:10.3732/ajb.93.7.972.

Leppanen, C., and D. Simberloff. 2016. "Implications of early production in an invasive forest pest." *Agricultural and Forest Entomology 19*: 217–224.

Liebhold, A. M., D. G. McCullough, L. M. Blackburn, S. J. Frankel, B. Von Holle, and J. E. Aukema. 2013. "A highly aggregated geographical distribution of forest pest invasions in the USA." *Diversity and Distributions 19*: 1208–1216.

Liendo, D., I. Biurrun, J. A. Campos, M. Herrera, J. Loidi, and I. García-Mijangos. 2015.

"Invasion patterns in riparian habitats: the role of anthropogenic pressure in temperate streams." *Plant Biosystems 149*: 289–297. doi:10.1080/11263504.2013.822434.

Lodge, D. M. 1993. "Biological invasions: lessons for ecology." *Trends in Ecology and Evolution 8*: 133–137. doi:10.1016/0169-5347(93)90025-K.

Lorimer, C. G. 1980. "Age structure and disturbance history of a southern Appalachian virgin forest." *Ecology 61*: 1169–1184.

Lovett, G. M., C. D. Canham, M. A. Arthur, K. C. Weathers, and R. D. Fitzhugh. 2006. "Forest ecosystem responses to exotic pests and pathogens in eastern North America." *Bioscience 56*: 395–405.

Lovett, G. M., M. Weiss, A. M. Liebhold, T. P. Holmes, B. Leung, K. F. Lambert, D. A. Orwig, et al. 2016. "Nonnative forest insects and pathogens in the United States: impacts and policy options." *Ecological Applications 26*: 1437–1455.

Martin, P. H., C. D. Canham, and P. L. Marks. 2009. "Why forests appear resistant to exotic plant invasions: intention introductions, stand dynamics and the role of shade tolerance." *Frontiers in Ecology and the Environment 7*: 142–148. doi:10.1890/070096.

McCarthy, B. C., C. J. Small, and D. L. Rubino. 2001. "Composition, structure and dynamics of Dysart Woods, an old-growth mixed mesophytic forest of southeastern Ohio." *Forest Ecology and Management 140*: 193–213. doi:10.1016/S0378-1127(00)00280-2.

McClure, M. S. 1987. "Biology and control of hemlock wooly adelgid." Bulletin 851. New Haven, CT: The Connecticut Agricultural Experiment Station.

McClure, M. S. 1990. "Role of wind, birds, deer, and humans in the dispersal of hemlock woolly adelgid (Homoptera: Adelgidae)." *Environmental Entomology 19*: 36–43.

McDonald, R. I., G. Motzkin, and D. R. Foster. 2008. "Assessing the influence of historical factors, contemporary processes, and environmental conditions on the distribution of invasive species." *Journal of the Torrey Botanical Society 135*: 260–271. doi 10.3159/08-Ra-012.1.

Merow, C., S. T. Bois, J. M. Allen, Y. Xie, and J. A. Silander. 2017. "Climate change both facilitates and inhibits invasive plant ranges in New England." *Proceedings of the National Academy of Sciences of the United States of America 114*: E3276–284. doi:10.1073/pnas.1609633114.

Oliver, C. D., and B. C. Larson. 1996. *Forest Stand Dynamics*. Updated Edition. New York: Wiley.

Orwig, D. A., A. A. B. Plotkin, E. A. Davidson, H. Lux, K. E. Savage, and A. M. Ellison. 2013. "Foundation species loss affects vegetation structure more than ecosystem function in a northeastern USA Forest." *PeerJ 1*: e41.

Orwig, D. A., J. R. Thompson, N. A. Povak, M. Manner, D. Niebyl, and D. R. Foster. 2012. "A foundation tree at the precipice: *Tsuga canadensis* health after the arrival of *Adelges tsugae* in central New England." *Ecosphere 3*: art10.

Parendes, L. A., and J. A. Jones. 2000. "Role of light availability and dispersal in exotic plant invasion along roads and streams in the H. J. Andrews Experimental Forest, Oregon." *Conservation Biology 14*: 64–75. doi:10.1046/j.1523-1739.2000.99089.x.

Poland, T. M., and D. G. McCullough. 2006. "Emerald ash borer: invasion of the urban forest and the threat to North America's ash resource." *Journal of Forestry 104*: 118–124.

Prati, D., and O. Bossdorf. 2004. "Allelopathic inhibition of germination by *Alliaria petiolata* (Brassicaceae)." *American Journal of Botany 91*: 285–288. doi:10.3732/ajb.91.2.285.

Rodewald, A. D., and P. Arcese. 2016. "Direct and indirect interactions between landscape structure and invasive or overabundant species." *Current Landscape Ecology Reports.* doi:10.1007/s40823-016-0004-y.

Schulz, B. K., and A. N. Gray. 2013. "The new flora of northeastern USA: quantifying introduced plant species occupancy in forest ecosystems." *Environmental Monitoring and Assessment 185*: 3931–3957. doi:10.1007/s10661-012-2841-4.

Skinner, M., B. L. Parker, S. Gouli, and T. Ashikaga. 2003. "Regional responses of hemlock woolly adelgid (Homoptera: Adelgidae) to low temperatures." *Environmental Entomology 32*: 523–528.

Smith, M. T., J. J. Turgeon, P. De Groot, and B. Gasman. 2009. "Asian longhorned beetle *Anoplophora glabripennis* (Motschulsky): lessons learned and opportunites to improve the process of eradication and management." *American Entomologist 55*: 21–25.

Snyder, C. D., J. A. Young, D. P. Lemarié, and D. R. Smith. 2002. "Influence of eastern hemlock (*Tsuga canadensis*) forests on aquatic invertebrate assemblages in headwater streams." *Canadian Journal of Fisheries and Aquatic Sciences 59*: 262–275.

Stotz, G. C., G. J. Pec, and J. F. Cahill. 2016. "Is biotic resistance to invaders dependent upon local environmental conditions or primary productivity? a meta-analysis." *Basic and Applied Ecology 17*: 377–837. doi:10.1016/j.baae.2016.04.001.

Tingley, M. W., D. A. Orwig, R. Field, and G. Motzkin. 2002. "Avian response to removal of a forest dominant: consequences of hemlock woolly adelgid infestations." *Journal of Biogeography 29*: 1505–1516.

Vilà, M., J. L. Espinar, M. Hejda, P. E. Hulme, V. Jarošík, J. L. Maron, J. Pergl, U. Schaffner, Y. Sun, and P. Pyšek. 2011. "Ecological impacts of invasive alien plants: a meta-analysis of their effects on species, communities and ccosystems." *Ecology Letters 14*: 702–708. doi:10.1111/j.1461-0248.2011.01628.x.

Wagner, D. L., K. J. Metzler, S. A. Leicht-Young, and G. Motzkin. 2014. "Vegetation composition along a New England transmission line corridor and its implications for other trophic levels." *Forest Ecology and Management 327*: 231–239. doi:10.1016/j.foreco.2014.04.026.

Ward, J. S., M. E. Montgomery, C. A. S-J. Cheah, B. P. Onken, and R. S. Cowles. 2004. "Eastern hemlock forests: guidelines to minimize the impacts of hemlock woolly adelgid." Technical Bulletin NA-TP-03-04. Northeastern Area State & Private Forestry. Morgantown, WV: USDA Forest Service.

Watkins, R. Z., J. Chen, J. Pickens, and K. D. Brosofske. 2003. "Effects of forest roads on understory plants in a managed hardwood landscape." *Conservation Biology 17*: 411–419. doi:10.1046/j.1523-1739.2003.01285.x.

# Chapter 13

# Silviculture for Eastern Old Growth in the Context of Global Change

*William S. Keeton, Craig G. Lorimer, Brian J. Palik,*
*and Frédérik Doyon*

When management for old-growth characteristics in eastern forests first began to be discussed in the late twentieth century, there was skepticism from some quarters as to whether it was a desirable or even a feasible idea. Old growth will recover on its own. Why not just let nature take its course? There were also those who saw little value in managing for old-growth features, perceiving this as a threat to more traditional management objectives (Puettmann et al. 2015). Since that time, concepts of managing for stand structural complexity, in ways that encourage some characteristics of old-growth forests, have caught on in a variety of contexts (Bauhus et al. 2009; Puettmann et al. 2009). In many ways this shift mirrors how the profession has grown to embrace multifunctional forestry broadly defined (Gustafsson et al. 2012). Old-growth silviculture increasingly has a place within this framework, filling the niche of enhancing the representation of late successional forests on landscapes where they are now vastly underrepresented relative to their abundance on landscapes prior to Euro-American settlement (Lorimer and White 2003; Rhemtulla et al. 2007). The working hypothesis is that this type of management will contribute to sustainable forest practices focused on providing a broad array of ecosystem goods and services, including those associated with late successional systems. And in recent decades there has been increasing interest in old-growth restoration more narrowly and management for older forest characteristics in working forests generally, both in terms of experimental research (e.g., Keeton 2006; Gronewold et al. 2010; Forrester et al. 2013; Palik et al. 2014) and practical applications (Hagenbuch et al. 2013; Fassnacht et al. 2015).

   Interest in structure-based silviculture has evolved from studies demonstrating the ecological significance of specific structural elements associated with late successional and old-growth forests (e.g., Tyrell and Crow 1994; McGee et al. 1999; Ziegler 2000; Després et al. 2014). Availability of these structures can be highly limited in forests managed under conventional even- and uneven-aged systems (Hale et al. 1999; McGee et al. 1999; Angers et al. 2005). But a central question remains: Is it better to let old growth recover passively or could silvicultural methods be used to restore or increase the representation of old-growth characteristics in secondary forests? Certainly it has become clear that humans have often negatively impacted forests, even in remote locations, in ways that nature cannot readily rehabilitate. Moreover, old distinctions between active forest management and natural-areas management have become blurred in recent decades. No longer are these distinctions viewed as a strict dichotomy (Keeton 2007). Agencies such as the National Park Service and organizations like The Nature Conservancy now routinely conduct prescribed burning, cut or thin undesirable trees, reestablish native species, and apply herbicides to control invasive plants (Schwartz et al. 2016). In the Anthropocene, the idea of just letting nature take its course no longer seems to be an adequate strategy if the goal is to maintain biologically diverse and healthy ecosystems that function in a similar way to forests in the past millennium. Especially in the eastern United States, where most forests even in reserves have been cutover one to several times, and the existing forest is increasingly subject to a blitzkrieg of novel stressors—invasive species, exotic earthworms, and airborne pollutants—passive management could result in depauperate ecosystems under chronic stress. Yet, it is clear that these novel stressors will make restoring, or even just maintaining, old-growth structural features technically challenging.

One reason for considering active management in eastern North American forests is an issue of time delay. Even if passive management could restore healthy old-growth forests, the process in the existing second growth would often involve a delay of another 60 to 150 years. Old growth is currently so rare that treatments in younger forests to accelerate old-growth structural features—large trees, canopy gaps, multilayered vegetation, standing snags, and fallen logs—have demonstrated ecological benefits (e.g., Bauhus et al. 2009; Dove and Keeton 2015). These benefits range from the late successional habitat values provided by such structures, to ecosystem service benefits like enhanced carbon and riparian functionality. The biodiversity argument is nuanced and focused on correcting the forest age class imbalances that are the legacy of the region's land-use history. In the Pacific Northwest, for example, over a thousand species were deemed strongly associated with late successional ecosystems (FEMAT 1993). But here in the East, the biodiversity value rests primarily on habitat quality and availability for species using late successional/old-growth habitats (figure 13-1; chapter 11). In our region only rarely does the biodiversity depend strictly upon old-growth habitats, although there are notable exceptions like some calicioid lichens and fungi (Selva 2003; chapter 11). But populations of often underrepresented late successional species would benefit from having more old-growth structures available in reserves, and more importantly, if these structures were better incorporated into managed landscapes comprising the majority of eastern forests.

We stress that management for late successional biodiversity is by no means mutually exclusive of early successional habitat management advocated for species such as disturbance-dependent birds (see, for example, Hunter et al. 2001). Early and late successional habitats share high biodiversity value (figure 13-1). The question is how best to optimize the mix of all habitat types at landscape scales. In other words, forest managers can provide both early and late seral habitats with careful planning and forest harvest scheduling. However, it will be important to consider the long-term implications for forest age class distributions of increased emphasis on early successional management. This is needed to avoid the unintended consequence of a landscape shifted over time by techniques like patch cutting into the stem-exclusion stage of development, which most often has the lowest biological diversity of any structural condition (figure 13-1).

In some ways, ecosystem service benefits are the most compelling reason to incorporate old-growth siviculture into holistic management. Often, forest managers and landowners in the eastern United States—spanning a huge range of ownership systems, parcel sizes, and degrees of timber emphasis—are most interested in sustainable forest management integrating multiple cobenefits. Chief among the ecosystem service values cited as an outcome of old-growth silviculture is climate change mitigation through enhanced carbon storage. Sequestered carbon in the form of biomass stored above- and belowground is many times higher in landscapes with abundant old-growth forests as compared to most managed landscapes (Rhemtulla et al. 2009; Keeton et al. 2011; McGarvey et al. 2015). Consequently, the types of high biomass structural conditions targeted by old-growth silviculture are precisely what are incentivized by rapidly developing compliance and voluntary carbon markets (Kerchner et al. 2015). It is no surprise, therefore, that old-growth silviculture often makes use of similar techniques (e.g., high degree of retention, extended rotations, enhancement of large tree and dead wood components) as those recommended for carbon forestry (Nunery and Keeton 2010; see chapter 14). Similarly, there is growing interest in the high

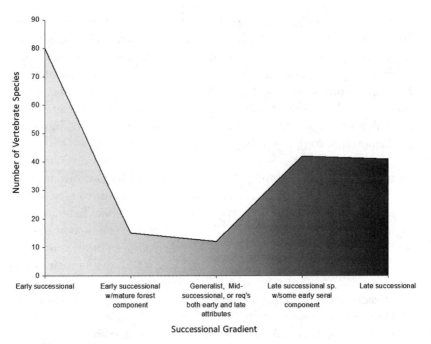

FIGURE 13-1. Number of vertebrate species in northern hardwood forests of New England, categorized by seral stage and habitat association. The data were extracted by K. Manaras-Smith, E. Travis, and W. S. Keeton (unpublished) from matrices published in DeGraaf and Yamasaki (2001). The categorizations are based on habitat preference and use scores developed by Manaras-Smith et al. from the various habitat uses, requirements, and associations presented in the matrices. When these scores where arrayed along a size class gradient (regenerating, pole-sized, sawtimber, large sawtimber, and uneven-aged), the use scores for large sawtimber and uneven-aged stands were more than twice that for regenerating and pole-sized stands; they were 70 percent greater for preference scores. However, when overlapping or multiple habitat uses (e.g., by the same species) were assessed, as displayed on the X axis in the figure above, the relationship took on a U-shaped form. This indicates the strong biodiversity value of both early and late successional habitats, as well as seral habitats with biological legacies and some attributes of both early and late successional forests.

degree of riparian functionality provided by structurally complex, late successional forests, particularly in terms of effects on stream habitats (Keeton et al. 2007; Warren et al. 2016) and flood resilience (Keeton et al. 2017).

At the same time, it should be recognized that old-growth management and retention of large legacy trees, while probably economically feasible, will likely reduce rates of stand production to some degree. This has already been reported in Douglas-fir forests of the Pacific Northwest (Zenner et al. 1998) and in southern Appalachian forests (Miller et al. 2006). In northern hardwoods, a simulation of 22 ecological forestry treatments with an individual-tree model suggested that reduced production could vary widely depending on the treatment. These ranged from a 9-percent decline with 7 reserve trees per hectare to a 55-percent decline with treatments retaining substantial coarse woody debris and a maximum residual DBH (diameter at breast height) target of 80 centimeters. Treatments designed to strike a balance between timber production and maintenance of old-growth stand structure involved losses of 27 to 30 percent compared to conventional single-tree selection (Hanson et al. 2012). In an experimental study in Vermont, a technique designed to accelerate the development of old-growth characteristics harvested only 60 percent of the merchantable volume produced by conventional selection harvests, but generated a moderate profit when market conditions, preharvest volumes, and trucking distances were favorable (Keeton and Troy 2006). With the advent of carbon markets providing financial incentives for forestry practices that maintain high stocking levels, we may see old-growth silviculture tapping into supplemental revenue streams in the future.

A major impediment to restoration of old-growth features—and one that has no easy solution in sight—is the recent cascade of novel and interacting environmental stressors, such as climate warming, exotic insects, diseases, earthworms, excessive deer densities, and invasive plants (see chapter 12). Insects, diseases, and drought often disproportionately affect large and old trees, hampering efforts to restore or maintain old-growth structure. The continued presence of old trees of susceptible species seems unlikely in the near future without some effective mechanism (e.g., biological control, backcrossing, genetic engineering—all contentious and fraught with challenges) to combat the attacking organisms. In addition to exotic pests and pathogens already well established in eastern North American forests, such as chestnut blight (*Cryphonectria parasitica*) and emerald ash borer (*Agrilus planipennis*), new ones, such as Asian long-horned beetle (*Anoplophora glabripennis*), mountain pine beetle (*Dendroctonus ponderosae*), and sudden oak death (*Phytophthora ramorum*), are on the horizon and may be capable of major devastation for multiple species. Given these potential realities, the principles of managing for old growth in this chapter should probably be regarded in the short term as interim guidelines for areas not yet heavily impacted by environmental stressors. Over the long-term, silviculture oriented toward old growth—like all forestry—will need to adapt to changing environmental conditions.

## Lessons from Old Growth

Traditionally, old growth was treated as a predictable and stable end point of forest development (Odum 1969; Bormann & Likens 1979). But the contemporary view rejects the idea of a stable "climax" stage, recognizing instead the potential for continuous change (termed "nonequilibrium dynamics") as old, complex stands are acted upon and respond to a range of natural disturbance types and intensities that occur chronically, periodically, or episodically (Ziegler 2002 ; D'Amato and Orwig 2008). Work in recent decades has shown that forest structure and composition continue to fluctuate after old-growth age thresholds are crossed (Spies 1997). In Pacific Northwest forests, varying degrees of old growth have been described (Franklin et al. 2002). In eastern forests, continuous changes in forest elements, such as gap fraction, coarse woody debris volume, biomass, and relative species abundance, have been observed across chronosequences, on permanent plots, and projected in simulations (Tyrrell and Crow 1994; Woods 2000; Keeton et al. 2011). Developmental stages in the live-tree population can be detected from systematic changes in the diameter distribution of old-growth stands, which tend to develop from unimodal to bimodal (compound) to descending monotonic in the absence of moderate or heavy disturbance (Lorimer and Halpin 2014). In the Great Lakes region, which has some extensive and rather pristine landscapes of old growth, all these stages are evident on the landscape as a patchwork, with roughly equal proportions of old growth in the different stages (Heinselman 1973; Lorimer and Halpin 2014).

In late successional forests, however, these stages do not appear to develop along a simple circular pathway from a young, even-aged forest to uneven-aged old growth, reverting to a young forest again after a stand-replacing disturbance (as in Bormann and Likens 1979). Rather, the pathways resemble a complex web, with stands moving backwards and forwards in erratic and unpredictable fashion among various stages in response to repeated mild and moderate disturbances (Hanson and Lorimer 2007; Halpin and Lorimer 2016; Meigs and Keeton 2018). Complicating things further are the effects of land-use history (e.g., wood harvesting, pests, pathogens, etc.) on long-term trajectories for forest development, resulting in a structure and composition that reflects human influences as much as fundamental stand dynamics (McLachlan et al. 2000). Thus, the contemporary view emphasizes the multiple pathways and rates by which old-growth ecosystems can develop (Donato et al. 2012; Halpin and Lorimer 2016; Urbano and Keeton 2017). Site productivity (soil fertility, local climate, aspect, etc.) and well as the type, intensity, and timing of disturbance events exert major influences on these developmental trajectories. This is a critical lesson for old-growth silviculture because the targets are not fixed: They are a continuous range of stand structural conditions, responding dynamically to their environment over time and space. Old-growth silviculture must work with, not against, these dynamics, manipulating stand development processes to achieve the desired ecosystem functions—like structurally complex habitats and carbon storage—but recognizing that continuous, sometimes unpredictable, change is certain (Fahey et al. 2018).

Consequently, there is no one-size-fits-all approach, but rather a range of possibilities that must be tailored to the stand or property-specific management objectives, disturbance regimes, land-use history, and site conditions. The idea of natural variability in old-growth structure and function was described further by Bauhus et al. (2009). They endorsed the concept of a range of "old-growthness," arguing that simple structural criteria or age thresholds fail to capture the dynamics of natural baselines or variation in the values of any one parameter, such as large tree density or downed coarse woody debris, in late successional forests. This conclusion was supported by the global review of old-growth structure performed by Burrascano et al. (2013). They found a wide range of variability around mean values for parameters, such as large tree density, downed woody debris volumes, and basal area, within both mature and old-growth age classes in temperate forests worldwide, including parts of the eastern United States. In some cases, structural characteristics typically associated with old growth may be equally, if not more, pronounced in mature forests, depending on site characteristics and disturbance history (Burrascano et al. 2013). This suggests a need to broaden our concept of late successional dynamics and to move away from overly narrow or rigid classification criteria. Therefore, it follows that targets for old-growth silviculture are a continuum of possibilities rather than a discrete or uniform set of objectives.

## Principles of Old-Growth Silviculture

Depending on landholding and management objectives, management for old growth may or may not involve the harvesting and removal of trees for wood products. In some cases, such as in natural areas or parklands, if any trees are felled, they are often left to decay on site. On other ownerships, silvicultural treatments in zones with active forest management may be selected to enhance or accelerate the development of old-growth structural features, with some trees being removed to help defray the costs of restoration. Another option for forest landscape managers is to designate "aging areas" where extended rotations or no-harvest inclusions might be employed to increase the variety of old-growth stand features (Mladenoff et al. 1994). In the eastern United States, these areas of active forest management are second- or third-growth forests, not remnant fragments of the original old growth in natural areas.

Given that old growth continuously changes in species and structure, managers should avoid establishing a static management target, reflecting the older conception of old growth as a single, homogeneous stage of development. In late successional forests of shade-tolerant tree species, such as spruce-fir and hemlock-northern hardwoods, foresters were long accustomed to establishing a structural goal for uneven-aged stands in which the residual size distribution after harvest resembles a negative exponential or "reverse J-shaped" curve (O'Hara 1998). This made sense as a means of ensuring sustainable harvests under the selection system (Crow et al. 1981). It would be easy to extend this concept to old-growth management simply by increasing the maximum residual diameter and stand basal area.

However, as has already been shown for conventional single-tree selection, this approach, if applied too broadly, lowers species diversity and stand heterogeneity because of the small and uniform gap sizes (Della-Bianca and Beck 1985; Leak and Sendak 2002; Schuler 2004). Applying this management target for old growth would imply that small gap dynamics are the only important disturbance process to mimic with old-growth silviculture. Although unforeseen natural disturbances in managed stands can maintain some of this heterogeneity, foresters should recognize that a wide range of structural conditions and disturbance histories are represented in old growth, including unimodal size distributions and two- or three-aged stands. On intact old-growth northern hardwood landscapes in Michigan, only about half of the old-growth stands approach a descending monotonic size distribution (i.e., vertically continuous canopies), with the remainder represented by unimodal or bimodal distributions (Lorimer and Halpin 2014). There is evidence that stands with a unimodal distribution (i.e., limited understory development) are ecologically important for some organisms. For example, the dense understory and numerous sapling gaps in the archetypal all-aged stand can be an obstruction when goshawks (*Accipiter gentilis*) forage beneath the canopy or when forest bats sense such environments as being filled with obstructions (Owen et al. 2004; Boal et al. 2005).

Multicohort management that maintains some uneven-aged stands with a variety of opening sizes can help create conditions that enable perpetuation of mid- and shade-intolerant tree species as a component of the stand, either through natural regeneration, planting, or both. Gap size is only one of several important factors affecting the recruitment and growth of the less shade-tolerant species. Therefore, ancillary treatments may be needed, especially on better sites affected by novel stressors (Bolton and D'Amato 2011; Kern et al. 2013; see further discussion below). Multiple-tree gaps are usually needed to ensure successful recruitment into the canopy before lateral closure occurs (Runkle and Yetter 1987; Leak 1999; Webster and Lorimer 2005). These gaps can be installed as group selection openings or as larger openings (up to roughly 0.5 hectares or more) in the context of an irregular, multicohort harvest removing about 40 to 50 percent of the stand basal area (Leak 1999; Hanson and Lorimer 2007; Hanson et al. 2012). Managers should be careful, however, not to overcompensate and create too many large openings. Heavier multicohort harvests probably need to be undertaken only once or twice during the entire 250-year life span of a cohort to mimic natural processes and accomplish ecological objectives. A temporary shelterwood overstory may also be needed in larger openings to hinder the development of dense shrub layers. Similar cautions about gap size and shrub competition apply to group selection (Kern et al. 2013; Walters et al. 2016). Furthermore, simulation studies have suggested that a high degree of smallscale forest fragmentation can occur if groups occupy 9 percent or more of the forest matrix in each cutting cycle (Gustafson and Crow 1996; Halpin and Lorimer 2017). Excessive lateral crown exposure could conceivably create additional physiological stress on residual trees under a warming climate.

Pioneer tree species, such as the pines (*Pinus* spp.), present much different management challenges compared to the shade-tolerant species. Maintaining an age-class mosaic is easier on public lands managed for multiple objectives, such as state and national forests in the United States or provincial forests in Canada, than in reserves, because conventional silvicultural methods (e.g., clear-cutting with reserves and irregular shelterwoods) can be used to establish the younger stands. However, the eventual replacement of old pioneer trees by shade-tolerant species will at some point require their conversion to younger stands if the proportion of old growth on the landscape is to remain stable at a level agreed upon in policy negotiations.

One of the greatest areas of interest in old-growth silviculture has been the idea of emulating the effects of natural disturbances on rates and pathways of forest development, with the intent of accelerating rates of old-growth structural and compositional development (North and Keeton 2008). Broadly defined, natural-disturbance based silviculture incorporates biological legacies (e.g., carryover of organically derived structures) into prescriptions. This emulates stand development processes, including small-scale disturbances, in intermediate treatments (i.e., thinning) and allows for appropriate intervals between regeneration harvests, known as *rotation periods* or *entry cycles*, for recovery of late successional habitats (Seymour et al. 2002; Franklin et al. 2007). Thinning, as a form of stand improvement, has been used for years as a method to more rapidly move stands into later stages of development. But, while conventional thinning increases growth increments on residual trees, it can also have the disadvantage for old-growth silviculture of simplifying some attributes of stand structure if employed too aggressively. This is particularly true if applied through spatially uniform timber marking and when less vigorous or defective trees are the main removals, which can substantially reduce the potential for dead tree recruitment. For this reason, concepts like variable density thinning—encouraging horizontal heterogeneity within stands—and deliberate retention of a component of dead, dying, and defective trees offer useful alternatives for intermediate treatments.

A variety of methods employed in conjunction with either intermediate treatments or regeneration harvests are also relevant. Management for old-growth conditions—often synonymous with management for structural complexity at the stand level in moist to wet temperate forests—can include a broad array of techniques used in tandem or individually, depending on stand conditions and the desired mix of cobenefits. Some foresters might even reject the notion that treatments must be strictly defined as either intermediate or regeneration harvest (Franklin et al. 1997). If natural disturbances can free up growing space, thereby enhancing growth in residual trees while regenerating a new cohort of trees, then so might hybrid, disturbance-based forestry practices (Keeton 2006). When variations of selection systems or multicohort management are used for old-growth objectives, oft-cited techniques include residual diameter distributions of variable forms, large tree retention, crown release around dominant and codominant trees, girdling to create dead trees (snags), enhancement of downed coarse woody debris, variable density thinning or tree selection, and creating variably sized, irregularly shaped harvest gaps (sometimes including within-gap legacy tree retention) (O'Hara 1998; Keeton 2006; North and Keeton 2008; Kern et al. 2016). For each, the forester must carefully consider both the baseline referenced in developing the silvicultural prescription, the dynamics of that baseline over time and space (since no baseline is truly stable), and how the technique will work with, and ideally accelerate, stand development processes, such as gap formation, understory reinitiation (i.e., regeneration) of mid- and shade-tolerant species, vertical differentiation of the canopy, tree growth across all canopy strata, and dead wood recruitment. Old-growth silviculture in eastern deciduous, coniferous, and mixed-wood forests, therefore, is like a tool box with multiple options the forester might consider, rather than a single approach. If there is a unifying theme, it is that old-growth silviculture should emulate and promote a range of stand development processes, including those induced by both self-thinning and natural disturbance effects.

## Experimentation: Can We Restore Elements of Old-Growth Structure and Function?

The idea of managing eastern forests for old-growth characteristics is no longer theoretical. Rather, experimental research over the last 20 years has clearly established the ability of carefully planned silvicultural interventions to enhance developmental rates for late successional structure, function, and composition. For example, to address the problems of low species diversity under single-tree selection, a number of field trials across the northern hardwood region have examined the effects of harvest-created openings on natural or planted seedlings of mid- or shade-intolerant tree species (Leak 1999; Webster and Lorimer 2005; Kern et al. 2012, 2013; Bolton and D'Amato 2011; Fahey and Lorimer 2013; D'Amato et al. 2015; Gauthier et al. 2016; Gottesman and Keeton 2017). Several broad conclusions have been reached. First, successful long-term recruitment into the canopy is not very likely in openings smaller than 180 to 200 square meters because of low light intensities and lateral crown encroachment from gap border trees. Also, in the case of white pine, openings may need to be considerably larger to reduce infection by white pine blister rust (*Cronartium ribicola*), which is aggravated by cold-air drainage in small openings. Second, success varies widely among habitats. On the richer habitats, competition with shrubs and taller saplings of shade-tolerant species is often so severe that survival of the mid-tolerant species is very low and growth of survivors is slow. Success is more likely on sites of moderate or below-average productivity. Success on better sites may require control of understory competition, scarification of the soil surface, and retention of fallen woody debris. Third, in areas of moderate to high deer densities, special measures may need to be taken to reduce losses from browsing, such as repeated application of deer repellants to terminal leaders. Fourth, preliminary evidence suggests that natural regeneration is often more vigorous and faster growing than planted seedlings, perhaps because of physiological stress in planted stock. Future improvements in planting stock, and perhaps improvements in containerized seedlings, may help remedy these problems.

Field studies have also evaluated the use of silvicultural treatments to hasten the development of old-growth structural features. A comparison of managed second-growth and unmanaged old-growth northern hardwood stands showed that the proportion of the canopy in openings and the gap size distribution were very similar in both management types. This was the case even though the harvests in second growth were conventional single-tree selections and were not intentionally designed to mimic old-growth processes (Goodburn 1996). In contrast, a comparison of canopy dynamics in Quebec, Canada, characterized with multitemporal LiDAR data, indicated the need for a recovery period of at least 20 years after selection cutting before canopy dynamics returned to something similar to the referenced old-growth deciduous forest (Senécal et al. 2018). The Quebec study concluded that, under a selection cutting cycle of 25 years, forest canopy dynamics will be distinctly different from old-growth forests for 80 percent of the harvest rotation period or entry cycle. Similarly, in a study of gap dynamics in old-growth hemlock-hardwood forests, Curzon and Keeton (2010) found that the canopy openings were not only larger, on average, than those documented for young and mature forests, but also irregularly shaped and variably sized, with ragged within-gap structure related to the abundance of legacy trees, both live and dead. A challenge for silviculture, therefore, is to emulate this tremendous variability in gap structure and canopy architecture associated with natural disturbance effects in old, unmanaged forests.

In principle, canopy gaps, snags and logs, and multilayered vegetation are elements of old growth that should be possible to restore fairly quickly in second-growth stands by silvicultural treatments. A study in Pennsylvanian oak forests, for example, showed that retention of cull trees and snags in stands during improvement cuts was successful in elevating bird abundance compared to stands with conventional treatment (Stribling et al. 1990). Designing structural enhancement treatments begins by comparing the habitat elements already provided in conventionally managed forests with those found in old-growth forests. For instance, total coarse woody debris volumes under conventional single-tree selection can range from about 50 to 70 percent of the volumes in unmanaged old growth (Hardt and Swank 1997; Goodburn and Lorimer 1999; McGee et al. 1999; Doyon et al. 2005). Harvest treatments designed to restore old-growth structural elements in Vermont northern hardwood-conifer forests resulted in downed log volumes about 88 percent of old-growth volumes and 50 to 100 percent greater than usually achieved in conventional single-tree selection (Keeton 2006). Mixed results have also been observed. In a recent largescale experiment, using commercial harvest to enhance old-growth features in Wisconsin, damage during harvesting and subsequent mortality reduced the retention of small preexisting snags. More surprising was that net density of standing snags larger than 25 centimeters DBH was not significantly increased five years after girdling 5 trees per hectare—a treatment that was intended to boost snag density (Fassnacht and Steele 2016).

The question of whether silvicultural treatments can accelerate the development of large trees was recognized at the outset as a more uncertain proposition. Canopy trees in second-growth stands are already vigorous, fast-growing trees receiving direct sunlight at the top of the crown and, in many cases, are 80 to 100 years old. Would thinning the canopy actually boost growth rates in vigorous and relatively old trees that are beyond the age at which thinning is usually conducted? At least for the shade-tolerant sugar maple (*Acer saccharum*), the answer may be yes. A retrospective field study of thinned stands indicated that substantial mean basal area growth increases of 74 to 107 percent occurred even in canopy trees more than 100 years old. The response, moreover, was proportional to the degree of crown perimeter exposure (Singer and Lorimer 1997). Simulations based on these field data suggested that the time required for a 77-year-old northern hardwood stand to reach minimum structural criteria of old growth could be reduced from 79 years to 36 years after a series of three moderate thinning treatments at 15-year intervals (Choi et al. 2007). Size distributions after 45 years resembled those of stands in the later stages of old growth. An important caveat was that, if the stand being treated was already uneven-aged with some large trees, aggressive thinning treatments could actually increase the amount of time required to reach minimum old-growth structural criteria, because some medium and large trees had to be cut to meet the targeted thinning intensity. Thinning caused little change to coarse woody debris volume when 20 percent of cut-tree volume was left on site and 80 percent was removed, but a higher proportion of coarse woody debris was in larger size classes on the treated sites.

In the northern hardwood region of the northeastern United States, one of the first experimental tests of old-growth silviculture evaluated an approach called Structural Complexity Enhancement (SCE) that was compared alongside conventional single-tree and group-selection harvests (figure 13-2; Keeton 2006). The study, started in 2001, is replicated at two research forests in northern Vermont and employs many of the individual

## Single-Tree Selection

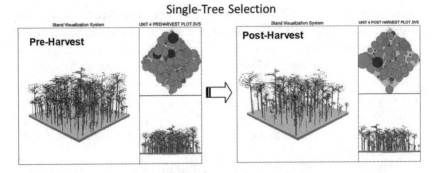

## Structural Complexity Enhancement

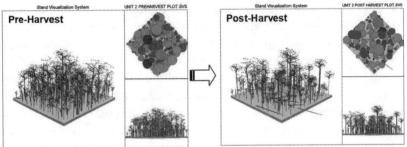

FIGURE 13-2. Visualizations using field data and contrasting 0.01-hectare sample plots in single-tree selection (above) and Structural Complexity Enhancement (SCE) treatments (below). The plots are part of an old-growth silvicultural experiment located on the Mt. Mansfield State Forest in Vermont. Tree coloration is species-specific, with plots seen from oblique, aerial, and side profile views. Differences between these immediately postharvest outcomes are subtle but note the high degree of postharvest structure (e.g., basal area and stem density), vertical complexity, and downed log densities in the SCE plot. Note also the somewhat lesser structural retention for single-tree selection, as well as the lack of downed logs and lower spatial variation in tree density.

techniques described above, but these methods were used jointly to accelerate development of multiple-stand structural characteristics (table 13-1). One of the innovative aspects of the study was a test of an alternative diameter distribution called a "rotated sigmoid" as the postharvest target. Sigmoidal form is one of several possible distributions in eastern US old-growth forests (Goodburn and Lorimer 1999). Although discussed in the theoretical literature (O'Hara 1998), the distribution had not previously been field tested as a way to shift growing-space allocation and, potentially, aid in the recruitment of large trees over time. To determine if SCE was

TABLE 13-1. Silvicultural prescriptions for treatments compared against an old-growth approach, called Structural Complexity Enhancement (SCE), developed for northern hardwood-conifer forests in northern New England, as described in Keeton (2006) and Ford and Keeton (2017). To make it a "fair" test, the study compared SCE against conventional selection systems modified to retain more structure, particularly in medium and larger tree size classes, than is typical in the region. The target basal area (postharvest for selection treatments; desired in the future for SCE), maximum tree diameter retained postharvest (selection treatments) or desired in the future (SCE), and q-factor define the shape of the residual (or postharvest) diameter distribution. For SCE, the variable q-factor approximated the rotated sigmoid distribution sometimes found in old-growth forests. The conventional treatments employed negative exponential distributions. Table modified from Ford and Keeton (2017).

| Target residual basal area $(m^2\ ha^{-1})$ | Max diameter (cm) | q-factor* | Structural objective | Silvicultural prescription |
|---|---|---|---|---|
| **Treatment: SINGLE-TREE SELECTION** | | | | |
| 18.4 | 60 | 1.3 | Structural retention | • Modified residual basal area and diameter distribution |
| | | | Vertically differentiated canopy | • Released advance regeneration<br>• Regenerated new cohort |
| **Treatment: GROUP SELECTION** | | | | |
| 18.4 | 60 | 1.3 | Structural retention | • Modified residual basal area and diameter distribution |
| | | | Vertically differentiated canopy | • Released advance regeneration<br>• Regenerated new cohort |
| | | | Horizontal diversification | • Spatially aggregated harvest (patches ~0.05 hectare) |
| **Treatment: STRCTURAL COMPLEXITY ENHANCEMENT** | | | | |
| 34.0 | 90 | 2.0/<br>1.1/<br>1.3 | Reallocated basal area to larger size class | • Rotated sigmoid diameter distributions<br>• High maximum diameter and target basal area<br>• Retained trees >60 centimeters DBH |
| | | | Vertically differentiated canopy | • Released advance regeneration<br>• Regenerated new cohort |

*continued on next page*

TABLE 13-1. *continued*

| Target residual basal area ($m^2 ha^{-1}$) | Max diameter (cm) | q-factor* | Structural objective | Silvicultural prescription |
|---|---|---|---|---|
| Treatment: STRCTURAL COMPLEXITY ENHANCEMENT | | | | |
| 34.0 | 90 | 2.0/ 1.1/ 1.3 | Horizontal diversi-fication | • Marked trees for variable density<br>• Created small gaps (~0.02 hectare) around crown-released trees |
| | | | Dampened growth decline in larger trees | • Enabled full (3- or 4-sided) and partial (2-sided) crown release |
| | | | Elevated coarse woody material volume and density | • Girdled trees to create snags<br>• Felled and left trees to create downed logs<br>• Pushed or pulled trees over to create tip-up mounds |

* q-factor is the ratio of the number of trees in each successively larger size class

successful, investigators tracked various indicators of stand development, financial viability, late successional biodiversity, tree regeneration, and aboveground carbon pools.

As confirmation of the effectiveness of crown release (see above) and possibly the rotated sigmoid residual diameter distribution, as techniques to enhance the representation of large trees, computer model simulations predict that the stands treated with SCE are likely to develop an average of 5 more large trees (greater than 50 centimeters DBH) per hectare than there would have been without treatment after 50 years. This contrasts with the conventional treatments, where simulations project 10 fewer large trees per hectare on average than would recruit in the absence of timber harvesting (Keeton 2006). Furthermore, an exciting outcome of this work has been finding that SCE can produce carbon storage cobenefits while simultaneously achieving late successional habitat objectives (Figure 13-3). In fact, ten years after harvest, measured aboveground carbon in SCE units had recovered to only 15.9 percent less than simulated no-harvest baselines, compared to 44.9 percent less in conventional treatments (Ford and Keeton 2017).

While all the treatments were generally successful in maintaining the overall richness and/or abundance of understory plants, terrestrial sala-

FIGURE 13-3. The outcome of Structural Complexity Enhancement at a northern hardwood forest 13 years after treatment (top) and in a mixed-woods forest 15 years after treatment (bottom). Both locations are in Vermont, USA. Note the successful development of large trees, canopy differentiation, high levels of aboveground biomass, and other elements of old-growth-like characteristics. Photo credits: W. S. Keeton (top), R. Aszalós (bottom).

manders, and fungi, the diversity of sensitive, late successional, herbaceous plants increased significantly under SCE and decreased significantly in the semiopen canopied conditions within group-selection treatments (McKenny et al. 2006; Smith et al. 2008). Importantly, the understory plant responses were strongly affected by overstory treatment and less influenced by soil chemistry and drought stress. Fungi and salamander responses were strongly associated with microsite characteristics, particularly silviculturally enhanced snag and downed log densities, and increased significantly under SCE, but showed no significant decrease in relatively small gaps (0.05 hectares in size on average) created by group selection (Dove and Keeton 2015). Fully established tree regeneration 13 years after both SCE and the selection harvests was spatially variable (or patchy) but sufficient to maintain desirable stocking and a diverse species mix. However, as many foresters in the northern hardwood region have experienced firsthand, the problem of competition between seedlings and saplings of more commercially desirable species and beech sprouts was a limitation of all these relatively low impact treatments, particularly group selection, and may require active beech control in the future (Gottesman and Keeton 2017).

Managing for old-growth characteristics takes on a different spin in forests with natural disturbance regimes best described as canopy-replacing. In these forests, the silvicultural approach of variable-retention harvesting (VRH) is being used to emulate a key aspect of natural disturbance, specifically retaining biological legacies from the pre- to post-disturbance forest (Franklin et al. 2007). Broadly speaking, biological legacies are defined as organisms, organic materials, and organically derived patterns in soil and vegetation that persist from the pre-disturbance ecosystem into the post-disturbance environment (Franklin et al. 2007).

In practice, legacy management often focuses on structures that enrich the new forest simply because they are large, including live and dead trees (snags) and logs on the forest floor, as well as patches (aggregates) of intact forest. VRH, when used for the purpose of legacy management, addresses the ecological principle of *continuity*—maintaining elements of structure, function, and biota from the original forest during regeneration harvesting (Palik and D'Amato 2017).

Beyond the obvious habitat value of large structures, there is the practical consideration that *large* generally equates to *old*; that is, these structures often take decades or centuries to develop. Thus, VRH, for structural enrichment, focuses on leaving some of the largest structures available at harvest, potentially in perpetuity in the case of live trees; that is, these trees are left to live out their natural life spans and ultimately transition into the snag and downed-log pool. VRH, for structural enrichment, is used globally (Gustafsson et al. 2012). In the eastern United States, it has been used most extensively in the northern Great Lakes region, particularly in pine- (*Pinus*) and aspen- (*Populus*) dominated forests (e.g., Palik and Zasada 2003; Klockow et al. 2013).

The first operational-scale application of VRH in the Lake States was in red pine forests of the Chippewa National Forest in northern Minnesota (Palik and Zasada 2003). This largescale management experiment was designed to evaluate ecosystem responses to dispersed and aggregated retention of pines, which were left in a harvested matrix that was regenerating to mixtures of pines, oaks (*Quercus*), and boreal hardwoods, a composition better reflective of the native ecosystem. An important finding of this work was that the spatial pattern of retention can be manipulated across stands so as to favor tree species of differing shade tolerances, depending on regeneration objectives. Moreover, reduction in resource availability to regenerating seedlings attributed to competition with retained pines can be minimized by concentrating reserve trees in groups surrounding openings. The practical application of these findings was that large, legacy trees can be maintained to structurally enrich the new forest, without greatly compromising new cohort growth (Montgomery et al. 2013). While VRH approaches are not old-growth silviculture in the traditional sense (nor as described earlier in this chapter), their use in forests actively managed for timber products, traditionally using even-aged methods, does represent a strategy for maintaining and enhancing structural elements characteristic of old forests.

## Adaptive Old-Growth Silviculture

A frequent criticism of old-growth forest restoration reflects the concern that this may be increasingly irrelevant in a rapidly changing world. There is the legitimate skepticism a) of whether historical baselines are still relevant in the context of global change, including climate change, biodiversity loss, invasive species, and atmospheric deposition of pollutants; and b) of whether late successional and old-growth forests will be adaptive to future shifts in disturbance regimes, potential loss of foundational species, and the needs of growing human populations. With species range shifts and the probable formation of novel species assemblages in the future—as occurred with past climate changes—managing for compositional baselines inferred from remaining primary forests or reconstructed through other methods (e.g., Cogbill et al. 2003) may no longer be appropriate. While managing for historical baseline composition overall is likely problematic, it still may be important to determine if any of the tree species that are part of the old-growth mix, even if occurring in low abundance, may be future-climate adaptive. This addresses the often cited first line of defense against climate change, specifically, maintaining tree species diversity in managed forests (e.g., Brang et al. 2014). In this context, one or more species in the ecosystem may be future-climate adaptive, so consideration should be given to increasing, maintaining, or restoring their abundance to provide the option to increase the numbers of these species in the future (Messier et al. 2013). Alternatively old-growth compositional objectives would need to track or anticipate range shifts, a difficult proposition at best. Furthermore, with the spread of invasive pests and pathogens and loss of foundational species such as the eastern hemlock (*Tsuga canadensis*) and species-specific structures like large, old American beech (*Fagus americana*), foresters may be working with increasingly simplified systems. And if disturbance intensities and frequencies increase (Diffenbaugh et al. 2013), old-growth structure may be particularly vulnerable on disturbance-prone sites. All these uncertainties beg the question, Are we trying to manage for the forest of the past rather than for the forest of the future?

And yet, it is reasonable to propose that managing for old-growth characteristics could also be a part of an adaptation strategy based on enhancing the resilience and the resistance properties of a landscape (D'Amato et al. 2011). Recent advances in adaptive silviculture have applied the science of complex systems to questions about how to manage for resilience to natural and anthropogenic stress. As pioneers in this field, Millar et al. (2007) proposed a three-pronged framework to guide adaptive silviculture: 1) resistance for forestalling impacts and protecting highly valued resources, 2) resilience for improving the capacity of ecosystems to return to desired conditions after disturbance, and 3) response for facilitating transition from current to new conditions. We wonder if elements of structure and function associated with old-growth may be more resilient to change than is species composition. Old-growth forests could certainly be considered complex adaptive systems (Messier et al. 2013; Fahey et al. 2018). For example, if shifting late successional species assemblages are nevertheless able to maintain high levels of carbon storage or to provide critical habitat elements, such as large snags, downed logs, gap environments, and tip-up mounds, then some structure-function relationships will be maintained. Biodiversity using such habitat elements will still need these structures in the future, even if the tree species providing these structures changes, assuming some degree of plasticity in these relationships, which is certainly not always the case. In this sense, managing for old-growth structure, could help functional processes recover from or accommodate compositional transitions. Moreover, the complex canopies of old-growth forests buffer below-canopy microclimates from fluctuations in climatic influences. This adds ecological resistance to changes occurring above the canopy and has been shown to help biodiversity persist in the face of warming conditions (Frey et al. 2016).

Managing for old-growth characteristics, as reflected in structural, compositional, and functional diversity of these stands, may well provide the flexibility needed to shift or transition forest development in different directions as evolving climate conditions warrant. This kind of diversification is akin to the concept of a diversified financial investment portfolio but applied to forest adaptation. A greater range of investment options better insures ability to adapt to changing market conditions. There is also interest in preferentially managing for certain groups of species based on functional traits that confer either resilience or resistance to disturbances and stress (Aubin et al. 2016). This approach complements old-growth silviculture quite well, as the latter typically entails variants of partial cutting that can incorporate preferential retention of species exhibiting adaptive functional traits.

In short, we remain optimistic that old-growth will not only have a place on the future landscape but will continue to provide important ecosystem services and essential habitat functions. But realizing this future will require a combination of protected-area strategies and silvicultural practices that are adaptive to change, particularly in terms of species composition, invasive species, and altered disturbance regimes.

## Conclusion

Old-growth silviculture in eastern North America has come a long way in the last couple of decades, developing from a largely theoretical proposition to an experimentally tested and operationally vetted endeavor. We have learned that old-growth baselines developed from primary (never cleared) forests are highly variable because of the formative role of disturbance histories and complex successional dynamics in shaping those systems. This means that desired future conditions may also encompass a range of possibilities. Moreover, there is no strict code for what it means to manage for old-growth characteristics. Targets, and the specific techniques employed to achieve them, can be tailored to site conditions and to suit landowner objectives and the desired mix of cobenefits, both timber and nontimber. For example, some degree of coarse woody debris retention or enhancement is easily incorporated into most harvesting systems with minimal trade-offs in terms of foregone timber yield. Other specifics, such as more complex prescriptions involving the form of postharvest diameter distribution or even tip-up mound creation, may or may not be used, depending on landowner interest or operational feasibility. Multicohort or irregular shelterwood systems, such as "expanding gap" systems incorporating permanent retention of legacy trees, are not only well tested (Raymond et al. 2009; Kern et al. 2016) but of increasing use operationally for a variety of objectives, including certain types of bird habitat (Hagenbuch et al. 2013). But these systems should comprise just one component of the overall efforts to diversity stand structures and enhance complexity at landscape scales since multicohort systems do not produce the full range of late successional forest conditions.

Given this inherent flexibility in silviculture oriented toward old-growth, a variety of practical applications present themselves. These range from full-fledged, old-growth restoration—for instance in nature reserves—to habitat provision within managed forests, to carbon forestry focused on promoting high stocking and high levels of biomass, to carefully planned, and probably more limited, interventions in riparian areas to enhance late successional, forest-stream interactions, including flood resilience. Choice of application will determine the appropriate rotation period or entry cycle, the number of harvest entries (e.g., one time only or on-going), and the degree of old-growth structure ultimately developed (see table 13-2). For instance, for restoration purposes in a nature reserve, managers might prefer a single entry, with stand development processes allowed to take over thereafter. In most "improved forest management" carbon applications, landowners generally favor on-going, active timber management, so long as it maintains the high levels of stocking encouraged by existing carbon market protocols. In the former, we might expect the full array of old-growth characteristics to develop over time, whereas in the latter, perhaps only some of these will develop and possibly to a lesser degree.

Clearly silviculture promoting the characteristics of old-growth forests can contribute to biodiversity conservation and terrestrial carbon storage in eastern forest ecosystems while providing both timber and nontimber economic opportunities. Active management for this purpose will complement strategies of protected-areas and can comprise one element of adaptive management (Keeton 2007). In the course of their professional careers, the authors of this chapter have seen growing accep-

TABLE 13-2. Examples of applications of old-growth silviculture, varying management objective, number of harvest entries, and degree of expected late successional structural development.

| Application | No. of Entries | Late Successional Structural Development | Carbon Storage Emphasis |
|---|---|---|---|
| Old-growth restoration | One or possibly two entries | High | High |
| Riparian management | Single or multiple | Moderate to high | Moderate to high |
| Timber emphasis | Multiple | Low to moderate | Low to moderate |

tance of old-growth silvicultural concepts and expanding application in a variety of contexts across a range of forestland ownerships—public and private. With continued experimentation and demonstration, the future looks bright for old-growth silviculture as a tool applicable to both restoration and integrated management on working forests, provided concepts of late successional forest structure and function can adapt to changing environmental-boundary conditions.

# References

Angers, V. A., C. Messier, M. Beaudet, and A. Leduc. 2005. "Comparing composition and structure in old-growth and harvested (selection and diameter-limit cuts) northern hardwood stands in Quebec." *Forest Ecology and Management 217*: 275–293.

Aubin, I., A. D. Munson, F. Cardou, P. J. Burton, N. Isabel, J. H. Pedlar, and C. Messier. 2016. "Traits to stay, traits to move: a review of functional traits to assess sensitivity and adaptive capacity of temperate and boreal trees to climate change." *Environmental Reviews 24*: 164–186.

Bauhus, J., K. Puettmann, and C. Messier. 2009. "Silviculture for old-growth attributes." *Forest Ecology and Management 4*: 525–537.

Boal, C. W., D. E. Andersen, and P. L. Kennedy. 2005. "Foraging and nesting habitat of breeding male northern goshawks in the Laurentian mixed forest province, Minnesota." *Journal of Wildlife Management 69*: 1516–1527.

Bolton, N. W., and A. W. D'Amato 2011. "Regeneration responses to gap size and coarse woody debris within natural disturbance-based silvicultural systems in northeastern Minnesota, USA." *Forest Ecology and Management 262*: 1215–1222.

Bormann, F. H., and G. E. Likens. 1979. *Pattern and Process in a Forested Ecosystem.* New York: Springer Verlag.

Brang, P., P. Spathelf, J. B. Larsen, J. Bauhus, A. Bončina, C. Chauvin, L. Drössler, et al. 2014. "Suitability of close-to-nature silviculture for adapting temperate European forests to climate change." *Forestry 87*: 492–503.

Burrascano, S., W. S. Keeton, F. M. Sabatini, and C. Blasi. 2013. "Commonality and variability in the structural attributes of moist temperate old-growth forests: A global review." *Forest Ecology and Management 291*: 458–479.

Choi, J., C. G. Lorimer, and J. M. Vanderwerker. 2007. "A simulation of the development and restoration of old-growth structural features in northern hardwoods." *Forest Ecology and Management 249*: 204–220.

Cogbill, C., J. Burk, and G. Motzkin. 2003. "The forests of presettlement New England, USA: spatial and compositional patterns based on town proprietor surveys." *Journal of Biogeography 29*: 1279–1304.

Crow, T. R., R. D. Jacobs, R. R. Oberg, and C. H. Tubbs. 1981. "Stocking and structure for maximum growth in sugar maple selection stands." *Research Paper NC-199*. North Central Forest Experiment Station. St. Paul, MN: USDA Forest Service.

Curzon, M. T., and W. S. Keeton. 2010. "Spatial characteristics of canopy disturbances in riparian old-growth hemlock-northern hardwood forests, Adirondack Mountains, New York, USA." *Canadian Journal of Forest Research 40*: 13–25.

D'Amato, A. W., J. B. Bradford, S. Fraver, and B. J. Palik. 2011. "Forest management for mitigation and adaptation to climate change: insights from long-term silviculture experiments." *Forest Ecology and Management 262*: 803–816.

D'Amato, A. W., P. F. Catanzaro, and L. S. Fletcher. 2015. "Early regeneration and structural responses to patch selection and structural retention in second-growth northern hardwoods." *Forest Science 61*: 183–189.

D'Amato, A. W., and D. A. Orwig. 2008. "Stand and landscape-level disturbance dynamics in old-growth forests in western Massachusetts." *Ecological Monographs 78*: 507–522.

DeGraaf, R. M., and M. Yamasaki. 2001. *New England Wildlife*. Hanover and Lebanon, NH: University Press of New England.

Della-Bianca, L., and D. E. Beck. 1985. "Selection management in southern Appalachian hardwoods." *Southern Journal of Applied Forestry 9*: 191–196.

Després, T., H. Asselin, F. Doyon, and Y. Bergeron. 2014. "Structural and spatial characteristics of old-growth temperate deciduous forests at their northern distribution limit." *Forest Science 60*: 871–880.

Diffenbaugh, N. S., M. Scherer, and R. J. Trapp. 2013. "Robust increases in severe thunderstorm environments in response to greenhouse forcing." *Proceedings of the National Academy of Sciences 110*: 16361–16366.

Donato, D. C., J. L. Campbell, and J. F. Franklin. 2012. "Multiple successional pathways and precocity in forest development: can some forests be born complex?" *Journal of Vegetation Science 23*: 576–584.

Dove, N. C., and W. S. Keeton. 2015. "Structural complexity enhancement increases fungi diversity in northern hardwood forests." *Fungal Ecology 13*: 181–192.

Doyon, F., D. Gagnon, and J. F. Giroux. 2005. "Effects of strip and single-tree selection cutting on birds and their habitat in a southwestern Quebec northern hardwood forest." *Forest Ecology and Management 209*: 101–116.

Fahey, R. T., B. Alveshere, J. I. Burton, A. D'Amato, Y. L. Dickinson, W. S. Keeton, C. C. Kern, et al. 2018. "Shifting conceptions of complexity in forest management and silviculture." *Forest Ecology and Management*. In Press.

Fahey, R. T., and C. G. Lorimer. 2013. "Restoring a midtolerant pine species as a component of late-successional forests: results of gap-based planting trials." *Forest Ecology and Management 292*: 139–149.

Fassnacht, K. S., D. R. Bronson, B. J. Palik, A. W. D'Amato, C. Lorimer, G., and K. J. Martin. 2015. "Accelerating the development of old-growth characteristics in second-growth northern hardwoods." General Technical Report NRS-144. Northern Research Station. Newtown Square, PA: USDA Forest Service.

Fassnacht, K. S., and T. W. Steele. 2016. "Snag dynamics in northern hardwood forests under different management scenarios." *Forest Ecology and Management 363*: 267–276.

FEMAT [Forest Ecosystem Management Assessment Team]. 1993. *Forest Ecosystem Management: An Ecological, Economic, and Social Assessment.* Portland, OR: USDA Forest Service.

Ford, S. E., and W. S. Keeton. 2017. "Enhanced carbon storage through management for old-growth characteristics in northern hardwoods." *Ecosphere 8*: 1–20.

Forrester, J. A., D. J. Mladenoff, and S. T. Gower. 2013. "Experimental manipulation of forest structure: near-term effects on gap and stand scale C dynamics." *Ecosystems 16*: 1455–1472.

Franklin, J. F., D. R. Berg, D. A. Thornburgh, and J. C. Tappeiner. 1997. "Alternative silvicultural approaches to timber harvesting: variable retention harvest system." In *Creating a Forestry for the 21st Century: The Science of Ecosystem Management*, edited by K. A. Kohm, and J. F. Franklin, 111–140. Washington, DC: Island Press.

Franklin, J. F., R. J. Mitchell, and B. Palik. 2007. "Natural disturbance and stand development principles for ecological forestry." General Technical Report NRS-19. Northern Research Station. Newton Square, PA: USDA Forest Service.

Franklin, J. F., T. A. Spies, R. Van Pelt, A. B. Carey, D. A. Thornburgh, D. R. Berg, D. B. Lindenmayer, M. E. Harmon, W. S. Keeton, D. C. Shaw, K. Bible, and J. Chen. 2002. "Disturbances and structural development of natural forest ecosystems with silvicultural implications, using Douglas-fir forests as an example." *Forest Ecology and Management 155*: 399–423.

Frey, S., A. Hadley, S. Johnson, M. Schulze, J. Jones, and M. Betts. 2016. "Spatial models reveal the microclimatic buffering capacity of old-growth forests." *Science Advances 22*: e1501392.

Gauthier, M-M., M-C. Lambert, and S. Bedard. 2016. "Effects of harvest gap size, soil scarification, and vegetation control on regeneration dynamics in sugar maple-yellow birch stands." *Forest Science 62*: 237–246.

Goodburn, J. M. 1996. "Comparison of forest habitat structure and composition in old-growth and managed northern hardwoods in Wisconsin and Michigan." Master's thesis, University of Wisconsin–Madison.

Goodburn, J. M., and C. G. Lorimer. 1999. "Population structure in old-growth and managed northern hardwoods: an examination of the balanced diameter distribution concept." *Forest Ecology and Management 118*: 11–29.

Gottesman, A., and W. S. Keeton. 2017. "Regeneration responses to management for old-growth characteristics in northern hardwood-conifer forests." *Forests 8*: 1–21.

Gronewald, C. A., A. W. D'Amato, and B. J. Palik. 2010. "The influence of cutting cycle and stocking level on the structure and composition of managed old-growth northern hardwoods." *Forest Ecology and Management 259*: 1151–1160.

Gustafson, E. J., and T. R. Crow. 1996. "Simulating the effects of alternative forest management strategies on landscape structure." *Journal of Environmental Management 47*: 77–94.

Gustafsson, L., S. C. Baker, J. Bauhus, W. J. Beese, A. Brodie, J. Kouki, D. B. Lindenmayer, et al. 2012. "Retention forestry to maintain multifunctional forests: a world perspective." *Bioscience 62*: 633–645.

Hagenbuch, S., K. Manaras, N. Patch, J. Shallow, K. Sharpless, M. Snyder, and K. Thompson 2013. *Managing Your Woods with Birds in Mind: A Vermont Landowner's Guide.* Huntington and Waterbury, VT: Audubon Vermont and Vermont Department of Forests, Parks, and Recreation.

Hale, C. M., J. Pastor, and K. A. Rusterholtz. 1999. "Comparison of structural and compositional characteristics in old-growth and mature, managed hardwood forests of Minnesota, U.S.A." *Canadian Journal of Forest Research 29*: 1479–1489.

Halpin, C. R., and C. G. Lorimer. 2016. "Trajectories and resilience of stand structure in response to variable disturbance severities in northern hardwoods." *Forest Ecology and Management 365*: 69–82.

Halpin, C. R., and C. G. Lorimer. 2017. "Predicted long-term effects of group selection on species composition and stand structure in northern hardwood forests." *Forest Ecology and Management 400*: 677–691.

Hanson, J. J., and C. G. Lorimer. 2007. "Forest structure and light regimes following moderate wind storms: Implications for multi-cohort management." *Ecological Applications 17*: 1325–1340.

Hanson, J. J., C. G. Lorimer, C. R. Halpin, and B. J. Palik. 2012. "Ecological forestry in an uneven-aged, late-successional forest: simulated effects of contrasting treatments on structure and yield." *Forest Ecology and Management 270*: 94–107.

Hardt, R. A., and W. T. Swank. 1997. "A comparison of structural and compositional characteristics of southern Appalachian young second-growth, maturing second-growth, and old-growth stands." *Natural Areas Journal 17*: 42–52.

Heinselman, M. L. 1973. "Fire in the virgin forests of the Boundary Waters Canoe Area, Minnesota." *Quaternary Research 3*: 329–382.

Hunter, W. C., D. A. Buechler, R. A. Canterbury, J. L. Confer, and P. B. Hamel. 2001. "Conservation of disturbance dependent birds in eastern North America." *Wildlife Society Bulletin 29*: 425–439.

Keeton, W. S. 2006. "Managing for late-successional/old-growth characteristics in northern hardwood-conifer forests." *Forest Ecology and Management 235*: 129–142.

Keeton, W. S. 2007. "Role of managed forestlands and models for sustainable forest management: perspectives from North America." *George Wright Forum 24*: 38–53.

Keeton, W. S., E. M. Copeland, and M. C. Watzin. 2017. "Towards flood resilience: exploring linkages between riparian forest structure and geomorphic condition in northeastern U.S. streams." *Canadian Journal of Forest Research 47*: 476–487.

Keeton, W. S., C. E. Kraft, and D. R. Warren. 2007. "Mature and old-growth riparian forests: structure, dynamics, and effects on Adirondack stream habitats." *Ecological Applications 17*: 852–868.

Keeton, W. S. and A. R. Troy. 2006. "Balancing ecological and economic objectives while managing for late-successional forest structure." In *Ecologisation of Economy as a Key Prerequisite for Sustainable Development. Proceedings of the International Conference*, Sept. 22–23, 2005, edited by L. Zahvoyska, 22–23. L'viv, Ukraine: Ukrainian National Forestry University.

Keeton, W. S., A. A. Whitman, G. G. McGee, and C. L. Goodale. 2011. "Late-successional biomass development in northern hardwood-conifer forests of the northeastern United States." *Forest Science 57*: 489–505.

Kerchner, C. and W. S. Keeton. 2015. "California's regulatory forest carbon market: panacea or Pandora's box for northeastern landowners?" *Forest Policy and Economics 50*: 70–81.

Kern, C. C., J. Burton, P. Raymond, A. D'Amato, W. S. Keeton, A. A. Royo, M. B. Walters, C. R. Webster, and J. L. Willis. 2016. "Challenges facing gap-based silviculture and possible solutions for mesic northern forests in North America." *Forestry 90*: 4–17.

Kern, C. C., A. W. D'Amato, and T. F. Strong. 2013. "Diversifying the composition and structure of managed, late-successional forests with harvest gaps: what is the optimal gap size?" *Forest Ecology and Management 304*: 110–120.

Kern, C. C., P. B. Reich, R. A. Montgomery, and T. F. Strong. 2012. "Do deer and shrubs override canopy gap size effects on growth and survival of yellow birch, northern red oak, eastern white pine, and eastern hemlock seedlings?" *Forest Ecology and Management 267*: 134–143.

Klockow, P. A., A. W. D'Amato, and J. B. Bradford. 2013. "Impacts of post-harvest slash and live-tree retention on biomass and nutrient stocks in *Populus tremuloides* Michx.-dominated forests, northern Minnesota USA." *Forest Ecology and Management 291*: 278–288.

Leak, W. B. 1999. "Species composition and structure of a northern hardwood stand after 61 years of group/patch selection." *Northern Journal of Applied Forestry 16*: 151–153.

Leak, W. B., and P. E. Sendak. 2002. "Changes in species, grade, and structure over 48 years in a managed New England northern hardwood stand." *Northern Journal of Applied Forestry 19*: 25–27.

Lorimer, C. G., and C. R. Halpin. 2014. "Classification and dynamics of developmental stages in late-successional temperate forests." *Forest Ecology and Management 334*: 344–357.

Lorimer, C. G., and A. S. White. 2003. "Scale and frequency of natural disturbances in the northeastern U.S.: Implications for early-successional forest habitats and regional age distributions." *Forest Ecology and Management 185*: 41–64.

McGarvey, J. C., J. R. Thompson, H. E. Epstein, and H. H. Shugart, Jr. 2015. "Carbon storage in old-growth forests of the Mid-Atlantic: toward better understanding the eastern forest carbon sink." *Ecology 96*: 311–317.

McGee, G. G., D. J. Leopold, and R. D. Nyland. 1999. "Structural characteristics of old-growth, maturing, and partially cut northern hardwood forests." *Ecological Applications 9*: 1316–1329.

McKenny, H. C., W. S. Keeton, and T. M. Donovan. 2006. "Effects of structural complexity enhancement on eastern red-backed salamander (*Plethodon cinereus*) populations in northern hardwood forests." *Forest Ecology and Management 230*: 186–196.

McLachlan, J. S., D. R. Foster, and F. Menalled. 2000. "Anthropogenic ties to late-successional structure and composition in four New England hemlock stands." *Ecology 81*: 717–733.

Meigs, G. W. and W. S. Keeton. 2018. "Intermediate-severity wind disturbance in mature temperate forests: effects on legacy structure, carbon storage, and stand dynamics." *Ecological Applications.* doi:10.1002/eap.1691.

Messier, C., K. Puettmann, and D. Coates. 2013. *Managing Forests as Complex Adaptive Systems: Building Resilience to the Challenge of Global Change.* New York, NY: Routledge.

Millar, C. I., N. L. Stephenson, and S. L. Stephens. 2007. "Climate change and forests of the future: managing in the face of uncertainty." *Ecological Applications 17*: 2145–2151.

Miller, G. W., J. N. Kochenderfer, and D. B. Fekedulegn. 2006. Influence of individual reserve trees on nearby reproduction in two-aged Appalachian hardwood stands. *Forest Ecology and Management 224*: 241–251.

Mladenoff, D. J., M. A. White, T. R. Crow, and J. Pastor. 1994. "Applying principles of landscape design and management to integrate old-growth forest enhancement and commodity use." *Conservation Biology 8*: 752–762.

Montgomery, R. A., B. J. Palik, S. B. Boyden, and P. B. Reich. 2013. "New cohort growth and survival in variable retention harvests of a pine ecosystem in Minnesota, USA." *Forest ecology and management 310*: 327–335.

North, M. P. and W. S. Keeton. 2008. "Emulating natural disturbance regimes: an emerging approach for sustainable forest management. In *Patterns and Processes in Forest Landscapes - Multiple Use and Sustainable Management*, edited by R. Lafortezza, J. Chen, G. Sanesi, and T. R. Crow, 341–372. The Netherlands: Springer Verlag.

Nunery, J. S. and W. S. Keeton. 2010. "Forest carbon storage in the northeastern United States: Net effects of harvesting frequency, post-harvest retention, and wood products." *Forest Ecology and Management 259*: 1363–1375.

Odum, E. P. 1969. "The strategy of ecosystem development." *Science 164*: 262 – 270.

O'Hara, K. L. 1998. "Silviculture for structural diversity: a new look at multi-aged systems." *Journal of Forestry 96*: 4–10.

Owen, S. F., M. A. Menzel, and J. W. Edwards. 2004. "Bat activity in harvested and intact forest stands in the Allegheny Mountains." *Northern Journal of Applied Forestry 21*: 154–159.

Palik, B. J., and A. W. D'Amato. 2017. "Ecological forestry: much more than retention harvesting." *Journal of Forestry 115*: 51.

Palik, B. J., R. A. Montgomery, P. B. Reich, and S. B. Boyden. 2014. "Biomass growth response to spatial pattern of variable-retention harvesting in a northern Minnesota pine ecosystem." *Ecological Applications 24*: 2078–2088.

Palik, B. J., and J. Zasada. 2003. "An ecological context for regenerating multi-cohort, mixed species red pine forests." *Research Note NC-382*. North Central Research Station. St. Paul, MN: USDA Forest Service.

Puettmann, K., D. Coates, and C. Messier. 2009. *A Critique of Silviculture: Managing for Complexity.* Washington, DC: Island Press.

Puettmann, K. J., S. M. Wilson, S. C. Baker, P. Donoso, L. Drossler, G. Amente, B. D. Harvey, et al. 2015. "Silvicultural alternatives to conventional even-aged forest management— What limits global adoption?" *Forest Ecosystems 2*: 1–16.

Raymond, P., S. Bédard, V. Roy, C. Larouche, and S. Tremblay. 2009. "The irregular shelterwood system: review, classification, and potential application to forests affected by partial disturbances." *Journal of Forestry 107*: 405–413.

Rhemtulla, J. M., D. J. Mladenoff, and M. K. Clayton. 2007. "Regional land-cover conversion in the US Upper Midwest: magnitude of change and limited recovery (1850–1935–1993)." *Landscape Ecology 22*: 57–75.

Rhemtulla, J. M., D. J. Mladenoff, and M. K. Clayton. 2009. "Historical forest baselines reveal potential for continued carbon sequestration." *Proceedings of the National Academy of Sciences 106*: 6082–6087.

Runkle, J. R., and T. C. Yetter. 1987. "Treefalls revisited: gap dynamics in the southern Appalachians." *Ecology 68*: 417–424.

Schuler, T. M. 2004. "Fifty years of partial harvesting in a mixed mesophytic forest: composition and productivity." *Canadian Journal of Forest Research 34*: 985–997.

Schwartz, N. B., D. L. Urban, P. S. White, A. Moody, and R. N. Klein. 2016. "Vegetation dynamics vary across topographic and fire severity gradients following prescribed burning in Great Smoky Mountains National Park." *Forest Ecology and Management 365*: 1–11.

Selva, S. B. 2003. "Using calicioid lichens and fungi to assess ecological continuity in the Acadian Forest Ecoregion of the Canadian Maritimes." *The Forestry Chronicle 79*: 550–558.

Senécal, J-F, F. Doyon, and C. Messier. 2018. "Management implications of varying gap detection height thresholds and other canopy dynamics processes in temperate deciduous forests." *Forest Ecology and Management*. In Press.

Seymour, R. S., A. S. White, and P. H. deMaynadier. 2002. "Natural disturbance regimes in northeastern North America: evaluating silvicultural systems using natural scales and frequencies." *Forest Ecology and Management 155*: 357–367.

Singer, M. T., and C. G. Lorimer. 1997. "Crown release as a potential old-growth restoration approach in northern hardwoods." *Canadian Journal of Forest Research 27*: 1222–1232.

Smith, K. J., W. S. Keeton, M. Twery, and D. Tobi. 2008. "Understory plant response to alternative forestry practices in northern hardwood-conifer forests." *Canadian Journal of Forest Research 38*: 1–17.

Spies, T. A. 1997. "Forest stand structure, composition, and function." In *Creating a Forestry for the 21st century,* edited by K. A. Kohm, and J. F. Franklin, 11–30. Washington, DC: Island Press.

Stribling, H. L., H. R. Smith, and R. H. Yahner. 1990. "Bird community response to timber stand improvement and snag retention." *Northern Journal of Applied Forestry 7*: 35–38.

Tyrrell, L. E., and T. R. Crow. 1994. "Structural characteristics of old-growth hemlock-hardwood forests in relation to age." *Ecology 75*: 370–386.

Urbano, A. R. and W. S. Keeton. 2017. "Forest structural development, carbon dynamics, and co-varying habitat characteristics as influenced by land-use history and reforestation approach." *Forest Ecology and Management 392*: 21–35.

Walters, M. B., E. J. Farinosi, J. L. Willis, and K. W. Gottschalk. 2016. "Managing for diversity: harvest gap size drives complex light, vegetation, and deer herbivory impacts on tree seedlings." *Ecosphere 7*: e01397.

Warren, D. R., W. S. Keeton, P. M. Kiffney, M. J. Kaylor, H. A. Bechtold, and J. Magee. 2016. "Changing forests-changing streams: riparian forest stand development and ecosystem function in temperate headwaters." *Ecosphere 7* (8). doi:10.1002/ecs2.1435.

Webster, C. R. and C. G. Lorimer. 2005. "Minimum opening sizes for canopy recruitment of midtolerant tree species: a retrospective approach." *Ecological Applications 15*: 1245–1262.

Woods, K. D. 2000. "Dynamics in late-successional hemlock-hardwood forests over three decades." *Ecology 81*: 110–126.

Zenner, E. K., S. A. Acker, and W. H. Emmingham. 1998. "Growth reduction in harvest-age, coniferous forests with residual trees in the western central Cascade Range of Oregon." *Forest Ecology and Management 102*: 75–88.

Ziegler, S. S. 2000. "A comparison of structural characteristics between old-growth and post-fire second-growth hemlock-hardwood in Adirondack Park, New York, U.S.A." *Global Ecology and Biogeography 9*: 373–389.

Ziegler, S. S. 2002. "Disturbance of hemlock-dominated old-growth in northern New York, USA." *Canadian Journal of Forest Research 32*: 2106–2115.

## Chapter 14

# Source or Sink? Carbon Dynamics in Eastern Old-Growth Forests and Their Role in Climate Change Mitigation

*William S. Keeton*

For decades forest scientists have thought that old-growth temperate forests were either carbon neutral or even carbon sources, emitting more greenhouse gases to the atmosphere through respiration and decomposition than they were absorbing through photosynthesis. However, recent research has questioned that assumption, showing that eastern old-growth forests may remain productive and have net positive carbon uptake later into succession and stand development than previously thought. These findings remain contentious and yet have profound implications for our understanding of the role of high-biomass, late successional forests in global carbon budgets. Emerging science strongly supports conservation of old-growth forests and management for old-growth structure as effective strategies in global efforts to reduce carbon dioxide emissions and moderate the intensity of future climate change (Luyssaert et al. 2008; Keith et al. 2009; Burrascano et al. 2013).

With forest ecosystems in eastern North America under increasing stress from rural sprawl and habitat fragmentation, climate disruption, and invasive species, protection of high-conservation-value forests, such as old growth, is more urgent than ever before. At the same time, many have wondered if conservation of remaining old-growth stands and active management to build old-growth characteristics into working forests would have climate mitigation benefits. This idea rests on the assumption that sequestered carbon is often stored at exceptionally high levels in old-growth forests, and thus represents "money in the bank" that might otherwise flux to the atmosphere if cleared or otherwise poorly managed. But the proposition has been bedeviled by conflicting views about carbon dynamics (i.e., rates of uptake, long-term storage capacity, flux rates between carbon pools, etc.) in late successional and old-growth forests, with some portraying old growth as a carbon source or at best carbon neutral. Set against this is the argument that maintaining carbon storage in high-biomass forests like old growth is a key element of climate mitigation and that the sink capacity of old growth has been significantly underestimated. Major advances in our scientific understanding of carbon dynamics, in relation to long-term forest development over roughly the last 30 years, have the potential to resolve this debate.

This chapter summarizes the intriguing and ongoing debate surrounding long-term carbon dynamics in older eastern forests. Carbon dynamics may change in the future as compounded, often interacting stressors and disturbances influence forest development, physiology, and distribution. Therefore, we must acknowledge that the role of old-growth forests as carbon sinks may shift in the future. The chapter explores a variety of options for strengthening the carbon sink capacity of eastern forested landscapes through conservation, restoration, and management for old-growth stands and structures.

## Early Ideas About Carbon Dynamics in Late Successional Forests

What happens to carbon uptake (or "sequestration") and accumulation (or "storage") in organic matter and free soil carbon as forests age? This is a question that ecologists have debated for decades. Forest carbon is sequestered—soaked up from the atmosphere—by plants (autotrophs) through photosynthesis, passed along through consumption to higher trophic levels (heterotrophs), stored in living and dead biomass, and eventually released back to the atmosphere in the form of carbon dioxide through plant and animal respiration and decomposition. These flux pathways—inputs and outputs of carbon from an ecosystem and among "pools" (e.g., live or dead biomass, whether standing or downed, whether above- or belowground, etc.) within the system—describe the forest carbon cycle and are at the crux of whether or not old-growth forests are carbon sinks (absorbing more carbon than they emit) or sources (emitting more carbon than they absorb).

To ecologists in the 1960s and earlier, it seemed reasonable that carbon dynamics would conform to those of largely theoretical systems, such as experimental microcosms, in which net growth was assumed to eventually plateau and even decline as a forest aged. This theory formed the basis of Eugene Odum's (1969) paper "The Strategy of Ecosystem Development," one of the most influential papers in successional ecology of its day. Odum postulated that, as forests age, rates of Net Primary Production (net carbon fixation through photosynthesis by plants after deducting autotrophic respiration and mortality) will peak after an initial period (e.g., decades to more than a century) of vigorous tree growth, decreasing afterwards as dominant trees reach their maximum sizes and growth rates decline. The slowdown should be a result both of the greater metabolic needs of older trees and, in some cases, as later research showed, hydrologic limitations and nutrient availability (Harmon 2001). At the same time, in this model, Net Primary Production is declining and heterotrophic respiration is increasing as more of the energy flow passes through the sequence of herbivores and carnivores and as dead organic material is processed by detritivores. Therefore, in a typical eastern deciduous or coniferous forest, for example, after the first one to two centuries of positive biomass accumulation, Net Ecosystem Productivity (NEP, the total amount of carbon added to a system after losses from autotrophic and heterotrophic respiration) should level off and ultimately stabilize close to zero.

This general model was widely applied and continues to underpin many projections of forest carbon budgets under climate change and land-use scenarios. In their seminal work, Bormann and Likens (1979) elaborated on the relationship between forest age and biomass, breaking these into four stages of development following stand-replacing disturbance or timber harvesting with even-aged regeneration: 1) reorganization, 2) aggradation, 3) transition, and 4) steady state. They based the biomass accumulation trends for the first two stages on empirical data collected in secondary forest stands located at the Hubbard Brook Experimental Forest in New Hampshire, and elsewhere. In contrast, the biomass dynamics, projected using a computer model for the latter stages of development, were largely theoretical but incorporated what was then the relatively new understanding of gap dynamics in late successional forests. This adjustment acknowledged the important role of density-independent mortality in late successional stands, driven, in this case, by finely scaled natural disturbances like windthrows and pathogens that kill small groups of trees (Runkle 1982). As a result, the Bormann and Likens model predicted that northern hardwood forests would maintain positive net productivity until peaking after about 170 years of stand development, followed by declining biomass in stands 200 to 350 years of age, and steady-state (i.e., shifting gap) biomass dynamics in stands older than 350 years of age. NEP was predicted to decline to zero "at full maturity" based on the earlier work by Odum (1969) and others (e.g. Whitaker et al. 1974).

An important qualifier in the following discussion is that positive NEP does not necessarily indicate carbon accumulation unless export and non-biological oxidation of carbon have been accounted for or shown to be negligible (Lovett 2006). Since the chapter refers to both carbon and biomass, the reader should be aware that carbon content is about one-half of biomass in live organic matter, declining with decomposition in dead organic matter.

## Old-Growth Forests as a Long-Term Carbon Sink

In the 1970s, the presumed lack of old-growth forests in the eastern United States was given as the reason for relying on theoretical projections for late successional biomass trends (Bormann and Likens 1979). But since that time, researchers have discovered and inventoried far more old growth in the eastern United States than was previously thought to exist (Davis 1996). These sites have included some primary forest blocks and landscapes that are not anomalous in terms of geophysical environment, as many previously mapped old-growth remnants were thought to be, surviving only because of inaccessibility or flukes of land-use history (Keeton et al. 2007; Rhemtulla et al. 2007; Fraver et al. 2009; Després et al. 2014; McGarvey et al. 2015). Examples include forests in northern Minnesota and Wisconsin, the Adirondack region of New York State, the White Mountains of New Hampshire, and parts of the southern Appalachian Mountains. Having places like these to work in gives researchers the opportunity to study carbon fluxes and other ecological processes in dynamic forest landscapes and across a range of site conditions, encompassing variations in disturbance history, successional pathway, soil productivity, landform, and local climate. New evidence from these investigations challenges the notion that NEP consistently declines to zero in old growth or that biomass peaks relatively early after just a couple of centuries of postdisturbance development.

This challenge has come mainly from two types of studies, those quantifying carbon storage (biomass) biometrically in old-growth forests, then comparing those measurements against younger forests, and those directly measuring gas exchange and carbon flux across forest canopies. To be clear, the point in contention is not whether old-growth forests store large amounts of carbon; on that point studies agree (Keith et al. 2009; Burrascano et al. 2013). Rather, the debate is whether NEP declines to zero in forests more than two centuries old (i.e., old growth as carbon neutral in terms of additional uptake capacity) or whether it continues to hover somewhere above zero for a significantly longer period (i.e., old-growth as a net sink, adding additional carbon to already high storage).

One of the first studies to provoke a reconsideration of earlier views was conducted in the US Pacific Northwest, renewing interest in the carbon sink potential of old-growth temperate forests generally and the negative effects on overall carbon balance—even accounting for sequestration rates and the life cycle of carbon in wood products—of converting old forests into young stands (Harmon et al. 1990). Since then a growing number of studies in the eastern United States and other temperate forest systems globally reached similar conclusions (Burrascano et al. 2013). These show consistently higher biomass or carbon density in old-growth forests (figure 14-1) compared to young and mature forests (Keith et al. 2009; Keeton et al. 2011; McGarvey et al. 2015), and, in some (but not all) cases, positive NEP values continuing later than expected in late successional and old-growth forests (Luyssaert et al. 2008; Gough et al. 2016). Many scientists now see a fundamental problem with generalizing about productivity from models grounded in observations of even-aged stands and extrapolating these assumptions to structurally complex, multi- or uneven-aged forests (Carey et al. 2001; Fahey et al. 2015).

Not all eastern old-growth forests are net carbon sinks, but many are. The idea that NEP might not decline to zero was put forth perhaps most notably in a paper by Luyssaert et al. (2008) using data points from 519 temperate (70 percent) and boreal (30 percent) study sites around the world. Reviewing these carbon balance studies, many of which used a method called "eddy covariance" to measure carbon dioxide flux above the canopy, the authors concluded that the vast majority of late successional forests were net carbon sinks, accumulating biomass much later into stand development and at much greater ages than previously thought. A limitation of large meta-analyses like Luyssaert et al. is the generalization of trends based on heterogeneous datasets, encompassing differing forest types, life spans, climates, and site qualities. However, other studies in eastern forests have supported this conclusion, though findings vary by forest type and site conditions (see, for example, Gower et al. 1996). For example, an analysis of deciduous and mixed deciduous-coniferous sites in the US Northeast and Midwest, compared also to similar forests in Europe and Asia, found sustained positive NEP assessed across the full range of late successional forest ages (up to 400 years; Gough et al. 2016). Similarly, high levels of net annual carbon accumulation (3.0 megagrams per hectare) were measured using the eddy covariance method in an approximately 200-year-old eastern-hemlock (*Tsuga canadensis*) stand in New England; about twice the typical

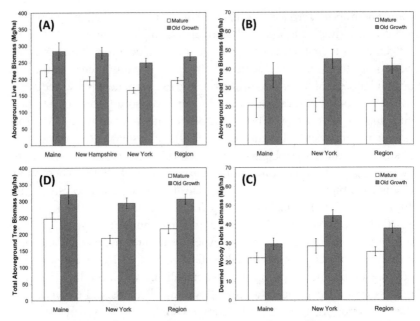

FIGURE 14-1. Aboveground biomass in live (A), dead (B), downed woody debris (C), and total tree (live + dead, D) for northern hardwood and hemlock-hardwood forests in Maine, New Hampshire, New York, and combined (Region). Only data for live tree biomass were available for New Hampshire. The figure compares biomass in mature stands (approximately 80 to 150 years of age) with primary old-growth stands (dominant tree ages older than 150 years). The datasets are described in Keeton et al. (2011) and were contributed by W. S. Keeton and G. McGee (Adirondack Mountains, New York), C. Goodale (White Mountains, New Hampshire), and A. Whitman (northern Maine). There were 94 sites in total (48 mature, 46 old-growth). Sample sizes are as follows: New York—mature n = 15, old-growth n = 21; New Hampshire—mature n = 10, old-growth = 5; Maine—mature n = 20, old-growth n = 20. Error bars are +/- one standard error of the mean.

rate for secondary forests in this region (Hadley and Schedlbauer 2002). Long-term work in a 110-year-old oak-maple (*Quercus-Acer*) stand on the Harvard Forest in Massachusetts indicates that, not only has aboveground biomass accumulated linearly over the preceding 42-year measurement period, but rates of accumulation are now ticking upwards. This does not mean that growth rates and NEP will not eventually decline as that forest continues to age, but it is curious that growth rates are actually increasing in the vigorous, dominant canopy trees as they emerge the "winners" from earlier periods of intense interstem competition. A similar phenomenon has been observed in temperate and boreal forests globally, suggesting that old-growth trees have the capacity for "renewed growth and physiological function postmaturity" (Philips et al. 2008: 1355), allowing their net growth rates to sometimes increase late in stand development contrary to expectation (Stephenson et al. 2014). While intriguing, more research is needed to substantiate how broadly applicable these findings are to eastern North America. For example, interannual climate variability can tilt the balance from positive to negative NEP (Pan et al. 2009). And furthermore, it will be important to separate these responses from likely growth increases, both current and future, due to carbon dioxide fertilization (caused by rising carbon dioxide levels in the atmosphere) and airborne nitrogen deposition (Fisk et al. 2002; Gough et al. 2016).

## Disturbance as a Key Driver of Carbon Sink Capacity

Natural disturbances help maintain positive carbon sink capacities in late successional forested landscapes. But disturbance intensity and extent influence carbon dynamics in fundamentally different ways. The finely scaled disturbances most common in eastern forested landscapes, such as high frequency, canopy gap-creating events, maintain and increase structural complexity in late successional forests (Runkle 1982; Kern et al. 2016). These regenerate new cohorts of trees, release already established regeneration in the understory, and free up growing space and access to light for midcanopy and codominant trees. As a consequence, positive growth can occur across multiple canopy strata; the combined effects of shifting gap dynamics at landscape scales can be positive NEP, as has been shown for the US Midwest (Gough et al. 2016). For every declining or dead tree in the upper canopy of an old-growth forest, there is a codominant or an intermediately positioned tree ready to take its place through a growth response triggered by newly available light and space. This was one of the reasons for the sustained carbon sink capacity of old forests cited by Luyssaert et al. (2008) in their global review. Other features of forests dominated by old growth, such as the buildup of carbon stored in woody debris as forests age (Harmon 2001; McGarvey et al. 2015) and large remnant old-growth trees persisting in postdisturbance stands (Keeton and Franklin 2005; D'Amato et al. 2017), contribute disproportionately greater shares of carbon storage (Brown et al. 1997; Keeton et al 2011; Urbano and Keeton 2017). In the Adirondack Mountains of New York State, for example, mature stands with scattered, large remnant old-growth trees have intermediate aboveground biomass at levels between mature (no remnants) and old-growth age classes (Keeton et al. 2007). Remnant or residual structure persisting after partial disturbances, like windthrows of moderate severity, can represent a significant carryover of carbon storage in multiaged stands (Meigs and Keeton 2018).

Another fundamental shift in old-growth ecology reflects the accumulating evidence for non- or "quasi-equilibrium" dynamics in late successional forest landscapes, attributable to partial disturbances, such as wind and ice storms of intermediate intensity (Hanson and Lorimer 2007; Woods 2004; Curzon and Keeton 2010; Meigs and Keeton 2018). These are likely to produce biomass trends inconsistent with predictions derived from earlier models (see Keeton et al. 2011; figure 14-2). In a forest landscape subjected to spatially heterogeneous, periodic, partial disturbances at variable frequencies, an early stage of true even-aged structural development, resulting from complete stand initiation, would rarely occur (Schulte and Mladenoff 2005; D'Amato and Orwig 2008). Instead, aboveground biomass would dip after each partial disturbance, with residual or multiaged structure maintaining a high degree of carbon stocking and relatively rapid recovery to predisturbance biomass levels (Woods 2004; Zeigler 2002). As a consequence, we would not expect to see a biomass peak related to optimal allocation of growing space as occurs through even-aged stand development. Rather, quasi-equilibrium dynamics would produce a logarithmic relationship between biomass accumulation and dominant tree age similar to some chronosequence studies (e.g. Keeton et al. 2011; Gough et al. 2016). Steady-state dynamics can occur if the interval between partial disturbances is long enough (Halpin and Lorimer 2016), but these can be periodically "upset" by partial disturbances of varying intensity (Ziegler 2002; Keeton et al. 2011), a possibility Bormann and Likens (1979) foresaw.

Disturbances and tree mortality from self-thinning are also important from a carbon standpoint because they flux carbon to both the standing and the downed woody-debris pools. But large woody debris as a carbon sink is a concept many people find hard to understand. It seems intuitive that, since detritus of all sorts is in various stages of decomposition, there must be a net flux of carbon out of this pool. Carbon flux from microbial respiration does increase as coarse woody debris accumulates, varying with moisture content, temperature, and species-specific decay rates (Gough et al. 2007; Forrester et al. 2013). Yet, so long as input rates of dead material equal or exceed outputs from decomposition, net carbon fluxes out of this pool can remain stable or even negative, meaning that overall storage increases (Harmon 2001). And since eastern old-growth forests can have exceptionally high carbon density in dead woody material as compared to

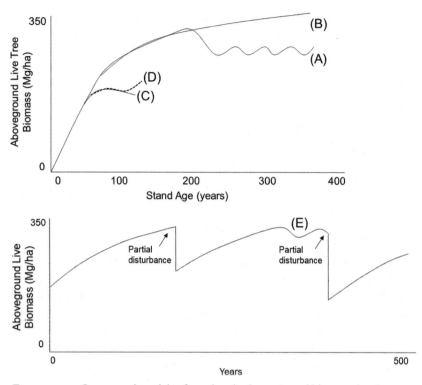

FIGURE 14-2. Conceptual models of stand-scale aboveground biomass development pathways for northern hardwood and mixed hardwood-conifer forests. (*top*) Pathway A depicts the widely cited Bormann and Likens (1979) model, based on empirical data for young stands and computer simulations for late successional stands. This pathway is consistent with recent empirical and modeling work by Halpin and Lorimer (2015). Pathway B is based on chronosequence data presented in Keeton (2011) and Gough et al. (2016), suggesting continued long-term biomass accumulation potential. Pathway C represents the potential for early biomass declines in secondary forests due to anthropogenic stresses, such as acid deposition (causing calcium depletion), invasive insects, and/or introduced diseases (Fahey et al. 2005; Cleavitt et al. 2018). Pathway D is an approximation of the potential to restore positive growth increments, as achieved in a sugar maple dominated stand through calcium additions (Battles et al. 2014). (*bottom*) Hypothetical aboveground biomass responses to partial disturbances, such as moderate severity windthrow and ice storms. Note that live-tree biomass declines following each disturbance, but it is hypothesized to recover due to maintenance of multiaged stands. If the interval between partial disturbances is long enough, steady-state dynamics driven by gap dynamics (E) may occur (Halpin and Lorimer 2015). Collectively the curves represent alternate pathways of biomass development described in the literature. Figure and caption modified from Keeton et al. (2011).

younger and managed forests, accumulated dead wood becomes a significant storage pool (Keeton et al. 2007; Keeton et al. 2011; Gunn et al. 2014). With dead wood representing a sizable carbon sink and looking over the long-term, net positive NEP in older stands could be attributable, at least in part, both to increases in the dead wood and to increases in the soil carbon pools (Harmon and Marks 2002).

Like all pools, the absolute amount or share of carbon stored in downed woody debris in an old-growth forest can vary widely (Burrascano et al. 2013), reflecting differences in pulse recruitment from disturbances, local climate, species composition, and other factors affecting microbial activity and decomposition rates (Fahey et al. 2010; Forrester et al. 2015). The proportion of total stand carbon represented by downed woody material (coarse and fine) can have a complex correlation with other aspects of stand structure, even showing an inverse relationship with live tree stocking (or relative density; Woodall et al. 2013). This relationship may indicate higher recruitment rates under those conditions. On the other hand, the absolute amount of carbon in downed woody material is probably highest in well-stocked, highly productive stands having at least some species with slower decay rates (Woodall et al. 2013). For northern hardwood and mixed hemlock-hardwood forests, Keeton et al. (2011) found biomass levels for downed coarse woody debris that were 33 percent higher in old growth as compared to mature stands but did not find a difference in the total proportion of aboveground carbon in this pool. In contrast, for a variety of forests dominated by mixed oak (*Quercus* spp.), eastern hemlock, and tulip popular (*Liriodendron tulipifera*) in the mid-Atlantic states, McGarvey et al. (2015) found carbon densities in old-growth stands that, for downed woody debris, were an astonishing 1,800 percent higher than surrounding younger forests. They attributed this difference in part to local disturbance history, including diseases like chestnut blight (*Cryphonectria parasitica*). Similar spikes in downed woody debris in late successional systems have been related to beech bark disease (*Neonectria faginata*) and hemlock wholly adelgid (*Adelges tsugae*) (Gunn et al. 2014; Case et al. 2017).

## Pathways of Biomass Accumulation

Late successional and old-growth eastern forests are carbon storage reservoirs, because they generally have substantially more biomass than managed and younger forests (Burrascano et al. 2013; Gunn et al. 2014; McGarvey et al. 2015); though not all studies support this conclusion (see, for example, Hoover et al. 2012). But does that mean that late successional forests continue to accumulate substantial amounts of biomass for centuries? Recent studies have reached differing conclusions on this question. Some, employing both "space for time substitutions" (chronosequences) and long-term monitoring data, have found overall trends of sustained positive biomass accumulation as eastern forests age and for longer periods than would have been predicted by earlier models (Keeton et al. 2011; Eisen and Plotkin 2015; Gough et al 2016). Aboveground biomass trends, for both live and dead trees, showed positive accumulations well past 300 years of age in the dominant canopy trees for the northern hardwood-conifer sites investigated by Keeton et al. (2007, 2011) in the US Northeast. Yet, this research was limited by the chronosequence approach used to infer long-term trends and which, of course, did not track or reconstruct the history of individual stands over time. Other limitations include the confounding influence of legacy structures and the difficulty inferring developmental history from the age of old, large trees alone. However, similar trends were observed by Gough et al. (2016) for forests in the US Midwest. In these chronosequence studies, total aboveground biomass declines towards an asymptote over time but does not show a distinct peak with a subsequent decline. This makes sense, because an asymptotic relationship with forest age is biologically plausible and consistent with there being a finite biomass carrying capacity.

But not all recent studies agree. Some show support for the Bormann and Likens (1979) hypothesis of a biomass peak or even early biomass declines in secondary forests. For example, Halpin and Lorimer (2016), working in northern hardwood stands of the upper Midwest, found biomass peaks after about 200 years of development, with a roughly15-percent decline in biomass by age 320 and steady-state dynamics thereafter. They predicted that continued aboveground biomass accumulation in stands older than 200 years was unlikely. Perhaps even more surprising was the finding that aboveground biomass accumulations in a control watershed at Hubbard Brook showed a much earlier than expected leveling-off after only 80 years of secondary stand development (Fahey et al. 2005). This decline was clearly linked to anthropogenic influences in subsequent work (Battles et al. 2014). These included beech bark disease, an introduced pathogen, that led to decline and mortality in American beech (*Fagus americana*). However, declines in accumulation of live tree biomass were most strongly related to calcium depletion caused by acid deposition (see also Cleavitt et al. 2018), which resulted in dramatic growth rate reductions in sugar maple (*Acer saccharum*). Experimental calcium additions in a paired watershed study elsewhere at Hubbard Brook restored a positive growth increment in sugar maple within just a few years (Battles et al. 2014).

Clearly there is a range of variability and multiple pathways along which biomass dynamics can proceed, varying with forest type, soil productivity, and disturbance history (Eisen and Plotkin 2015). Both sustained biomass accumulation well into old-growth development and somewhat earlier biomass peaks and modest declines are plausible scenarios. There is considerable variation around the averages for biomass in eastern old-growth forests; in some cases, highly productive mature stands may have biomass levels equal to or even exceeding those in less productive old-growth stands (Burrascano et al. 2013). But the generally higher levels of carbon storage in old growth, as compared to young and mature stands on comparable sites (Burrascano et al. 2013; Gunn et al. 2014; McGarvey et al. 2016), remain a key difference regardless of long-term (e.g., longer than 200 years) biomass accumulation dynamics. For this reason, the mounting evidence strongly supports the idea that conserving remaining old-growth forests and managing for old-growth characteristics in working forests would have carbon value, comprising part of a holistic forest carbon management approach (McKinley et al 2011; Ford and Keeton 2017). Allowing forest biomass to fully recover in secondary forests on some portion of the northeastern landscape has the potential to increase in situ carbon storage in those stands by a factor of 2.3 to 4.2, depending on site-specific variability (Keeton et al. 2011). This would sequester an additional 72 to 172 megagrams per hectare of carbon over the current levels typical of young to mature secondary hardwoods. Similar recovery potentials have been estimated for forested landscapes in the upper Midwest based on reconstructions prior to Euro-American settlement forest carbon storage there and comparisons against current stocking levels (Rhemtulla et al. 2009).

## What Does the Future Hold?

There is good reason to be concerned that the carbon sink and storage capacities of late successional and old-growth forests may change into the future. Symptoms of global change are ubiquitous: Shifts in precipitation and temperature regimes, greater frequency of extreme weather events, and spread of invasive pests and pathogens, for example, are already altering the dynamics of eastern forests (Janowiak et al. 2018). When these combine with other anthropogenic stressors, such as ground level ozone and habitat fragmentation from exurban and rural sprawl, the result is an increasingly vulnerable landscape (Ollinger et al. 2008). Climate change and loss of foundational species will alter both successional dynamics and rates and pathways of future biomass development (Ellison et al. 2005; Iverson et al. 2008).

Particularly troubling is the incremental loss of important late successional tree species from eastern forests, such as hemlock in areas infested by hemlock wholly adelgid and ash (*Fraxinus* spp.) succumbing to emerald ash borer (*Agrilus planipennis*), two invasive insects (chapter 12). These threats are superimposed on the functional extirpation of American chestnut (*Castanea dentata*) and American elm (*Ulmus americana*) that occurred during the twentieth century, although efforts are underway to introduce disease resistant varieties of both. Loss of mature American beech trees due to beech bark disease—and the resulting development of dense understories of beech sprouts that inhibit regeneration of other species (Gottesman and Keeton 2017)—has already fundamentally altered the structure and successional dynamics of many northern hardwood stands. When these threats interact with others that result in growth declines in sugar maple due to calcium depletion and the Asian long-horned beetle (*Anoplophora glabripennis*)—which, in North America, preferentially infest maples (*Acer* spp.) but also other genera—it can seem as though the majority of late successional species or species-specific structures (like large beech) are at risk. This leaves the composition of future late successional forests uncertain, even accounting for the species range shifts predicted for the next century (Janowiak et al. 2018). A simplified ecosystem is a likely scenario, necessitating active efforts to improve resilience to change, for instance by maintaining high diversity of functional traits within the species mix in managed forests.

Whether climate change will result in negative or positive effects on late successional forest carbon storage will depend on many factors, including the intensity of local climate disruption, effects on tree growth, extent of species range shifts, formation of novel species assemblages, and interactions with other stressors, such as disturbances and land use (Ollinger et al. 2002; Beckage et al. 2008). For example, if disturbance frequencies increase, effectively reducing the available time interval for development of late successional structure and biomass, net carbon storage capacity at landscape scales would decline (Seidl et al. 2014; Thom and Seidl 2016). In some cases, stands with old-growth structural features, such as tall or less structurally sound large trees, may be more susceptible to some kinds of disturbances, like windthrow, with associated implications for carbon flux, although this remains in debate (Gardiner et al. 2016). On the other hand, more frequent disturbances may accelerate rates at which species assemblages shift and adapt compositionally in response to future climate conditions (Thom et al. 2017).

The question of how tree growth rates will respond to elevated concentrations of carbon dioxide in the air (termed "carbon dioxide fertilization") is another critical uncertainty, with researchers disagreeing about the extent to which growth responses are likely (Foster et al. 2010; McMahon et al. 2010). A related question is whether increased summer moisture stress may ultimately overwhelm enhanced water-use efficiency in plants conferred by carbon dioxide fertilization; soil nitrogen availability also limits the potential for growth increases. Sure to affect growth and phenology is the observed lengthening of growing seasons, though this preferences some species (sometimes including invasives) over others by altering the competitive playing field. Most likely, there will be variable growth responses, some transient and others longer lasting. Thus, while old-growth reference stands suggest an inherent carbon storage capacity within the system, future dynamics are likely to depart from historic baselines as environmental boundary conditions are altered by global change (Ollinger et al. 2008; Seidl et al. 2008).

## Conclusion

As we approach the third decade of the twenty-first century, attitudes have shifted dramatically since the latter twentieth century, and it is now well accepted that conserving remaining old-growth forests provides not only habitat for biodiversity (Lindenmayer and Franklin 2002; see chapter 11) but also cobenefits like carbon storage. Equally promising, however, is the potential of active management for old-growth characteristics to confer carbon benefits. And in eastern forests that historically developed multi- or uneven-aged structure, silviculture can be used to direct or accelerate the development of stand structural complexity and associated high levels of carbon storage (Keeton 2006; Ford and Keeton 2017). Research has shown that those two traits—stand structural complexity and carbon storage (figure 14-3)—often covary, increasing simultaneously as secondary forests redevelop toward a late successional condition (Urbano and Keeton 2017). And for that reason, many of the silvicultural approaches that are of interest to landowners for a range of other reasons related to structural complexity (habitat, water, aesthetics, nontimber forest products, etc.) also have utility for carbon forestry (Fahey et al. 2018). Since old-growth structural and compositional characteristics encompass wide variability

(Burrascano et al. 2013), foresters can build a range of "old growthness" into managed forests rather than expecting them to conform to a strict set of targets (Bauhus et al. 2009). This has the added benefit of providing greater flexibility for integrating carbon and late successional habitat with other management objectives.

Rapidly expanding compliance and voluntary carbon markets, both in North America and internationally, now emphasize maintenance of high carbon stocking in working forests registering to generate greenhouse gas emissions offsets. Carbon markets present significant financial opportunities for eastern forest landowners, providing incentives for sustainable forest management and conservation of habitat and open spaces (Kerchner and Keeton 2015). But to participate in "improved forest management" carbon projects under current market protocols, landowners generally must modify their management systems to increase carbon stocking over time. And that is precisely where old-growth silviculture fits in as one potential option—no longer focused just on ecological objectives but rather as a new opportunity for revenue generation (chapter 13).

As secondary forested landscapes mature and recover from historic clearing and nineteenth century agricultural uses, a range of new options becomes available for managers of working forests to integrate timber and other commodities with carbon objectives. It is absolutely feasible to maintain and/or increase carbon storage on working forests through a variety of silvicultural approaches (Ducey et al. 2013; Fassnacht et al. 2015) that incorporate or enhance elements of late successional forest structure, such as large legacy trees, coarse woody debris, and disturbance gaps (Forrester et al. 2015; Kern et al. 2016; Ford and Keeton 2017; Fahey et al. 2018). Carbon forestry approaches include retention and multicohort systems (Gustafsson et al. 2012), extended rotations or entry cycles (Gronewold et al. 2010; Silver et al. 2013), variants of selection systems (single tree and group) that maintain high postharvest structure (Chen et al. 2015; Ford and Keeton 2015), and optimized timber harvest scheduling (Hoover and Heath 2011; Peckham et al. 2012; chapter 13).

Silvicultural approaches incorporating elevated postharvest structural retention and extended rotations have been shown to increase net carbon storage in northern hardwood forests. This was true even when accounting for the life cycle of harvested carbon transferred to wood products (Nunery and Keeton 2010). Adaptive silvicultural systems—in this case focusing on species diversity and representation of functional traits—have similarly been shown to maintain or increase aboveground biomass under high emissions scenarios for the Great

FIGURE 14-3. An old-growth hemlock-hardwood forest in the Adirondack Mountains of New York State. The photo was taken in the autumn, and there is a light dusting of snow. Note the high degree of structural complexity in terms of variation of tree ages and sizes, vertically continuous canopy, and accumulations of dead wood both standing and downed. Research has shown a close correlation between structural complexity like this and carbon storage in eastern old-growth forests (e.g., Urbano and Keeton 2017). Photo credit: W. S. Keeton.

Lakes states (Duveneck et al. 2014). Research on the effects of alternate forest management approaches on net carbon balance has differed in the emphasis placed on sequestration rates (e.g., Davis et al. 2009; Peckham et al. 2012) versus storage (e.g. Harmon and Marks 2002; Nunery and Keeton 2010; Duveneck et al. 2014), sometimes reaching conflicting conclusions as a result. This highlights the need to reconcile disparate forest carbon accounting frameworks (Fahey et al. 2010), as carbon markets and the Intergovernmental Panel on Climate Change have endeavored to do over the last decade.

Of course, there are tradeoffs between carbon forestry approaches, the yield of harvested timber, and other objectives (Schwenk et al. 2012), and, therefore, any landowner will need to carefully evaluate these. But management for old-growth characteristics does not preclude other approaches across larger scales. Rather, it provides one particular climate mitigation option—reduced net greenhouse gas emissions through enhanced carbon storage in the terrestrial biosphere. Yet, there are other options, including those that increase sequestration rates in somewhat more intensively managed forests, where high growth rates result in greater uptake of carbon. But this approaches carries risks because it banks on the idea that, once a stand is cut, the carbon flux from biomass removals and the foregone carbon storage, it would have developed as an older, more structurally complex forest (i.e., the "opportunity cost"), are offset by rapid sequestration rates and transfer the carbon to wood-product pools. However, a long series of papers has established that, if the residency time (or "life cycle") of carbon in wood products is accounted for, the more intensively managed forest rarely catches up with the less intensively managed one in terms of net atmospheric carbon dioxide reductions (Harmon and Franklin 1990; Harmon and Marks 2002; Nunery et al. 2010). There remains disagreement on this point, depending on the carbon accounting framework and specifics of the harvesting regime and wood-products mix (as an example, see Peckham et al. 2012).

However, innovations within the forest sector can greatly enhance the carbon benefits of actively managed forests. One of these innovations is to increase the emphasis on management for durable wood products with longer life spans, and a second innovation is to substitute wood for materials, such as concrete and steel, that have high carbon footprints associated with their production (Eriksson et al. 2007; Malmsheimer et al. 2008). Once these "substitution effects" or avoided emissions are considered, wood products become a significant contributor to the reduction of emissions. Hence, the excitement around wood products like cross-laminated wooden beams that are increasingly used for construction of large, timber-framed structures.

With this in mind, carbon forestry at the landscape scale becomes an exercise in integrating and optimizing the mix of strategies, producing a full range of ecosystem goods, services, and values (Ray et al. 2009; McKinley et al 2011). Conservation of remaining old-growth forests, together with active management for old-growth characteristics on some portion of the landscape, are essential elements of a holistic carbon strategy for eastern North America. Conserved old-growth forests provide important yet underrepresented habitat features, while enhancing overall carbon storage and helping to mitigate fluxes from more intensively managed stands. Old-growth forests and carbon, therefore, go hand in hand as part of an integrated approach to climate change mitigation.

## Acknowledgements

The author would like to thank Anthea Lavallee, Hubbard Brook Research Foundation, and Craig Lorimer, University of Wisconsin-Madison, for helpful reviews of this chapter.

## References

Battles J. J., T. J. Fahey, C. T. Driscoll, J. D. Blum, and C. E. Johnson. 2014. "Restoring soil calcium reverses forest decline." *Environmental Science and Technology Letters 1*: 15–19.

Bauhus, J., K. Puettmann, and C. Messier. 2009. "Silviculture for old-growth attributes." *Forest Ecology and Management 258*: 525–537.

Beckage, B., B. Osborne, D. G. Gavin, C. Pucko, T. Siccama, and T. Perkins. 2008. "A rapid upward shift of a forest ecotone during 40 years of warming in the Green Mountains of Vermont." *Proceedings of the National Academy of Sciences 105*: 4197–4202.

Bormann, F. H., and G. E. Likens. 1979. *Pattern and Process in a Forested Ecosystem*. New York: Springer-Verlag.

Brown, S., P. Schroeder, and R. Birdsey. 1997. "Aboveground biomass distribution of US eastern hardwood forests and the use of large trees as an indicator of forest development." *Forest Ecology and Management 96*: 37–47.

Burrascano, S., W. S. Keeton, F. M. Sabatini, and C. Blasi. 2013. "Commonality and variability in the structural attributes of moist temperate old-growth forests: A global review." *Forest Ecology and Management 291*: 458–479.

Carey, E. V., A. Sala, R. Keane, and R. M. Callaway. 2001. "Are old forests underestimated as global carbon sinks?" *Global Change Biology 7*: 339–344.

Case, B. S., H. L. Buckley, A. Barker-Plotkin, D. A. Orwig, A. M. Ellison. 2017. "When a foundation crumbles: forecasting forest dynamics following the decline of the foundation species *Tsuga canadensis*." *Ecosphere 8*: e01893.

Chen, J., J. Xu, R. Jensen, and J. Kabrick. 2015. "Changes in aboveground biomass following alternative harvesting in oak-hickory forests in the Eastern USA." *iForest - Biogeosciences and Forestry 8*: 622–630.

Cleavitt, N. L., J. J. Battles, C. E. Johnson, and T. J. Fahey. 2018. "Long-term decline of sugar maple following forest harvest, Hubbard Brook Experimental Forest, New Hampshire." *Canadian Journal of Forest Research 48*: 23–31.

Curzon, M. T., and W. S. Keeton. 2010. "Spatial characteristics of canopy disturbances in riparian old-growth hemlock-northern hardwood forests, Adirondack Mountains, New York, USA." *Canadian Journal of Forest Research 40*: 13–25.

D'Amato, A. W., and D. A. Orwig. 2008. "Stand and landscape-level disturbance dynamics in old-growth forests in western Massachusetts." *Ecological Monographs 78*: 507–522.

D'Amato, A. W., D. A. Orwig, D. R. Foster, A. Barker Plotkin, P. K. Schoonmaker, and M. R. Wagner. 2017. "Long-term structural and biomass dynamics of virgin *Tsuga canadensis– Pinus strobus* forests after hurricane disturbance." *Ecology 98*: 721–733.

Davis, M. B. 1996. "Extent and location." In *Eastern Old-Growth Forests: Prospects for Rediscovery and Recovery*, edited by M. B. Davis., 18–34. Washington, DC: Island Press.

Davis, S. C., A. E. Hessl, C. J. Scott, M. B. Adams, R. B. Thomas. 2009. "Forest carbon sequestration changes in response to timber harvest." *Forest Ecology and Management 258*: 2101–2109.

Després, T., H. Asselin, F. Doyon, and Y. Bergeron. 2014. "Structural and spatial characteristics of old-growth temperate deciduous forests at their northern distribution limit." *Forest Science 60*: 871–880.

Ducey, M. J., J. S. Gunn, and A. A. Whitman. 2013. "Late-successional and old-growth forests in the Northeastern United States: structure, dynamics, and prospects for restoration." *Forests 4*: 1055–1086.

Duveneck, M. J., R. M. Scheller, and M. A. White. 2014. "Effects of alternative forest management on biomass and species diversity in the face of climate change in the northern Great Lakes region (USA)." *Canadian Journal of Forest Research 44*: 700–710.

Eriksson, E., A. R. Gillespie, L. Gustavsson, O. Langvall, M. Olsson, R. Sathre, and J. Stendahl. 2007. "Integrated carbon analysis of forest management practices and wood substitution." *Canadian Journal of Forest Research 37*: 671–681.

Eisen, K., and A. B. Plotkin. 2015. "Forty years of forest measurements support steadily increasing aboveground biomass in a maturing, *Quercus*-dominant northeastern forest." *Journal of the Torrey Botanical Society 142*: 97–112.

Ellison, A. M., M. S. Bank, B. D. Clinton, E. A. Colburn, K. Elliott, C. R. Ford. D. R. Foster, et al. 2005. "Loss of foundation species: consequences for the structure and dynamics of forested ecosystems." *Frontiers in Ecology and Environment 3*: 479–486.

Fahey, R. T., B. C. Alvesherea, J. I. Burton, A. W. D'Amato, Y. L. Dickinson, W. S. Keeton, C. C. Kerne, et al. 2018. "Shifting conceptions of complexity in forest management and silviculture." *Forest Ecology and Management*. In Press.

Fahey, R. T., A. T. Fotis, and K. D. Woods. 2015. "Quantifying canopy complexity and effects on productivity and resilience in late-successional hemlock-hardwood forests." *Ecological Applications 25*: 834–847.

Fahey, T. J., T. G. Siccama, C. T. Driscoll, G. E. Likens, J. Campbell, C. E. Johnson, J. J. Battles, et al. 2005. "The biogeochemistry of carbon at Hubbard Brook." *Biogeochemistry 75*: 109–176.

Fahey, T. J., P. B. Woodbury, J. J. Battles, C. L. Goodale, S. P. Hamburg, S. V. Ollinger, and C. W. Woodall. 2010. "Forest carbon storage: ecology, management, and policy." *Frontiers in Ecology and Environment 8:* 245–252.

Fassnacht, K. S., D. R. Bronson, B. J. Palik, A. W. D'Amato, C. G. Lorimer, and K. J. Martin. 2015. "Accelerating the development of old-growth characteristics in second-growth northern hardwoods." General Techical Report NRS-144. North Central Research Station. Newtown Square, PA: USDA Forest Service.

Fisk, M. C., D. R. Zak, and T. R. Crow. 2002. "Nitrogen storage and cycling in old- and second-growth northern hardwood forests." *Ecology 83*: 73–87.

Ford, S. E., and W. S. Keeton. 2017. "Enhanced carbon storage through management for old-growth characteristics in northern hardwoods." *Ecosphere 8*: 1–20.

Forrester, J. A., D. J. Mladenoff, A. W. D'Amato, S. Fraver, D. L. Lindner, N. J. Brazee, M. K. Clayton, and S. T. Gower. 2015. "Temporal trends and sources of variation in carbon flux from coarse woody debris in experimental forest canopy openings." *Oecologia 179*: 889–900.

Forrester, J. A., D. J. Mladenoff, and S. T. Gower. 2013. "Experimental manipulation of forest structure: near term effects on gap and stand scale C dynamics." *Ecosystems 16*: 1455–1472.

Foster, J. R., J. I. Burton, J. A. Forrester, F. Liu, J. D. Muss, F. M. Sabatini, R. M. Scheller, and D. J. Mladenoff. 2010. "Evidence for a recent increase in forest growth is questionable." *Proceedings of the National Academy of Sciences 107*: E86–E87.

Fraver, S., A. S. White, and R. S. Seymour. 2009. "Natural disturbance in an old-growth landscape of northern Maine, USA." *Journal of Ecology 97*: 289–298.

Gardiner, B., P. Berry, and B. Moulia. 2016. "Review: Wind impacts on plant growth, mechanics and damage." *Plant Science 245*: 94–118.

Gough, C. M., P. S. Curtis, B. S. Hardiman, C. M. Scheuermann, and B. Bond-Lamberty. 2016. "Disturbance, complexity, and succession of net ecosystem production in North America's temperate deciduous forests." *Ecosphere 7*: e01375. doi:10.1002/ecs2.1375.

Gough C. M., C. S. Vogel, C. Kazanski, L. Nagel, C. E. Flower, and P. S. Curtis. 2007. "Coarse woody debris and the carbon balance of a north temperate forest." *Forest Ecology and Management 244*: 60–67.

Gottesman, A. J., and W. S. Keeton. 2017. "Regeneration responses to management for old-growth characteristics in northern hardwood-conifer forests." *Forests 8*: 1–21.

Gower, S. T., R. E. McMurtrie, and D. Murty. 1996. "Aboveground net primary production decline with stand age: potential causes." *Trends in Ecology and Evolution 11*: 378–382.

Gronewold, C. A., A. W. D'Amato, and B. J. Palik. 2010. "The influence of cutting cycle and stocking level on the structure and composition of managed old-growth northern hardwoods." *Forest Ecology and Management 259*: 1151–1160.

Gunn, J. S., M. J. Ducey, and A. A. Whitman. 2014. "Late-successional and old-growth forest carbon temporal dynamics in the Northern Forest (Northeastern USA)." *Forest Ecology and Management 312*: 40–46.

Gustafsson, L., S. C. Baker, J. Bauhus, W. J. Beese, A. Brodie, J. Kouki, D. B. Lindenmayer, et al. 2012. "Retention forestry to maintain multifunctional forests: A world perspective." *BioScience 62*: 633–645.

Hadley, J. L., and J. L. Schedlbauer. 2002. "Carbon exchange of an old-growth eastern hemlock (*Tsuga canadensis*) forest in central New England." *Tree Physiology 22*: 1079-1092.

Halpin, R. C., and C. Lorimer. 2016. "Long-term trends in biomass and tree demography in northern hardwoods: An integrated field and simulation study." *Ecological Monographs 86:* 78–93.

Hanson, J. J., and C. G. Lorimer. 2007. "Forest structure and light regimes following moderate wind storms: implications for multi-cohort management." *Ecological Applications 17*: 1325–1340.

Harmon, M. E. 2001. "Carbon sequestration in forests: addressing the scale question." *Journal of Forestry 99*: 24–29.

Harmon, M. E., W. K. Ferrell, and J. F. Franklin. 1990. "Effects on carbon storage of conversion of old-growth forests to young forests." *Science 247*: 699–702.

Harmon, M. E., and B. Marks. 2002. "Effects of silvicultural practices on carbon stores in Douglas-fir-western hemlock forests in the Pacific Northwest, USA: results from a simulation model." *Canadian Journal of Forest Research 32*: 863–877.

Hoover, C. M., and L. S. Heath. 2011. "Potential gains in C storage on productive forestlands in the Northeastern United States through stocking management." *Ecological Applications 21*: 1154–1161.

Hoover, C. M., W. B. Leak, and B. G. Keel. 2012. "Benchmark carbon stocks from old-growth forests in northern New England, USA." *Forest Ecology and Management 266*: 108–114.

Iverson, L., A. Prasad, and S. Matthews. 2008. "Modeling potential climate change impacts on the trees of the northeastern United States." *Mitigation and Adaptation Strategies for Global Change 13*: 487–516.

Janowiak, M. K.; A. W. D'Amato, C. W. Swanston, L. Iverson, F. R. Thompson III, W. D. Dijak, S. Matthews, et al. 2018. "New England and northern New York forest ecosystem vulnerability assessment and synthesis: a report from the New England Climate Change Response Framework project." General Technical Report NRS-173. Northern Research Station. Newtown Square, PA: USDA Forest Service.

Keeton, W. S. 2006. "Managing for late-successional/old-growth characteristics in northern hardwood-conifer forests." *Forest Ecology and Management 235*: 129–142.

Keeton, W. S., and J. F. Franklin. 2005. "Do remnant old-growth trees accelerate rates of succession in mature Douglas-fir forests?" *Ecological Monographs 75*: 103–118.

Keeton, W. S., C. E. Kraft, and D. R. Warren. 2007. "Mature and old-growth riparian forests: Structure, dynamics, and effects on Adirondack stream habitats." *Ecological Applications 17*: 852–868.

Keeton, W. S., A. A. Whitman, G. C. McGee, and C. L. Goodale. 2011. "Late-successional biomass development in northern hardwood-conifer forests of the Northeastern United States." *Forest Science 57*: 489–505.

Keith, H., B. G. Mackey, and D. B. Lindenmayer. 2009. "Re-evaluation of forest biomass carbon stocks and lessons from the world's most carbon-dense forests." *Proceedings of the National Academy of Sciences 106*: 11635–11640.

Kerchner, C. D., and W. S. Keeton. 2015. "California's regulatory forest carbon market: Viability for northeast landowners." *Forest Policy and Economics 50*: 70–81.

Kern, C. C., J. Burton, P. Raymond, A. D'Amato, W. S. Keeton, A. A. Royo, M. B. Walters, C. R. Webster, and J. L. Willis. 2016. "Challenges facing gap-based silviculture and possible solutions for mesic northern forests in North America." *Forestry 90*: 4–17.

Lindenmayer, D. B., and J. F. Franklin. 2002. *Conserving Forest Biodiversity: A Comprehensive Multiscaled Approach*. Washington, DC: Island Press.

Lovett, G. M., J. J. Cole, and M. L. Pace. 2006. "Is net ecosystem production equal to ecosystem carbon accumulation?" *Ecosystems 9*: 1–4.

Luyssaert, S., E-D. Schulze, A. Borner, A. Knohl, D. Hessenmoller, B. E. Law, P. Ciais, and J. Grace. 2008. "Old-growth forests as global carbon sinks." *Nature 455*: 213–215.

Malmsheimer, R. W., P. Heffernan, S. Brink, D. Crandall, F. Deneke, C. Galik, E. Gee, et al. 2008. "Forest management solutions for mitigating climate change in the United States." *Journal of Forestry 106*: 115–171.

McGarvey, J. C., J. R. Thompson, H. E. Epstein, and H. H. Shugart. 2015. "Carbon storage in old-growth forests of the Mid-Atlantic: toward better understanding of the eastern forest carbon sink." *Ecology 96*: 311–317.

McKinley, D. C., M. G. Ryan, R. A. Birdsey, C. P. Giardina, M. E. Harmon, L. S. Heath, R. A. Houghton, et al. 2011. "A synthesis of current knowledge on forests and carbon storage in the United States." *Ecological Applications 21*: 1902–1924.

McMahon, S. M., G. G. Parker, and D. R. Miller. 2010. "Evidence for a recent increase in forest growth." *Proceedings of the National Academy of Sciences 107*: 3611–3615.

Meigs, G. W., and W. S. Keeton. 2018. "Intermediate-severity wind disturbance in mature temperate forests: effects on legacy structure, carbon storage, and stand dynamics. *Ecological Applications*. In Press.

Nunery, J., and W. S. Keeton. 2010. "Forest carbon storage in the northeastern United States: net effects of harvesting frequency, post-harvest retention, and wood products." *Forest Ecology and Management 259*: 1363–1375.

Odum, E. P. 1969. "The strategy of ecosystem development." *Science 164*: 262–270.

Ollinger, S. V., J. D. Aber, P. B. Reich, and R. J. Freuder. 2002. "Interactive effects of nitrogen deposition, tropospheric ozone, elevated CO2, and land use history on the carbon dynamics of northern hardwood forests." *Global Change Biology 8*: 545–562.

Ollinger, S. V., C. L. Goodale, K. Hayhoe, and J. P. Jenkins. 2008. "Potential effects of climate change and rising CO2 on ecosystem processes in northeastern U.S. forests." *Mitigation and Adaptation Strategies for Global Change 13*: 467–485.

Pan, Y., R. Birdsey, J. Hom, and K. McCullough. 2009. "Separating effects of changes in atmospheric composition, climate and land-use on carbon sequestration of U.S. Mid-Atlantic temperate forests." *Forest Ecology and Management 259*: 151–164.

Peckham, S. D., S. T Gower, and J. Buongiorno. 2012. "Estimating the carbon budget and maximizing future carbon uptake for a temperate forest region in the U.S." *Carbon Balance and Mangaement 7*: 6.

Phillips, N. G., T. N. Buckley, and D. T. Tissue. 2008. "Capacity of old trees to respond to environmental Change." *Journal of Integrative Plant Biology 50*: 1355–1364.

Ray, D. G., R. S. Seymour, N. S. Scott, and W. S. Keeton. 2009. "Mitigating climate change with managed forests: balancing expectations, opportunity, and risk." *Journal of Forestry, January/February*: 50–51.

Rhemtulla, J. M., D. J. Mladenoff, and M. K. Clayton. 2007. "Regional land-cover conversion in the US upper Midwest: magnitude of change and limited recovery (1850–1935–1993)." *Landscape Ecology 22*: 57–75.

Rhemtullaa, J. M., D. J. Mladenoff, and M. K. Clayton. 2009. "Historical forest baselines reveal potential for continued carbon sequestration." *Proceedings of the National Academy of Sciences 106*: 6082–6087.

Runkle, J. R. 1982. "Patterns of disturbance in some old-growth mesic forests of eastern North America." *Ecology 62*: 1041–1051.

Schulte, L. A., and D. J. Mladenoff. 2005. "Severe wind and fire regimes in northern forests: historical variability at the regional scale." *Ecology 86*: 431–445.

Schwenk, W. S., T. M. Donovan, W. S. Keeton, and J. S. Nunery. 2012. "Carbon storage, timber production, and biodiversity: comparing ecosystem services with multi-criteria decision analysis." *Ecological Society of America 22*: 1612–1627.

Seidl, R., W. Rammer, P. Lasch, F. W. Badeck, and M. J. Lexer. 2008. "Does conversion of even-aged, secondary coniferous forests affect carbon sequestration? A simulation study under changing environmental conditions." *Silva Fennica 42*: 369–386.

Seidl, R., M. J. Schelhaas, W. Rammer, and P. J. Verkerk. 2014. "Increasing forest disturbances in Europe and their impact on carbon storage." *Nature Climate Change 4*: 806–810.

Silver, E. J., A. W. D'Amato, S. Fraver, B. J. Palik, and J. B. Bradford. 2013. "Structure and development of old-growth, unmanaged second-growth, and extended rotation *Pinus resinosa* forests in Minnesota, USA." *Forest Ecology and Management 291*: 110–118.

Stephenson, N. L., A. J. Das, R. Condit, S. E. Russo, P. J. Baker, N. G. Beckman, D. A. Coomes, et al. 2014. "Rate of tree carbon accumulation increases continuously with tree size." *Nature 507*: 90–93.

Thom, D., W. Rammer, and R. Seidl. 2017. "Disturbances catalyze the adaptation of forest ecosystems to changing climate conditions." *Global Change Biology 23*: 269–282.

Thom, D. and R. Seidl. 2016. "Natural disturbance impacts on ecosystem services and biodiversity in temperate and boreal forests." *Biological Reviews 91*: 760–781.

Urbano, A. R., and W. S. Keeton. 2017. "Forest structural development, carbon dynamics, and co-varying habitat characteristics as influenced by land-use history and reforestation approach." *Forest Ecology and Management 392*: 21–35.

Whittaker, R. H., F. H. Bormann, G. E. Likens, and T. G. Siccama. 1974. "The Hubbard Brook ecosystem study: forest biomass and production." *Ecological Monographs 44*: 233–252.

Woodall, C. W., B. F. Walters, S. N. Oswalt, G. M. Domke, C. Toney, and A. N. Gray. 2013. "Biomass and carbon attributes of downed woody materials in forests of the United States." *Forest Ecology and Management 305*: 48–59.

Woods, K. D. 2004. "Intermediate disturbance in a late-successional hemlock-northern hardwood forest." *Journal of Ecology 92*: 464–476.

Ziegler, S. S. 2002. "Disturbance regimes of hemlock-dominated old-growth forests in northern New York, U.S.A." *Canadian Journal of Forest Research 32*: 2106–2115.

# Chapter 15

# Conclusion: Past, Present, and Future of Old-Growth Forests in the East

*William S. Keeton and Andrew M. Barton*

*Old growth*—the term evokes something deeply rooted in the human psyche. We imagine the forest primeval, something timeless from our distant collective memory or perhaps nostalgia for what we imagine might once have been. Maybe we long for a time when life was less complicated, when the trappings of modern civilization were not so pervasive. There is the romanticism of the precolonial landscape "in which a squirrel could travel tree-to-tree from Georgia to Maine without ever touching the ground," a legend that somehow manages to leave out millions of indigenous peoples who influenced that landscape for millennia (Mann 2006). We ground this image in reassuring ideas like "equilibrium dynamics," in which primary or "virgin" forests are envisioned as stable and unchanging for centuries and natural disturbances are unfortunate events external to the system. Inputs equal outputs, everything in perfect balance.

But contemporary old-growth ecology paints a different picture, one no less mysterious and compelling, yet challenging because it requires a fundamental shift in thinking, an acceptance that change and dynamism themselves are intrinsic to healthy, well-functioning forest ecosystems. We have to recalibrate. The retrained eye sees the messy consequence of disturbances and death as *complexity*, creating niches that foster biodiversity and as evidence of an old-growth system functioning as it should, as long as it operates within a range of variability not profoundly altered by global change (Aplet and Keeton 1999). If there is one core lesson we have learned in more than a half century of research on eastern old-growth forests, it is that a dynamic system is a healthy one: Late successional forests are shaped by, and, in fact, dependent on, drivers of change, such as natural disturbances, successional dynamics, climate variability, and competition among organisms. And it is the outcome of these dynamics—forest stand development, carbon fixation, and hydrologic regulation, for example—that provide ecosystem services and habitat. This unifying theme runs through the chapters in this book and is the starting point for anyone beginning their study of old-growth forest ecology.

To be clear, the old-growth scholar or seeker is no less fascinated by the past, present, and future prospects of late successional ecosystems than the romantic searching for "unspoiled nature." But each has a different expectation for what they will find: The scholar understands that precolonial landscapes were highly complex and that old-growth forests themselves encompassed a wide range of structural and site conditions, types, and disturbance histories. Landscape patterns were complex mosaics with tremendous variation in age, shape, and composition, in which the full range of successional stages would have been represented, the preponderance (i.e., 70 to 80 percent) in a late successional condition because of the infrequency of large, high-intensity disturbances (Lorimer and White 2003). Indigenous influences on forests were also important, with widely variable impacts across eastern North America. In the Northeast, for example, burning by Native Americans and First Nations peoples was concentrated mostly around settlements and along trade routes but was less prevalent elsewhere (Russell 1983). A time traveler to that long-ago landscape might be surprised by the range of forest conditions encountered there.

Because of this variability across the landscape, what we think of as "old growth"—primary (i.e., never cleared) forests in a late stage of structural development, dominated by late successional species—might have taken many forms. It was not always the cathedral forest of our imagination. In some cases, the full complement of archetypical structural features—many large trees, living and dead; very high accumulations of large downed logs; big tip-up mounds; complex canopies, etc.—would have been present. But in other cases, some of those features may have been poorly developed. This would have depended, for instance, on how productive or fertile the site was and on the complex history of disturbances (e.g., type, timing, intensity, and sequencing), often unique to individual sites, as discussed further below.

Based on these diverse patterns, the authors of this book express differing views on how to define old growth. Some (e.g., chapters 10, 11, and 13) argue that old growth is a structural and compositional condition that develops late in succession and is a subset of the range of successional stages and structural conditions found in primary forest patch mosaics. This approach recognizes that, in a landscape shaped not just by small, gap-forming disturbances, but also by less frequent larger disturbances, there might be continuous variation in legacy structure within multiaged stands and multiple pathways of stand structural development (Lorimer and Halpin 2014; Meigs and Keeton 2018). These could produce a diversity of stand structural conditions with continuously varying degrees of "old growthness" (Bauhus et al. 2009). Others (e.g., chapters 3 and 4) argue for a broader definition of old growth that encompasses ecosystems without such structural conditions or successional stages and pathways. We encourage readers to once again inspect how chapter 4 attempts to resolve this definitional challenge by advocating for multiple dimensions of old growth: human influence, structure and age, and forest continuity. As we argued in chapter 1, at the core of this conundrum is the immense complexity of nature (Wirth et al 2009) and the intertwined use of "old growth" in scientific and cultural spheres (Pesklevits et al. 2011). Undoubtedly, the debate over how to conceive of and best classify old-growth forests will continue.

Fundamentally, however, we conclude that as a functional condition, old growth does not have to be restricted to primary forests, but rather has the potential to redevelop over time in anthropogenically influenced secondary forests as the legacy of more severe human disturbances (like clearing) fades. If a big part of what truly matters are the functions that old-growth ecosystems perform and the services those processes provide to nature and humanity, then our chief concern should be recovery of these functions and adaptation to future stresses that might compromise those services. A wide variety of structural conditions and developmental pathways is likely to yield different mixes of functions and services. Therefore, what we classify as old growth, from a functional perspective, need not constitute a single archetype or necessarily conform to a single definition, particularly a definition that demands an increasingly unrealistic prerequisite of zero human influence. From this standpoint, we see recovery of functional old-growth as not just possible, but probable on eastern landscapes.

We are challenged to temper our expectations every time we walk into an old-growth forest, understanding that each stand may present something a little different than what we have seen before, and yet, equally compelling. Those that seek out and venture into old-growth forests may still find tranquility, respite, and maybe even a glimpse of a world that seems to have faded away. But they should expect the unexpected.

## Old-Growth Discoveries

The last three decades have been a golden age of discovery of old-growth tracts in eastern North America (Davis 1996; Kershner and Leverett 2004; Blozan 2011). These newly found and already known tracts of old growth have been cataloged or summarized in three remarkable publications by Mary Byrd Davis (1993, 1996, 2003; see also Tyrrell et al. 1998). It is a testament to the efforts of these many prospectors that, due to the sheer numbers of old-growth tracts across eastern North America, updating this compilation is now a monumental task.

Never-cut, old stands have been discovered from Florida and Arkansas to Ontario and Quebec, located along steep cliffs, in river swamps, on isolated islands, in remote woods, and even in plain view. In many cases, the parcels are tiny, but that's not always the case. The Southern Appalachian Forest Coalition, for example, mapped over 45,000 hectares of old growth on national forest lands in that region, much more than recognized by the US Forest Service (Krajick 2003; Southern Appalachian Forest Coalition 2018). Similarly, old-growth tracts, thousands of hectares in size, have been identified in wilderness areas within Adirondack State Park in New York State (Keeton et al. 2007). The ages of some of the newly discovered stands have been astonishing, altering the conventional wisdom about eastern old growth: Bald cypress over 1,600 years old in swamps on the Black River of North Carolina (Stahle et al. 1988) and cedars that date back to the year 1,000 C.E. on the cliff faces of the Niagara Escarpment (Kelly and Larson 2007).

New methods are being tested to help determine how much old growth might be left. For example, the Wisconsin Department of Natural Resources used Forest Inventory and Analysis data to estimate that approximately 31,000 hectares of old growth may remain in that state. Others are using light detection and ranging (LiDAR) to map canopy complexity across broad landscapes. The intent is to aid old-growth mapping efforts, an excellent example of this being work across the southern Appalachian region (Hansen et al. 2014). And, of course, a rich and growing literature has reconstructed presettlement forest age, composition, and structure using a variety of documentary surveys, in effect "rediscovering" the historic baselines for how much and what types of old-growth forests may have occurred on eastern landscapes (Cogbill et al. 2003; Rhemtulla et al. 2009; Thompson et al. 2013; chapter 7). These approaches are helping scientists understand not only what might once have been, but also how the dynamics of land-use history have profoundly altered forest distribution, structure, and function. They also offer further insights into novel associations of species that occurred in the past and might develop in the future (Goring et al. 2016).

Stimulated initially by the emerging old-growth controversy in the Pacific Northwest, old-growth sleuths have included scientists but, also, many amateur naturalists with a passion for forests with big, old trees. The discovery of new old-growth forests is important for their conservation value alone, and in some cases, the stands are examples of rare communities (Kelly and Larson 2007). But these new tracts have also expanded and altered our understanding of the ecology of old forests, especially those that have pushed back the record of climate and tree responses by centuries (Stahle et al. 1988; Thompson et al. 2013). In addition, larger blocks of old-growth and primary forests give researchers a unique opportunity to observe and understand the interaction of succession, natural disturbances, climate, and other sources of variation that play out across broader landscapes and across a range of site conditions that are representative of geophysical diversity more generally (Keeton et al. 2011; McGarvey et al. 2015). These old-growth discoveries have been critical to the development of contemporary models, stressing the multiple pathways by which the compositional, structural, and functional characteristics of old growth can develop (Lorimer and Halpin 2014; Urbano and Keeton 2017).

There is little doubt that undiscovered old growth remains in eastern North America. New finds, such as two 300-year-old red pines in old growth in Nepisiguit Protected Natural Area in New Brunswick in 2015 (New Brunswick Museum 2015), continue to make the news. Discovery, mapping, inventory, and description are incomplete in most regions of eastern North America, leaving an open field for young ecologists and amateur naturalists alike.

## Frameworks for Understanding Variation in Old Growth across Eastern North America

The field of ecology is grounded in describing and understanding the variation of nature across the Earth. The old-growth forests of eastern North America examined in this book reveal a remarkable diversity of natural expression, from continental to local spatial scales. As shown in the map in plate 1, these ecosystems are distributed across a latitudinal gradient more than 4,000 kilometers long, from subtropical to eastern temperate to northern to boreal zones. These vast ecological sectors are further delineated into multiple ecoregions at two finer levels, illustrated in the map. The chapters also reveal striking variation in old-growth communities at much smaller scales, across local environmental gradients: from moister to drier sites, lower to higher elevations, north- to south-facing slopes, and infertile to fertile soils, to name a few (see especially chapters 3, 4, 5, and 7).

What are the key, measurable ways that we can make sense of this vast diversity of old-growth forests in eastern North America? Five stand-level attributes, which form the basis of most descriptions of old growth in this book, are especially important:

- *stand basal area*—a small or a large amount of tree biomass per area?
- *maximum size of trees*—little or big trees dominant in the canopy?
- *maximum age of trees*—all younger or a large number of old trees?
- *vertical structural complexity*—few or many forest layers and plant-growth forms?
- *amount and relative sizes of woody debris*—small and fine or voluminous and large?

At a broader, landscape-level scale, primary forest ecosystems also vary in their patch dynamics (Fahey et al. 2012). Many exhibit a forest mosaic with patches spanning a continuous range of ages, from newly initiated to old stands, while others are more homogeneous, with mainly mature forests (Picket and White 1985). This also means that primary forest landscapes vary in the relative amount of truly old versus younger forests, as well as in the presence of biological legacies (Foster et al. 2003; Pederson et al. 2014).

Table 15-1 profiles 11 old-growth and primary forest types discussed in this book for these six key attributes. The values are qualitative and relative, and they obscure variability within each type. Nevertheless, collectively they provide a sample of the remarkable diversity of old growth across eastern North America. For example, cove hardwood forests in the southern Appalachian Mountains (chapter 4) fit the conventional perception of old growth with large, old trees; high levels of basal area, structural complexity, and coarse woody debris; and little variation in patch ages across the landscape (i.e., mainly old forest). In contrast, primary boreal forests (chapter 8) exhibit smaller, younger trees; low levels of structural complexity; moderate stand basal area and woody debris; and a multitude of patch ages (with relatively little old forest).

What explains the striking contrasts revealed in table 15-1? Which natural forces, varying at local and continental scales, could lead to such

TABLE 15-1. Key differences among old-growth forest types across eastern North America discussed in this book. *Ch*—chapter in book; *Max Size*—maximum tree size; *Max Age*—maximum tree age; *Stand Basal*—stand basal area; *CWD*—coarse woody debris; *Complexity*—physical structural complexity; *Patch % Old*—landscape percent of primary forest in old growth under natural conditions.

| Forest Type | Ch | Max Size | Max Age | Stand Basal | CWD | Complex-ity | Patch % Old |
|---|---|---|---|---|---|---|---|
| Bottomland and Swamp Hardwoods (Southeastern US) | 2 | Inter | Inter | Inter | High | Inter | Inter |
| Pine Savanna (Southeastern US) | 3 | Inter | Old | Low | Low | Low | Very High |
| Xeric Pine-Oak (Southern Appalachians) | 4* | Small | Inter | Inter | Low | Low | Inter |
| Cove Hardwoods-Hemlock (Southern Appalachians) | 4 | Very Large | Very old | High | High | High | High |
| Mixed Mesophytic (Central Appalachians) | 5 | Large | Very old | High | High | High | High |
| Jack Pine (US-Canada Border) | 6 | Small | Young | Inter | Inter | Low | Low |
| Red Pine (US-Canada Border) | 6 | Inter | Old | Inter | Inter | Inter | Inter |
| Mixedwoods (conifers & hardwoods) (New England and Southeastern Canada) | 6 | Inter | Old | High | High | High | High |
| Northern Hardwoods (Northeastern US and Southeastern Canada) | 6 | Large | Old | High | High | High | High |
| Northern Hardwoods-Hemlock (Northern Lake States) | 7 | Large | Old | High | High | High | High |
| Boreal Forest (Canada) | 8 | Inter | Inter | High | High | Inter | Low |

*Information also from Flatley et al. 2013 and Grissino-Mayer 2015.

different expressions of old-growth forests? The short answer is variation in natural disturbance regimes; that is, differences from place to place in the intrinsic types, sizes, severities, and frequencies of disturbances (see Glossary). The longer answer involves the complex relationships among climate, disturbance, and the functional traits of tree species. Let's start with a comparison of natural disturbance regimes, which is provided in table 15-2 for the 11 old-growth forest types from table 15-1.

The chief dichotomy in natural disturbance regimes involves the occurrence of fire (table 15-2). Wildfire requires a spark for ignition (e.g., lightning), conditions dry enough (during some season or year) to carry fire, and a sufficient amount and a continuity of fuel to spread fire. Some locations meet these requirements (e.g., pine savannas); others do not. Northern hardwoods, for example, are almost always too moist (chapter 6). Fire-prone environments themselves differ in the particulars of their fire regimes. In boreal forests (chapter 8), conditions favoring fire occur infrequently enough that high levels of live and dead fuels build up during those intervals. Thus, when they do occur, fires tend to burn at high-severity into the forest crown, killing most or all trees and initiating a new stand. Jack pine (*Pinus banksiana*) stands work similarly (chapter 6). In southern pine savannas (chapter 3), in contrast, consistently dry conditions during the growing season promote frequent fires. Because this activity maintains low levels of woody debris (fuel), fires tend to burn at low severity on the surface, killing smaller but not larger stems. Climate conditions usually support a fire regime with both infrequent and surface stand-replacing fires in typical red pine (*P. resinosa*) ecosystems (chapter 6), a dynamic apparent in southern xeric pine stands as well (chapter 4; Flatley et al 2013). These three contrasting fire regimes, then, produce mature forest stands with very different characteristics. They also result in contrasting primary forest landscapes, with relatively little old forest in boreal and jack pine areas, high percentages in pine savannas, and intermediate amounts in red pine and xeric southern pine complexes.

The constituent tree species in these communities play important feedback roles in reinforcing their intrinsic fire regimes. For example, longleaf pine (*P. palustris*) is well adapted to surviving surface fires with its thick bark and self-pruned trunk, and its needles actually promote fire spread; trees in boreal and jack pine stands largely lack such fire-resistant traits and are readily killed by fire; red pine and some of the xeric pines of the southern Appalachians fall somewhere in between. All of these species also possess traits that enhance the probability of their regeneration in postfire conditions: serotinous cones in black spruce (*Picea mariana*), jack pine, and ta-

TABLE 15-2. Differences among old-growth forest types (see table 15-1) in natural disturbance regimes, climate, and latitudinal distribution. The chapter (*Ch*) most relevant to each forest type is provided.

| Forest Type | Ch | Natural Disturbance | Climate | Latitude |
|---|---|---|---|---|
| Bottomland & Swamp Hardwoods (Southeast) | 2 | 1⁰: Flooding 2⁰: Wind, fire | Warm, mesic, seasonally hydric | 300–350 N |
| Pine Savanna (Southeast) | 3 | 1⁰: Frequent surface fire 2⁰: Wind | Warm, seasonally xeric | 250–350 N |
| Xeric Pine & Oak* (s. Appalachians) | 4 | 1⁰: Mixed-severity fire 2⁰: Wind, pests | Warm/Cold, xeric | 340–400 N |
| Cove Hardwoods & Hemlock (s. Appalachians) | 4 | 1⁰: Small gaps 2⁰: Moderate-severity wind | Warm/Cold, mesic | 350–380 N |
| Mixed Mesophytic (c. Appalachians) | 5 | 1⁰: Small gaps 2⁰: Moderate-severity wind, fire | Warm/Cold, mesic | 350–400 N |
| Jack Pine (US/Canada Border) | 6 | 1⁰: Infrequent stand-replacing fire 2⁰: Wind | Cold, xeric | 450–550 N |
| Red Pine (US/Canada Border) | 6 | 1⁰: Mixed-severity fire 2⁰: Wind | Cold, xeric | 450–550 N |
| Mixedwoods (conifers & hardwoods) (New England & Canada) | 6 | 1⁰: Small gaps, pests 2⁰: Moderate-severity wind, ice, fire | Cold, mesic | 400–500 N |
| Northern Hardwoods (NE US & Canada) | 6 | 1⁰: Small gaps 2⁰: Moderate-severity wind, ice, fire | Cold, mesic | 350–500 N |
| Northern Hardwoods-Hemlock (Northern Lake States) | 7 | 1⁰: Small gaps 2⁰: Moderate-severity wind, ice, fire | Cold, mesic | 420–500 N |
| Boreal Forest (Canada) | 8 | 1⁰: Infrequent stand-replacing fire 2⁰: Wind, pests, ice | Very cold, short growing season, periodically xeric | 500–650 N |

*Information also from Flatley et al. 2013 and Grissino-Mayer 2015.

ble mountain pine (*Pinus pungens*); grass-stage seedlings in longleaf pine; resprouting in oaks (*Quercus* spp.) and pitch (*Pinus rigida*) and shortleaf pines (*Pinus echinata*); clonal spread in aspen (*Populus tremuloides*); and fast juvenile growth in many species.

Most of the old-growth ecosystems in table 15-2 that are not fire-prone are governed primarily by a gap-phase natural disturbance regime, in other words, small openings in the canopy caused by the death of single trees or small groups of trees (chapters 4–7). Stand-replacing disturbances occur rarely. Intermediate-intensity disturbances occurring at intervals of several hundred years, on the other hand, may play an important role in engendering heterogeneity and persistence of midtolerant species that otherwise would be outcompeted (chapter 6 and 7). Tree species typically found in these old-growth ecosystems possess traits that allow them to tolerate typical intact forest conditions (shade tolerance, resilience to wind) and to take advantage of temporary resource pulses caused by canopy openings (e.g., rapid growth release, resprouting). Because of the rarity of more severe disturbances, gap-phase natural disturbance regimes engender primary forest landscapes composed largely of older forest. Higher frequencies of intermediate- and high-severity disturbances result in less old-growth area and more variation across patches in age and other associated stand characteristics.

This discussion raises three key points. First, the natural disturbance regime of any old-growth (or younger) ecosystem results from an interplay of climate, disturbance types, and the functional traits of plant species (see chapters 6 and 8). The chief climate drivers in this simple conceptual framework ultimately are themselves the product of geography: latitude, proximity to major bodies of water, and location with respect to major recurring air masses. At local levels, physiography—elevation, slope direction, and slope steepness—plays a similar role in governing microclimate and the probabilities of different disturbances and the species capable of persisting. (Soil heterogeneity can similarly control these patterns at both large and small scales.) Next, natural disturbances are clearly not external events imposed on these forests, but instead are intrinsic elements of these systems (Holling et al. 2002; Lorimer and White 2003). Disturbances and living organisms form a complex evolutionary and ecological feedback system that confers unique properties on each old-growth ecosystem. Finally, differences in these climate-disturbance-tree species systems (table 15-2), rooted in underlying geography and physiography, appear to explain a large portion of the tremendous variation across eastern North America in both stand- and forest-level attributes (table 15-1).

## Pristine Forests, the Anthropocene, and the Future

"For the past three centuries, the effects of humans on the global environment have escalated. Because of these anthropogenic emissions of carbon dioxide, global climate may depart significantly from natural behaviour for many millennia to come. It seems appropriate to assign the term 'Anthropocene' to the present, in many ways human-dominated, geological epoch..."                *Nature 415: 23*

So began Nobel Prize-winner Paul Crutzen's 2002 proposal titled "Geology of mankind." Sixteen years later, the Anthropocene has continued apace. Atmospheric carbon dioxide concentrations were at 407 parts per million in 2017, up from 374 in 2002. Seventeen of the warmest 18 years on record have occurred since 2001. Arctic sea ice shrank to a record low in 2012. Sea level rose 54 millimeters from 2002 to 2017 (NASA 2018). Heavy metals from industrial emissions are found in the most remote places (Beltcheva et al. 2011), and five trillion pieces of plastic float on the world's oceans (Eriksen et al. 2014).

Like every other corner of the Earth, old growth in eastern North America has not escaped the imprint of human civilization. The relationship between humans and old forests has a complex and fraught history. The first contemporary conceptions of old-growth ecosystems in the twentieth century were rooted in the Romantic era philosophy of forests as primeval sanctuaries isolated from humanity, where one could seek "sublimity and uplift in the contemplation of nature, linking nature and the divine" (Cox 1985, 156). There was gradual recognition, however, that, even before contemporary times, indigenous people exploited and shaped forests (Mann 2006), with varying impacts across landscapes (see Bush and Silman 2007), including eastern North America (chapter 1). As an example, aboriginal populations were small and their impacts light in northern New England, whereas in the Appalachian Mountains to the south, Indian-controlled fires apparently contributed to the abundance of mixed-oak forests observed at the beginning of Euro-American settlement (Brose et al. 2001).

Modern industrial times are another issue altogether. Old growth emerged as a compelling issue as a result of more than two centuries of land clearing and harvesting up to a point at which these forests could truly be described as remnant. Mary Byrd Davis' 1996 *Eastern Old-Growth Forests: Prospects for Discovery and Recovery* broadly recognized that even those enduring remnants bore a human footprint, regardless of their size or apparent remoteness. Top predators, such as the eastern timber wolf (*Canis lupus*) and the eastern mountain lion (*Puma concolor*), which played important regulatory roles in these communities, were missing. Fire suppression had altered many ecosystems—mixed-oak communities and pine savannas, for example. The loss of the American chestnut (*Castanea dentate*), a foundation species, was still reverberating through communities many decades after its demise. Habitat fragmentation was constraining both wildlife viability and the natural dynamics and connectivity of metapopulations.

The chapters of our book depict an assemblage of old-growth tracts with an even heavier human imprint. Chapter 3 details the long history of misguided fire suppression and habitat fragmentation in pine savannas in the South. Chapter 4 presents a conceptual model for taking into account a wide range of intensities of past and ongoing human impact on old growth in Great Smoky Mountains National Park. The basis of chapter 5 is that a century of fire suppression in the mixed mesophytic forests of eastern Kentucky is leading to the decline of oaks in favor of more mesophytic species such as maples—part of the so-called mesophication of eastern forests (Nowacki and Abrams 2008). Chapter 7 documents changes in old-growth northern hardwood-hemlock forests of the northern Lake States, pointing to altered fire regimes and greatly magnified deer densities, among other pressures. As discussed in chapter 8, one of the chief challenges of maintaining large areas of old-growth boreal forest is the inexorable pressure of timber harvesting.

These are extensions and intensifications of the anthropogenic impacts described in Davis (1996). New pressures have emerged, however, of perhaps more long-lasting import. Chapter 12 portrays the growing constellation of nonnative, invasive species disrupting forests across eastern North America, including the hemlock woolly adelgid, the emerald ash borer, and the Asian longhorned beetle. Like the chestnut blight, these invaders are in some places leading to wholesale extirpation or decline of major tree species, with no end in sight.

The most pervasive threat to old-growth forests is climate change. This issue was raised in Davis (1996) just after the release of the Second Assessment of Climate Change in 1995 by the Intergovernmental Panel on Climate Change. The Fifth Assessment was released in 2013 and 2014, and we know far more about the subject today. Ecologists know that forests will change, probably dramatically, but there remains considerable uncertainty about the details of these alterations. It is significant that there is not a chapter in the book dedicated to that topic. That is because current and future changes in the climate permeate each and every aspect of the ecology and conservation of old-growth ecosystems, including those in each of the chapters of this book. Altered dynamics due to climate change is thus a cross-cutting theme. Increases in temperature and growing-season length, for example, will likely exacerbate the impacts of invasive species (chapter 12). Climate warming may lead to higher burn rates in the boreal forest, which would mean reduced abundance of old forest, even in the absence of harvesting (chapter 8). On a positive note, there's reason to believe that pine savannas might be resilient to the effects of climate change, especially if appropriate fire regimes and connectivity can be reestablished (chapter 3).

What is clear from this book is that any sense of old growth as pristine, remote from human influence, has been largely banished from ecological science and conservation. This realistic stance has led the authors of this book to place more, not less, value on old growth, which, despite anthropogenic disruptions, provides benefits to people and the planet even more compelling, given the environmental state of the Earth. Crucially, these challenges have also motivated the contributors to think creatively and put to the test new ideas for recovering extant and creating new old growth for the future. At the end of *Eastern Old-Growth Forests* (Davis 1996), Bill McKibben eloquently captures this forward-looking sentiment:

> "When we look at a hemlock on a slope above the Cold River in the Berkshires, or a towering white pine south of Cranberry Lake in the Adirondacks, or a massive tulip poplar in a cove in the Smokies, we must not imagine that its glory devalues the second- and third-growth birch and beech a quarter-mile distant. Instead, the majesty of the ancient forest makes this tentative wildness all the more valuable, for it shows what it might become one day."

## Old Growth and Resilience in a Changing World

The chapters in this book tell a sobering story of a rapidly changing world in which ecological baselines are shifting and the future is uncertain. It seems reasonable to ask, in the Anthropocene, Is there still a place for old-growth forests in eastern North America? We would argue that yes, there most definitely is. The reasons are manifold. As Aldo Leopold reminded us, *"To keep every cog and wheel is the first precaution of intelligent tinkering."* (Leopold 1949). This precautionary approach takes on new meaning in the context of global environmental change, because the imperative now lies in maintaining the capacity of ecosystems to adapt and remain resilient to those changes (Millar et al. 2007). As discussed in chapter 13, old-growth forests contribute to adaptive capacity through the diversity of species that inhabit them, and the reservoir of potentially adaptive functional traits expressed from that diversity. It is also possible that aspects of old-growth structure may prove more resilient to climate change than will the species assemblages we have today, which are likely to change, though this remains a working hypothesis. However, potential losses of foundational species, due to continued range expansions and introductions of invasive pests and pathogens, seriously jeopardize that resilience by altering forest structure and many other impacts (Lovett et al. 2016). Maintaining and expanding the presence of old growth on the landscape will contribute to adaptive forest management and conservation by enhancing the potential for ecological transitions even as environmental boundary conditions change.

Idiosyncratically, the highly dynamic nature of late successional forests may itself add adaptive capacity to eastern landscapes. The dynamics of these systems will certainly shift in the future as disturbance regimes change and potentially novel pathways of forest development play out. However, complex interactions between late successional forests and disturbances will provide subsidies to the system, even when disturbance effects are severe and undesirable for economic objectives. These interactions, such as disturbance openings and multiple pathways of redevelopment, will create opportunities for adaptive shifts in species composition (Thom et al. 2017), while maintaining ecological memory in the form of biological legacies from previous stands incorporated into recovering postdisturbance stands (Johnstone et al. 2016).

There are other good reasons to believe old growth will have a place on the future landscape. Chief among these is that the habitat requirements of species using old-growth forests are not going away (chapters 3 and 11), and those species will benefit from old-growth structures under the future climate just as they do today, though ranges, competitive dynamics, and assemblages will change. Equally important are the ecosystem services that late successional and old-growth forests will provide on future landscapes. The science is clear that these services include high levels of carbon storage (chapter 14), as well as riparian functionality and flood resilience (chapter 9). For these reasons, active management for old-growth features and structural complexity at both stand and landscape scales is increasingly seen as an important silvicultural option in secondary forests, an approach that could expand the area of forest with old-growth characteristics (chapter 13). Silvicultural systems emulating the full array of disturbance dynamics in eastern forests and resulting in multiaged stand structures will have particular utility in this regard (chapter 6). From this standpoint, managing secondary forests

FIGURE 15-1. Old-growth eastern hemlock-northern hardwood forest in the Adirondack State Park, New York State. Photo credit: W. S. Keeton.

for old-growth characteristics will be of direct value to human communities in our efforts to moderate climate change (Ford and Keeton 2017).

In some ways old-growth forests are timeless, on the one hand increasingly displaced from their past, yet having the opportunity to adapt to the future. Where old growth has survived in the East, it is a reminder of our shared history and of the complexity, messiness, and grandeur of the forests that were once here. Yet, the contributors to this book make a compelling case that old growth is not only still relevant but is, in fact, vital in our changing world, and that prospects are high for restoring or actively managing for old-growth characteristics on a portion of secondary forest landscapes. This endeavor has a place within an integrated, holistic approach to adaptive forest management and conservation, ensuring the sustainability of a broad array of forest ecosystem benefits to humans and other organisms alike. With care and attention, future generations will be able to have the experience of walking into an eastern old-growth forest, and perhaps they too will catch a glimpse of something timeless, mysterious, and compelling (figure 15-1).

# References

Aplet, G. H., and W. S. Keeton. 1999. "Application of historical range of variability concepts to biodiversity conservation." In *Practical Approaches to the Conservation of Biological Diversity*, edited by R. Baydack, H. Campa, and J. Haufler, 71–86. Washington, DC: Island Press.

Bauhus, J., K. Puettmann, and C. Messier. 2009. "Silviculture for old-growth attributes." *Forest Ecology and Management 258*: 525–537.

Beltcheva, M., R. Metcheva, V. Peneva, M. Marinova, Y. Yankov, V. Chikova. 2011. "Heavy metals in Antarctic notothenioid fish from South Bay, Livingston Island, South Shetlands (Antarctica)." *Biol Trace Elem Res. 141*: 150–158. doi:10.1007/s12011-010-8739-5.

Blozan, W. 2011. "The Last of the Giants: Documenting and saving the largest eastern hemlocks." *American Forests*, Spring. Accessed February 8, 2018. http://www.americanforests.org/magazine/article/the-last-of-the-giants.

Brose, P., T. Schuler, D. Van Lear, and J. Berst. 2001. "Bringing fire back: the changing regimes of the Appalachian mixed-oak forests." *Journal of Forestry 99*: 30–35.

Bush, M. B. and M. R. Silman. 2007. "Amazonian exploitation revisited: ecological asymmetry and the policy pendulum." *Frontiers in Ecology and the Environment 5*: 457–465.

Cogbill, C., J. Burk, and G. Motzkin. 2003. "The forests of presettlement New England, USA: spatial and compositional patterns based on town proprietor surveys." *Journal of Biogeography 29*: 10–11.

Cox, T. 1985. "Americans and their forests: romanticism, progress, and science in the late nineteenth century." *Journal of Forest History 29*: 156-168. doi:10.2307/4004710.

Crutzen, P. J. 2002. "Geology of mankind." *Nature 415*: 23.

Davis, M. B. 1993. *Old Growth in the East: A Survey*. Richmond, VT: Cenozoic Society.

Davis, M. B., ed. 1996. *Eastern Old-Growth Forests: Prospects for Rediscovery and Recovery*. Washington, DC: Island Press.

Davis, M. B. 1996. "Extent and location." In *Eastern Old-Growth Forests: Prospects for Rediscovery and Recovery*, edited by M. B. Davis, 18–32. Washington, DC: Island Press.

Davis, M. B. 2003. *Old Growth in the East: A Survey*. Revised edition. Georgetown, KY: Appalachia-Science in the Public Interest.

Eriksen M., L. C. M. Lebreton, H. S. Carson, M. Thiel, C. J. Moore, J. C. Borerro, F. Galgani, et al., 2014. "Plastic Pollution in the World's Oceans: More than 5 Trillion Plastic Pieces Weighing over 250,000 Tons Afloat at Sea." *PLoS ONE 9* (12): e111913. https://doi.org/10.1371/journal.pone.0111913.

Fahey R. T., C. G. Lorimer, and D. J. Mladenoff. 2012. "Habitat heterogeneity and life-history traits influence presettlement distributions of early-successional tree species in a late-successional, hemlock-hardwood landscape." *Landscape Ecology 27*: 999–1013.

Flatley, W. T., C. W. Lafon, H. S. Grissino-Mayer, and L. B. LaForest. 2013. "Fire history, related to climate and land use in three southern Appalachian landscapes in the eastern United States." *Ecological Applications 23*: 1250–1266.

Ford, S. E., and W. S. Keeton. 2017. "Enhanced carbon storage through management for old-growth characteristics in northern hardwoods." *Ecosphere 8*: 1–20.

Foster, D., F. Swanson, J. Aber, I. Burke, N. Brokaw, D. Tilman, and A. Knapp. 2003. "The importance of land-use legacies to ecology and conservation." *BioScience 53*: 77–88.

Goring, S. J., D. J. Mladenoff, C. V. Cogbill, S. Record, C. J. Paciorek, S. T. Jackson, M. C. Dietze, et al. 2016. "Novel and lost forests in the upper midwestern United States, from new estimates of settlement-era composition, stem density, and biomass." *PLOS One 11* (12): eoisi935. doi:10.1371/journal.pone.0151935.

Grissino-Mayer, H. D. 2015. "Fire as a once-dominant disturbance process in the yellow pine and mixed pine-hardwood forests of the Appalachian Mountains." In *Natural Dis-

*turbances and Historic Range of Variation: Type, Frequency, Severity, and Post-Disturbance Structure in Central Hardwood Forests USA*, edited by C. H. Greenberg, and B. S. Collins, 123–146. Heidelberg, Germany: Springer.

Hansen, A. J., L. B. Phillips, R. O. Dubayah, S. Goetz, and M. Hofton. 2014. "Regional-scale application of LIDAR: variation in forest canopy structure across the southeastern US." *Forest Ecology and Management 329*: 214–226.

Holling, C. S., L. H. Gunderson, and D. Ludwig. 2002. "In quest of a theory of adaptive change." In *Panarchy: Understanding Transformations in Human and Natural Systems*, edited by L. H. Gunderson, and C. S. Holling, 3–22. Washington, DC: Island Press.

Johnstone, J. F., C. D Allen, J. F. Franklin, L. E. Frelich, B. J. Harvey, P. E. Higuera, M. C. Mack, et al. 2016. "Changing disturbance regimes, ecological memory, and forest resilience." *Frontiers in Ecology and the Environment 14*: 369–378.

Keeton, W. S., C. E. Kraft, and D. R. Warren. 2007. "Mature and old-growth riparian forests: Structure, dynamics, and effects on Adirondack stream habitats." *Ecological Applications 17*: 852–868.

Keeton, W. S., A. A. Whitman, G. C. McGee, and C. L. Goodale. 2011. "Late-successional biomass development in northern hardwood-conifer forests of the Northeastern United States." *Forest Science 57*: 489–505.

Kelly, P. E., and D. W. Larson. 2007. *The Last Stand: a Journey Through the Ancient Cliff-face Forest of the Niagara Escarpment*. Toronto: Dundurn.

Kershner, B., and R. Leverett, 2004. *The Sierra Club Guide to the Ancient Forests of the Northeast*. San Francisco: Sierra Club Books.

Krajick, K. 2003. "Methuselahs in our midst." *Science 302*: 768–769. doi:10.1126/science .302.5646.768.

Leopold, A. 1949. *A Sand County Almanac, and Sketches Here and There*. New York: Oxford University Press.

Lorimer, C. G., and C. R. Halpin. 2014. "Classification and dynamics of developmental stages in late-successional temperate forests." *Forest Ecology and Management 334*: 344–357.

Lorimer, C. G., and A. S. White. 2003. "Scale and frequency of natural disturbances in the northeastern US: implications for early successional forest habitats and regional age distributions." *Forest Ecology and Management 185*: 41–64.

Lovett, G. M., M. Weiss, A. M. Liebhold, T. P. Holmes, B. Leung, K. F. Lambert, D. A. Orwig, et al. 2016. "Nonnative forest insects and pathogens in the United States: Impacts and policy options." *Ecological Applications 26*: 1437–1455.

Mann, C. 2006. *1491: New Revelations of the Americas before Columbus*. New York: Vintage Books.

McGarvey, J. C., J. R. Thompson, H. E. Epstein, and H. H. Shugart. 2015. "Carbon storage in old-growth forests of the Mid-Atlantic: Toward better understanding of the eastern forest carbon sink." *Ecology 96*: 311–317.

McKibben, B. 1996. "Afterword: Future old growth." In *Eastern Old-Growth Forests: Prospects for Rediscovery and Recovery*, edited by M. B. Davis, 359–363. Washington, DC: Island Press.

Meigs, G. W., and W. S. Keeton. 2018. "Intermediate-severity wind disturbance in mature temperate forests: effects on legacy structure, carbon storage, and stand dynamics." *Ecological Applications*. doi:10.1002/eap.1691.

Millar, C. I., N. L. Stephenson, and S. L. Stephens. 2007. "Climate change and forests of the future: managing in the face of uncertainty." *Ecological Applications 17*: 2145–2151.

NASA. Global Climate Change. Vital Signs of the Planet. Accessed February 7, 2018. https://climate.nasa.gov.

New Brunswick Museum. 2015. "Oldest red pines discovered in NB protected area." Accessed February 8, 2018. http://www.fundy-biosphere.ca/en/news/atlantic-canada-s-oldest-red -pines-discovered-in-nb-protected-area.html.

Nowacki, G. J., and Abrams, M. D. 2008. "The demise of fire and "mesophication" of forests in the eastern United States." *BioScience 58*: 123–138.

Pederson N., J. M. Dyer, R. W. McEwan, A. E. Hessl, C. J. Mock, D. A. Orwig, H. E. Rieder, and B. I. Cook. 2014. "The legacy of episodic climatic events in shaping temperate, broadleaf forests." *Ecological Monographs 84*: 599–620.

Pesklevits, A., P. Duinker, and P. Bush. 2011. "Old-growth forests: anatomy of a wicked problem." *Forests 2*: 343–356. doi:10.3390/f2010343.

Pickett, S. T. A., and P. S. White. 1985. *The Ecology of Natural Disturbance and Patch Dynamics.* San Diego: Academic Press.

Rhemtulla, J. M., D. J. Mladenoff, and M. K. Clayton. 2009. "Historical forest baselines reveal potential for continued carbon sequestration." *Proceedings of the National Academy of Sciences 106*: 6082–6087.

Russell, E. W. B. 1983. "Indian-set fires in the forests of the northeastern United States." *Ecology 64*: 78–88.

Southern Appalachian Forest Coalition, 2018. Accessed February 8, 2018. http://www.safc.org.

Stahle, D. W., M. K. Cleaveland, and J. G. Hehr. 1988. "North Carolina climate Changes reconstructed from tree rings: A.D. 372 to 1985." *Science 240*: 1517–1519. doi:10.1126/science.240.4858.1517.

Thom, D., W. Rammer, and R. Seidl. 2017. "Disturbances catalyze the adaptation of forest ecosystems to changing climate conditions." *Global Change Biology 23*: 269–282.

Thompson J. R., D. N. Carpenter, C. V. Cogbill, and D. R. Foster. 2013. "Four centuries of change in northeastern United States forests." *PloS One 8*: e72540. doi:10.1371/journal.pone.0072540.

Tyrrell, L. E., G. J. Nowacki, T. R. Crow, D. S. Buckley, E. A. Nauertz, J. N. Niese, J. L. Rollinger, and J. C. Zasada. 1998. "Information about old growth for selected forest Type groups in the eastern United States". General Technical Report NC-197. North Central Forest Experiment Station. St. Paul, MN: US Department of Agriculture, Forest Service.

Urbano, A. R., and W.S. Keeton. 2017. "Forest structural development, carbon dynamics, and co-varying habitat characteristics as influenced by land-use history and reforestation approach." *Forest Ecology and Management 392*: 21–35.

Wirth, C., C. Messier, Y. Bergeron, D. Frank, and A. Kahl. 2009. "Old-Growth Forest Definitions: a Pragmatic View." In *Old-Growth Forests: Fate, Function, and Value*, edited by C. Wirth, G. Gleixner, and M. Heimann, 11–33. Berlin: Springer-Verlag.

*Note:* Bold terms within definitions are also in the glossary. Italicized terms are not in the glossary, but their meaning is indicated by context.

**Abiotic** – Referring to the nonliving portion of the environment, such as water, rocks, and temperature, in contrast to **biotic**, which refers to living elements of the environment.

**Acid rain** – The pH of normal rain is typically approximately 5.6, slightly acidic, due to the natural formation of carbonic acid in the atmosphere. Rainfall more acidic than that level, resulting from human-caused pollution, is considered acid rain. The term is also used as shorthand for all forms of *acid precipitation*.

**Adaptive management** – A decision-making process that involves iteratively improving management activities by monitoring outcomes.

**Age-class distribution** – The relative numbers of individuals in groupings of ages within a population (e.g., the numbers within 0-10, 11-20, and 21-30 years old or within seedling, sapling, pole, and mature tree classes).

**Age of culmination** – The age at which a stand of trees grown for harvest reaches its highest average growth rate. Age of culmination is considered the time to harvest in order to maximize wood volume over time.

**Allelopathy** – The release of substances by a plant into soil, resulting in reduced germination, establishment, growth, or reproduction of neighboring plants.

**Anthropocene** – The Age of Humans. A proposed new geological epoch, not yet accepted by professional geological societies, beginning at the time of significant human ecological impact on the Earth.

**Asymmetric competition** – Competition between two plants (or other organisms), usually of very different sizes, such that the process results in a large difference between the two in access to an essential resource, such as light.

**Autotrophic** – Obtaining food from inorganic substances (light energy or chemicals).

**Basal area** – The cross-sectional area of a tree trunk at 1.4 meters (4.5 feet) height off the ground (DBH). Basal area can be summed for all trees to calculate *stand basal area*, which is generally reported in square feet per acre or square meters per hectare. Basal area, which is easy to measure, is used as a surrogate for tree volume or biomass.

**Belowground competition** – Competition among neighboring plants for essential resources, such as water and nutrients, in the soil. Contrast with competition aboveground for light.

**Benthic** – Referring to the bottom of a body of water, such as a streambed.

**Biodiversity** – The variety of ecosystems, species, and genes. Also called *biological diversity*.

**Biodiversity hot spot** – Regions that support particularly high levels of biodiversity (and often **endemism**).

**Biogeographic** – Pertaining to patterns of geography of species, ecosystem types, or ecoregions.

**Biomass** – The mass of living matter in a given place, sometimes specified for a particular species or group of species, which is generally reported in mass (weight) per unit area.

**Biometric** – Relating to the use of measurements to address biological questions.

**Biotic** – Referring to living organisms, in contrast to **abiotic**, which refers to nonliving elements of the environment.

**$C_4$ plant** – Plant species using the $C_4$ photosynthesis pathway, as opposed to the more common $C_3$ pathway. Termed $C_4$ because the first product is a four-carbon compound. Many grasses are $C_4$ plants.

**Cambium** – The living tissue at the juncture of bark and wood that produces wood (primarily *xylem*, which transports water and minerals) to the inside and **phloem** (which transports food produced by photosynthesis) to the outside.

**Carbon budget** – Sum of carbon inflow and outflow for a plant, ecosystem, or land area.

**Carbon dynamics** – The *flux* of carbon into and out of a system (a plant, forest, area, or country).

**Carbon markets** – A market aimed at reducing greenhouse gas emissions by companies and countries, in which carbon credits from public and private landholders can be bought and sold.

**Carbon neutral** – Carbon uptake equals carbon release. In the case of a tree or forest, uptake is through photosynthesis; release is through cellular **respiration**.

**Carbon sink** – A tree or an area for which carbon uptake is larger than carbon release, which leads to **carbon storage**.

**Carbon source** – A tree or area for which carbon uptake is less than carbon release, leading to net emission of carbon to the atmosphere.

**Carbon storage** – The amount of carbon stored within an organism, population, or entire ecosystem in a given place. Carbon storage occurs when the uptake of carbon is larger than the release to the air. Also known as *carbon sequestration*. An ecosystem or area that stores carbon is known as a **carbon sink** (chapter 14).

**Cladogram** – A branching diagram, which shows the evolutionary relatedness of species (and sometimes the times of past species divergences), developed with the use of cladistics.

**Clear-cutting** – A forest management or **silvicultural** system in which all or nearly all trees in an area are harvested.

**Clementsian** – A reference to Frederic Clements, an influential early twentieth-century American ecologist, who maintained that plant communities develop like superorganisms after a disturbance (e.g., fire) in a unidirectional manner toward a particular endpoint best suited to local conditions, termed a *climatic climax*. Disturbances should, according to Clements, be thought of as a process that causes a community to deviate from the ideal. An antonym of Clementsian is *Gleasonian*, named after the plant taxonomist and ecologist Henry Gleason, who in the 1910s and 1920s refuted some of Clements's ideas.

**Clonal** – The production of offspring via asexual, vegetative reproduction.

**Coarse woody debris** (CWD) – Dead standing (*snags*) and downed, decaying trees. The threshold diameter is arbitrary and reflects objectives of a given study, but often CWD is defined as material having diameters greater than 10 or 25 centimeters. Smaller material (twigs, branches) is considered *fine woody debris*.

**Cohort** – Individuals of a similar age, often establishing after a disturbance. *Multicohort* refers to more than one cohort, as in multicohort harvesting.

**Competitive exclusion** – The exclusion of a species from a particular place as a result of competition with another species or group of species.

**Composition** – See **Species composition**.

**Congener** – A member of the same genus. All species within the oak genus (*Quercus*) are congeners, for example.

**Conservation easement** – A voluntary legal agreement between a land-owner and an entity, such as a land trust, in which restrictions are placed on use of the land in an effort to promote conservation goals.

**Demography** – The statistics of births, deaths, and growth into older age classes that reveal the patterns and changes in populations.

**De novo ecosystems** – See **no-analog communities**.

**Dendrochronology** – The science of dating events, growth, and past environments, using the patterns in the annual rings of trees.

**Dendroecology** – The use of dendrochronology to address ecological questions.

**Density-dependent mortality** – Mortality resulting from processes, such as competition, that increases with density among organisms. In forests, density-dependent mortality leads to lower density and larger trees.

**Density-independent mortality** – Mortality resulting from agents, such as natural disturbance, unrelated to the density of organisms.

**Depauperate** – In ecology, meaning relatively low diversity of species.

**Derecho** – A long-lived, straight-line windstorm, producing strong winds and covering a large area.

**Detritivore** – A decomposer organism that feeds on dead organic matter, breaking it into smaller pieces. Worms are detritivores.

**Diameter at breast height (DBH)** – By convention, the diameter of a tree is measured at "breast height" (approximately 1.4 meters above the ground) to eliminate the influence of root buttressing on the diameter measurement. DBH is used to determine **basal area**.

**Diameter-limit harvesting** – The harvesting of all trees above a certain size.

**Disturbance** – According to chapter 6, it is "any relatively discrete event in time that disrupts ecosystem, community, or population structure and changes resources, substrate availability, or the physical environment." *Disturbance processes* refer to the frequency, intensity, size, and effects of disturbances on communities and ecosystems

**Dynamics** – Literally, *changes*. In ecology, dynamics refers to the ways that natural communities and ecosystems change over time. It can refer to a directional, cyclic, or random change over time of an entire ecosystem or to continuous changes in patches within an ecosystem despite little change to the total system in terms of species, structure, and functions.

**Ecosystem goods and services** – Benefits provided by ecosystems to all organisms, although emphasis is generally placed on humans as beneficiaries. Examples include wood, clean water, air purification, crop pollination, and **carbon storage**.

**Ecotone** – A location where two types of habitats, ecosystems, or ecoregions meet.

**Eddy covariance** – A method of measuring the exchange rates of gases, such as carbon dioxide, from natural ecosystem or human landscapes.

**Emergent** – A tree whose canopy sticks out above the rest of the forest. Very tall white pines are good examples.

**Endemism** – The occurrence of a species only in a defined geographic location (habitat type, forest, preserve, state, country, etc.). Species exhibiting endemism are termed *endemic*.

**Environmental gradients** – A continuous change across space in an important environmental characteristic, such as moisture, light, and soil. Mountains exhibit elevation gradients, from low to high elevation.

**Equilibrium** – In ecology, when species composition, physical structure, and processes of an ecosystem stay roughly the same over time, even if changes occur in small patches of the ecosystem. Also known as *steady state*. See **gap dynamics**.

**Evapotranspiration** – Movement of water from the land to the air as a result of evaporation and transpiration (water loss from inside leaves through stomates, which are openings that allow for gas exchange).

**Feedbacks** – *Positive feedbacks* amplify the original process, while *negative feedbacks* diminish it. In climate science, for example, a positive feedback of climate warming will cause further warming, whereas a negative feedback will lead to less warming.

**Fire cycle** – The cycle from the occurrence of a fire at a site, through vegetation recovery, to the occurrence of another fire. Fire cycles vary from short (annual fires) to long (hundreds of years or more).

**Fire suppression** – Human actions aimed at reducing the frequency, intensity, or size of fires, usually by extinguishing them but also by habitat manipulations or land use (e.g., livestock grazing).

**Fragmentation** – The fracturing of previously continuous habitat into smaller, more isolated tracts.

**Function** – A reference to natural processes that occur in ecosystems, such as nutrient cycling, soil development, natural disturbance, respiration in living organisms, and carbon storage.

**Gap dynamics** – Changes within patches in forests induced by the death of plants, opening up space, and increasing resources. Many forests contain manifold patches, each at a different age or stage of recovery from disturbance, creating an internally dynamic forest.

**Gap phase dynamics** – See **gap dynamics**.

**Geographic turnover** – Changes in species composition (the species comprising a community) across space or time.

**Geomorphology** – The shape of the land with respect to geological features.

**Gradient** – See **environmental gradient**.

**Headwater** – Streams near the source of rivers or other waterways.

**Heterotrophic** – Obtaining food by consuming other organisms.

**Historic range and variability** – The range of ecosystem processes, especially the types, frequencies, severities, and sizes of natural disturbances in the past before extensive human impact.

**Hydrologic** – Referring to the movement of water with respect to the land.

**Hypsithermal** – A period from about 9,000 to 5,000 years before the present when, as a result of astronomical forces involving the Earth and the Sun, conditions were warmer than before or after this period.

**Individualistic** – A species behaving or occurring in a way that is unique and independent of other species.

**Inorganic** – Nonliving and not containing carbon.

**Intermediate treatments** – Interventions in managed forests after tree regeneration aimed at improving stand conditions and eventual economic and forest quality outcomes.

**Intraspecific competition** – Competition for resources, such as water and nutrients, among members of the same species.

**Invasive** – A nonnative species that establishes, spreads, and thrives in an area, leading to negative consequences for native species.

**LiDAR** – *Light detection and ranging* is a remote sensing method that uses pulsed light from lasers to detect the distances to objects. Combined with other data, it is used to produce precise three-dimensional images of the shape of the surface of the Earth, including forests.

**Life history** – The life cycle attributes of a species, especially the number and size of offspring, growth pattern, length of life, and age and size at maturity, reproduction, and mortality. *Life history theory* attempts to understand how the combination of these characteristics or life history strategies adapt species to their particular niche and environment.

**Low-gradient stream** – A slow-moving stream along an area with a gradual change in elevation. In contrast to a *high-gradient stream*, which moves quickly along a steep elevation grade.

**Macroenvironment** – The environmental characteristics of large areas. Includes *macroclimate*. Contrasts with **microenvironment**.

**Macroinvertebrates** – Large invertebrates (animals without backbones), visible with the naked eye.

**Megafauna** – Literally, *large animals,* but often associated with mammals (e.g., ground sloth, giant camel, mastodon, cave bear), larger than their modern counterparts, that lived during the late Pleistocene as recently as 12,000 years ago and went abruptly extinct. Evidence suggests a role for human overhunting and possibly climate change in these extinctions.

**Mesic** – Sites with a moderate amount of moisture. **Xeric** refers to dry and *hydric* to wet. *Submesic* is less than moderate moisture but not xeric.

**Mesophytic** – Technically, plants needing a moderate amount of moisture. The *mixed mesophytic forest* region, described by the influential ecologist E. Lucy Braun, refers to the unglaciated plateaus centrally located within the eastern deciduous forest with a very high diversity of tree species.

**Metapopulation** – A group of populations of a species separated by space but interacting, as individuals move from one population to another.

**Microbial** – Organisms too small to be seen with the naked eye.

**Microburst** – A potentially damaging localized downdraft within a thunderstorm.

**Microenvironment** – The environmental characteristics of a small, local area. Also referred to as *microhabitat* and includes *microclimate*. Contrasts with **macroenvironment** and *mesoenvironment*, the latter of which refers to an area intermediate in size.

**Microsite** – Small, localized site with environmental conditions differing from its surroundings.

**Mixedwood** – Containing both hardwoods (angiosperm dicots such as maple, oak, ash, and birch) and softwoods (coniferous tree species, such as spruce and pine).

**Mycorrhiza** – A symbiotic (living together) and mutualistic (mutually beneficial) relationship between a fungus and the roots of a vascular plant (those with a set of tissues for conducting food, water, and nutrients). *Mycorrhizae* play an essential role in the success of many plant species.

**Natural disturbance regime** – The types, severities, sizes, and frequencies of natural disturbances intrinsic to an area (excluding direct anthropogenic causes). *Types* include fire, blowdown from wind, flooding, insect or pathogen outbreak, ice damage, and others. *Severity* is the degree to which an ecosystem is affected by a disturbance, in other words, tree mortality, loss of living biomass, formation of openings, and similar occurrences. *Size* is the amount of area affected. *Frequency* is expressed as return interval, that is, the typical time between events at any one place for a disturbance type.

**Nitrification** – The process of conversion by bacteria in the soil of one form of nitrogen (usually ammonia or ammonium) to another (first nitrite and then nitrate).

**Nitrogen deposition** – The deposition of nitrogen compounds, above natural levels, onto ecosystems as a result of the burning of fossil fuels and high-intensity agriculture. **Acid rain** is a major contributor to nitrogen deposition.

**Nitrogen (N) mineralization** – The conversion by bacteria of nitrogen in soil organic matter to forms available to plants.

**No-analog community** – A natural community of species occurring in the past or expected to occur in the future that does not exist today. In other words, the species in a no-analog community do not occur together in modern communities.

**Paleoecology** – Literally, *older or ancient ecology*. The study of the interactions of organisms and their environment and with each other over geological time scales.

**Partial cuts** – Harvests that remove only part of a forest stand. **Shelterwood** and **selection harvesting** are partial cuts.

**Patch dynamics** – This term captures the idea that ecosystems are composed of patches created at different times by disturbances. As a result, these patches differ in age, physical structure, and species composition. The full idea of patch dynamics encompasses processes, such as natural disturbance, that create the patches, the changes over time (the dynamics) of those patches as they recover from disturbance, and the interactions among patches.

**Phenology** – The study of the timing of life cycle events in organisms on daily to seasonal time scales (e.g., flowering times).

**Phloem** – Living cells in plants that transport the food made through photosynthesis to all parts of the plants. In tree trunks, active phloem is produced by the **cambium** in a very thin layer at the inner part of the bark.

**Physiognomy** – The predominate physical shape of vegetation (e.g., shrublands, grasslands, multilayered forest).

**Physiography** – The physical shape of land, including elevation, steepness, slope direction, and degree of dissection.

**Phytophagous** – Plant-eating.

**Pioneer species** – The species that first colonize a site after a disturbance because they are adapted to the conditions created by that event.

**Prescribed fire** – A planned fire set in order to achieve specific management goals. Contrasts with lightning-caused fire and arson-caused fire. The prescription considers safety, weather, topography, vegetation, and goals.

**Primary forest** – A forest that has never been harvested. Contrast with *secondary forest*, one that has grown after one or more cycles of harvesting.

**Processes** – In ecology, a general term for the chemical and biological flows, cycles, and events that connect organisms to their environment and each other. Examples are the uptake of carbon (through photosynthesis), decomposition (of dead wood), cycling of nutrients (into and from plants, soils, streams, and the air), and natural disturbances (e.g., fire, wind). Also called **functions**.

**Productivity** – The amount of living mass produced in a given area, usually expressed as mass per unit surface. *Primary productivity* is the biomass generated by producers (**autotrophs**).

**Propagule** – Tissues that can lead to reproduction of an organism, whether sexually or asexually. In plants, this includes seeds, spores, gemmae (asexual outgrowths in mosses), and stems that, when separated from the mother plant, can root.

**Pyrodiversity** – Spatial variation in physical conditions that promote the occurrence, frequency, and intensity of fire. Spatial heterogeneity in flammable leaves creates pyrodiversity, for example.

**Pyrogenic** – Literally, *causing fire*.

**Recruitment** – The process by which juveniles successfully join a population by birth, seed production, germination, growth, and survival. Immigration of new individuals is also included in recruitment.

**Regeneration** – Establishment of a species or a plant community through the germination of seeds and spores or by asexual reproduction. *Regeneration harvest* is a logging operation designed to simultaneously extract wood and regenerate a stand.

**Resilience** – In ecology, the capacity of an organism, population, or ecosystem to recover quickly after being perturbed.

**Resistance** – In ecology, the capacity of an organism, population, or ecosystem to exhibit little change in the face of external stress (e.g., disturbance).

**Respiration (cellular)** – Metabolic processes within a cell that break down organic material, harvesting energy for the cell's activities in the process releasing carbon dioxide.

**Retention** – In forestry, a harvest that preserves elements of the previous stand to provide continuity in physical structures and species diversity.

**Reverse J-shaped distributions** – A distribution of sizes or ages within a tree population with many young or small individuals and increasingly fewer as size or age rises. The term comes from the similarity to a reversed letter J when such a distribution is graphed. Many **shade tolerant** tree species in natural conditions exhibit such distributions. Contrasts with an *even-aged distribution* where nearly all individuals are of similar size or age.

**Rotation** – The period of time between harvests (or a type of natural disturbance, such as fire) on a given site. *Extended rotation* refers to longer periods between harvests.

**Saproxylic** – Referring to invertebrates (insects, mites, nematodes, worms, etc.) dependent on dead or decomposing wood.

**Savanna** – An ecosystem with predominately scattered trees among grasses and forbs.

**Scarify** – Physical disturbance (scraping, for example) of the soil surface to promote germination and establishment of seedlings. Also, physical, chemical, or thermal weakening of the coat of a seed to encourage germination. Some seeds require such action before germinating.

**Seedling establishment** – The process of successful germination, growth, and survival of juvenile plants.

**Selection system** – A forest management or **silvicultural** system in which single trees or small groups of trees are harvested evenly across a stand, resulting in a forest with trees of a variety of ages and a largely intact canopy. This approach provides for long-term, sustained growth and production of the residual trees to allow periodic entries about every 15 to 25 years.

**Sequestration** – The accumulation of a substance (e.g., carbon, organic matter) in live or dead tissues.

**Seral** – Referring to **succession**.

**Serotiny** – Cones or fruits that release seeds only after they are heated to high temperatures, usually by fire. Serotiny is an adaptation of plants to areas with crown fires.

**Shade tolerance** – The degree to which a plant can grow in shaded conditions, such as in a forest understory. Plants are generally classified as being *shade tolerant* or *shade intolerant*, with intermediate groupings classified as *midtolerant*.

**Shannon Diversity Index** – A formula for calculating *species diversity* in a given area. The index includes both the number of species (**species richness**) and the relative numbers across species (*evenness*). Thus, a site with three species, each with similar abundance, is more diverse than a site with three species, two of which are rare. And a site with four species is more diverse than one with only three, if evenness is the same between the two sites.

**Shelterwood system** – A forest harvest management or **silvicultural** system in which mature trees are cut in two or three stages over time, allowing the establishment of new individuals before all mature trees are removed. Contrasts with **clear-cutting**, in which all mature trees are cut at once, and **selection**, in which single trees or small groups of trees are cut.

**Silviculture** – According to the US Forest Service, it is "the art and science of controlling the establishment, growth, composition, health, and quality of forests and woodlands to meet the diverse needs and values of landowners and society, such as wildlife habitat, timber, water resources, restoration, and recreation on a sustainable basis." *Ecological silviculture* emphasizes basing management actions on an understanding of the ecology of tree species and forests, especially using natural disturbance dynamics as a guide for harvesting systems (chapter 6).

**Single-tree selection** – A form of selection logging in which scattered single trees are harvested. See **selection**.

**Soil respiration** – The respiration (breakdown of food in cells to generate energy) of organisms residing in the soil.

**Soil texture** – The relative amount of sand (large particles), silt (medium-sized particles), and clay (small particles) in soil, which confers a perceptible "feel" (gritty, smooth, and sticky, respectively) and influences nutrient and moisture availability.

**Species composition** – The suite of species (or sometimes types of species) occurring in a particular place. Also called **composition**.

**Species richness** – The number of species in a given place. Species richness is one part of the formula for calculating *species diversity*, which also includes the *evenness* of species numbers, that is, the relative numbers across species. See **Shannon Diversity Index**.

**Sprouts** – New stems arising (from stumps or roots, for example) after the death of the above-ground trunk of a tree. Also called **resprouts**.

**Stand-replacing** – A disturbance (usually natural) that destroys the present stand of trees and replaces it with another through regeneration after the disturbance.

**Stem-exclusion stage** – An early stage in successional development of a forest stand after disturbance in which regeneration of new stems ceases and the death of some less successful older stems commences.

**Structure** – Referring either to the *physical structure* of a natural community or the patterns of occurrence of the sizes and types of species in a community.

**Succession** – Also known as *secondary succession*, which is the process of change in an ecological community after disturbance, both in terms of environmental factors (e.g., light availability), physical structure, and species composition. Succession in an area that has not been previously occupied by a community (e.g., retreat of glaciers) is known as *primary succession*.

**Surface fire** – Fire that occurs only on the land surface and does not move into upper layers of the forest canopy.

**Taiga** – From the Russian for a cold, northern coniferous forest. Also known as *boreal forest*.

**Thinning** – In forestry, the removal of trees to increase the growth of the remaining ones.

**Tip-up mounds** – Mounds in forests caused by uprooted trees, usually accompanied by adjacent depressions where the root mass occurred previous to uprooting. Tip-up mounds create uneven topography and serve as key sites for tree establishment.

**Treefall gaps** – Openings in the canopy of a forest caused by the death and fall of one or more trees. Light availability and sometimes other essential resources are elevated at these sites until canopy reclosure.

**Uneven-aged management** – Harvesting (e.g., **selection system**) that results in residual trees of different ages, in contrast to *even-aged management* that leads to regeneration of a single cohort of trees of the same age (e.g., from **clear-cutting**).

**Windthrow** – Wind-caused death of trees, usually referring to smaller, less intense events as opposed to hurricanes and tornados.

**Xeric** – Dry conditions. **Mesic** refers to a moderate amount of moisture and *hydric* to wet.

CONTRIBUTORS

**LORETTA L. BATTAGLIA** is an associate professor of plant biology at Southern Illinois University–Carbondale. Her research interests include invasive species, wetland ecology, and the effects of climate change on coastal ecosystems.

**HEATHER A. BECHTOLD** is an assistant professor in the Department of Biological Sciences at Lock Haven University of Pennsylvania. Her research interests investigate the interconnectedness of terrestrial and aquatic landscapes and how humans change ecosystem processes. She and her family live along the West Branch of the Susquehanna River.

**YAN BOULANGER** has been a research scientist in forest ecology at Natural Resources Canada since 2013. His main areas of expertise are related to the projection of forest landscapes and natural disturbances under climate change. He is also an associate professor at Université du Québec à Rimouski.

**PHILIP J. BURTON** is a professor in the Ecosystem Science and Management Program at the University of Northern British Columbia. His research interests include ecological succession and disturbance, forest regeneration, ecosystem restoration, and factors contributing to forest resilience.

**JULIA I. CHAPMAN** is a doctoral candidate in the Department of Biology at the University of Dayton. Her research focuses on plant community assembly and the spatiotemporal dynamics of biodiversity, especially in old-growth temperate forests.

**BEVERLY S. COLLINS** is a professor in the Department of Biology at Western Carolina University, Cullowhee, North Carolina. She specializes in plant community ecology and dynamics.

**WILLIAM H. CONNER** is professor at Clemson University's Baruch Institute of Coastal Ecology and Forest Science. His research interests include freshwater forested wetlands, wetland management, wetland creation and restoration, the effects of humans and nature on natural environments, the use of wetlands for wastewater treatment, estuarine-upland connections, and changing land-use impacts on natural systems.

**JENNIFER K. COSTANZA** is a research assistant professor in the Department of Forestry and Environmental Resources at North Carolina State University. Her research interests are in landscape ecology and conservation. Her research aims to build a body of knowledge about climate and land-use change threats to ecosystems and landscapes.

**ANTHONY W. D'AMATO** is an associate professor and director of the forestry program in the Rubenstein School of Environment and Natural Resources at the University of Vermont. He specializes in silviculture and applied forest ecology, particularly in the context of global change.

**FRÉDÉRIK DOYON** is a professor in the Department of Natural Resources at the Université du Québec en Outaouais in Gatineau, Canada and a member of the Institut des Sciences de la Forêt tempérée. He is a forest and landscape ecologist. His research interests include landscape dynamics of managed and unmanaged forests under global change and climate-adaptive silviculture

**TIMOTHY J. FAHEY** is Liberty Hyde Bailey Professor of Natural Resources at Cornell University. He has worked for many years at the Hubbard Brook Experimental Forest with a particular emphasis on nutrient cycling and fine root dynamics.

**JODI A. FORRESTER**, a former research scientist at the University of Wisconsin–Madison, is an assistant professor in the Department of Forestry and Environmental Resources at North Carolina State University. Her research examines how forest ecosystems respond to disturbances.

**SHAWN FRAVER** is an associate professor of forest ecology in the School of Forest Resources at the University of Maine. His research addresses how disturbances influence forest composition, structure, and development.

**SYLVIE GAUTHIER** has been working as a research scientist with the Canadian Forest Service of Natural Resources Canada since 1993. Her research aims to characterize disturbance regimes and their effects on forest dynamics. She is also working on translating ecological knowledge into sustainable forests management and climate change adaptation strategies.

**LOUIS DE GRANDPRÉ** is a research scientist at the Canadian Forest Service in Quebec City. His research interests include community ecology, biodiversity and the impact of disturbances on boreal forest dynamics. He also works to integrate this knowledge in a forest ecosystem management framework.

**JOHN S. GUNN** is a research assistant professor of forest management with the New Hampshire Agricultural Experiment Station and UNH Cooperative Extension at the University of New Hampshire in Durham. His research interests include greenhouse gas accounting of the forest products sector and invasive plant impacts on forests.

**DANIEL KNEESHAW** is a member of the Center for Forest Research and professor in forest ecology and environmental sciences at the University of Quebec in Montréal. He's worked on boreal old growth, stand dynamics following disturbances, and the effect on tree growth and mortality of interactions between moisture stress and defoliation.

**CLIFFORD E. KRAFT** is a professor in the Department of Natural Resources at Cornell University. His research focuses on ecosystem interactions that influence freshwater fish populations in northern temperate lakes and rivers, with an emphasis on waters in the Adirondack region of New York State.

**CRAIG G. LORIMER** is a professor emeritus in the Department of Forest and Wildlife Ecology, University of Wisconsin–Madison. His research interests include forest stand dynamics, natural disturbance regimes, ecology of old-growth forests, and ecological forestry practices.

**RYAN W. MCEWAN** is an associate professor and environmental biology program director in the Department of Biology at the University of Dayton. His research focuses on the causes and consequences of forest dynamics, ecological invasion, and biodiversity in tropical, temperate, and boreal forests.

**GREGORY G. MCGEE** is an associate professor in the Department of Environmental and Forest Biology at the State University of New York College of Environmental Science and Forestry, Syracuse. His research interests include forest ecosystem management, forest restoration, and science education.

**DAVID J. MLADENOFF** is a professor in the Department of Forest and Wildlife Ecology at the University of Wisconsin–Madison. His research focuses on forest ecosystem and landscape ecology. He recently studied the effects on ecosystem processes of restoring old-growth structure to young forests and using broadscale historical data to understand regional disturbance legacies.

**DAVID A. ORWIG** is a forest ecologist at the Harvard Forest in Petersham, Massachusetts. Using community ecology and tree-ring studies, he has spent the last two decades studying the ecology of old-growth forests and the complex interactions between invasive pests and pathogens and dominant host tree species.

**BRIAN J. PALIK** is a research ecologist and science leader for applied ecology with the USDA Forest Service, Northern Research Station. He specializes in research on ecological approaches to sustaining managed forests.

**ROBERT K. PEET** is a professor of biology at the University of North Carolina. He has studied plant communities across the breadth of southeastern ecosystems. Among his special interests are community dynamics, species diversity, and the use of large datasets to understand biogeographic patterns in species distribution and community assembly.

**WILLIAM J. PLATT** is a professor of biology at Louisiana State University and Henry Beadel Fellow at Tall Timbers Research Station. He has studied the Southeastern Coastal Plain hot spot of biodiversity and endemism for more than four decades with a focus on fire and hurricane disturbance ecology in sites with old-growth attributes.

**PATRICIA RAYMOND** is a research forester at the Direction de la recherche forestière—Forest Research Branch of the government of Québec, Canada. Her research interests are in forest ecology and silviculture.

**JULIE P. TUTTLE** is a doctoral candidate in ecology at the University of North Carolina at Chapel Hill. Her research interests include southern Appalachian forests, landscape vegetation pattern and dynamics, species distribution modeling, and the influence of human activities, natural disturbance, and finescale environmental gradients on forest change.

**DANA R. WARREN** is an assistant professor at Oregon State University with an appointment in both the Department of Forest Ecosystems and Society and the Department of Fisheries and Wildlife. Dana studies stream and riparian ecosystems and the ecology of aquatic biota.

**PETER S. WHITE** is a professor at the University of North Carolina at Chapel Hill. He teaches ecology and conservation biology. His research interests include ecosystem dynamics, patterns of species diversity, and conservation ethics. He and his students have worked in Great Smoky Mountains National Park for over three decades.

# ABOUT THE EDITORS

## ANDREW M. BARTON

Raised in the southern Appalachians of western North Carolina, *Andrew (Drew) M. Barton* is a forest ecologist, science writer, and professor of biology at the University of Maine at Farmington (UMF). His research focuses on fire-adapted pine communities in Maine and changing climate, wildfire, and forests in the Sky Islands of the American Southwest. He is the author of the award-winning book, *The Changing Nature of the Maine Woods*. Drew cofounded the Michigan National Forest Watch and the UMF Sustainable Campus Coalition and was a key player in the Mt. Blue-Tumbledown Conservation Alliance, which protected 30,000 acres of forestland in western Maine. He received a BA in biology from Brown University, an MS in zoology from the University of Florida, and a PhD in ecology from the University of Michigan.

## WILLIAM S. KEETON

*William (Bill) S. Keeton* spent his early days tromping through the woods of the Finger Lakes Region in upstate New York, becoming a life-long forest enthusiast. In the mid-1990s, while working on public lands issues in the Pacific Northwest, he became fascinated by old-growth forests and has worked on them ever since. His research focuses on forest dynamics, including disturbance ecology, carbon functions, forest-stream interactions, and old-growth silviculture. Bill Keeton is a professor of Forest Ecology and Forestry at the University of Vermont (UVM), where he also serves as a fellow in the Gund Institute for Environment. He directs the UVM Carbon Dynamics Laboratory and previously chaired the undergraduate forestry program. Bill is a member of the Board of Trustees for the Vermont Land Trust and on the advisory board for Science for the Carpathians. He is currently chairing the International Union of Forest Research Organization's Working Group on Old-Growth Forests and Reserves. He received a BS in natural resources from Cornell University, an MES in conservation biology and policy from Yale University, and a PhD in forest ecology from the University of Washington.

# INDEX

*Note:* Page numbers followed by the letter f or t indicate figures and tables, respectively.